About Island Press

Island Press is the only nonprofit organization in the United States whose principal purpose is the publication of books on environmental issues and natural resource management. We provide solutions-oriented information to professionals, public officials, business and community leaders, and concerned citizens who are shaping responses to environmental problems.

In 2002, Island Press celebrates its eighteenth anniversary as the leading provider of timely and practical books that take a multidisciplinary approach to critical environmental concerns. Our growing list of titles reflects our commitment to bringing the best of an expanding body of literature to the environmental community throughout North America and the world.

Support for Island Press is provided by The Nathan Cummings Foundation, Geraldine R. Dodge Foundation, Doris Duke Charitable Foundation, Educational Foundation of America, The Charles Engelhard Foundation, The Ford Foundation, The George Gund Foundation, The Vira I. Heinz Endowment, The William and Flora Hewlett Foundation, Henry Luce Foundation, The John D. and Catherine T. MacArthur Foundation, The Andrew W. Mellon Foundation, The Moriah Fund, The Curtis and Edith Munson Foundation, National Fish and Wildlife Foundation, The New-Land Foundation, Oak Foundation, The Overbrook Foundation, The David and Lucile Packard Foundation, The Pew Charitable Trusts, The Rockefeller Foundation, The Winslow Foundation, and other generous donors.

The opinions expressed in this book are those of the author(s) and do not necessarily reflect the views of these foundations.

CONSERVATION IN
THE INTERNET AGE

Conservation in the Internet Age

THREATS AND OPPORTUNITIES

Edited by
James N. Levitt

ISLAND PRESS
Washington • Covelo • London

ISLAND PRESS is a trademark of The Center for Resource Economics.

Library of Congress Cataloging-in-Publication Data
Conservation in the Internet age: threats and opportunities / edited by James N. Levitt.

   p.   cm.

Includes bibliographical references (p.).

   ISBN 1-55963-913-X (pbk. : alk. paper)

   1. Nature conservation. 2. Internet. I. Levitt, James N.

QH75 .C677 2002

333.7'2—dc21

                        2002005947

British Cataloguing-in-Publication Data available.

CONTENTS

PART II
THE POTENTIAL IMPACT OF NEW NETWORKS ON LAND USE AND BIODIVERSITY 61

PART III
HARNESSING THE POWER OF NEW NETWORKS TO ACHIEVE CONSERVATION OBJECTIVES 141

For links to vivid, interactive Web sites that supplement the chapters of this book, please visit:

www.islandpress.org/internetage

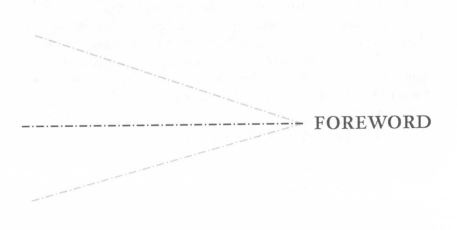

FOREWORD

Thomas J. Vilsack, Governor of Iowa

Teddy Roosevelt once said that "nine-tenths of wisdom is being wise in time." As we begin a new century, we have a unique opportunity to create a dialogue on conservation that encompasses the realities of the Internet. This global network, dazzling in its ability to reach sparsely populated and unspoiled lands as easily as it reaches urban centers, is a double-edged sword. The Internet connects far-flung populations, yet it has the potential to alter our country's physical landscape profoundly as it enables people to live and work wherever they desire. The Net is one of several substantial and growing threats to our open spaces and natural resources that demand our timely concern and action.

It is fitting that I quote Teddy Roosevelt, because in 1908 he called the nation's governors together for the first time in history. His urgent mission for this meeting was to conserve public land and natural resources. Out of that gathering, and his myriad other conservation initiatives, came the creation of many of the state and national forests, parks, and wildlife reserves that we enjoy today. Another result of that historic meeting was the birth of the National Governors' Association. Recently, as governor of Iowa, I chaired the association's Natural Resources Committee. As chair, I emphasized that, nearly 100 years after Teddy Roosevelt sounded the call for conserving public land, it is time to look at the untapped potential of conserving private land. Most of the land in our nation is privately owned. If significant improvements in land conservation across our country are to occur, government will have to provide the incentives and resources that encourage people to do the right thing for the land. Coincidentally, one of the vehicles for encouraging good private land stewardship will be the Internet.

It is encouraging that public discourse on the intersection of technology and conservation has begun in earnest. In June 2000, a group of government policy makers, nonprofit leaders, private entrepreneurs, and researchers met at Harvard University's John F. Kennedy School of Government to consider the challenges of conservation in the Internet Age. That meeting, at which I was the keynote speaker, became the genesis for this book. The thoughtful essays contained here give perspective on the far-reaching impacts of communication and transportation networks throughout history and explain how digital networks bring both positive and negative effects to our natural and built environments.

The time for a prescient dialogue on conserving open spaces is now. As governor of a largely rural state, I am keenly aware of the possibilities and pitfalls that rural development brings. The trend for companies to locate outside heavily congested areas to provide the goods and services our burgeoning world population demands is beginning to affect prime Iowa farmland. Consider that in 1846 the state of Iowa contained 95,000 people. Within the first decade of our existence, we grew to 195,000 people, and within the first twenty years, we had reached our first 1 million people. By 1900, we had 2.2 million people living in our state—more than in California or Florida. Iowa was the tenth most populated state in the Union. We had thirteen members of Congress. We were a significant force in the nation.

Today Iowa has yet to reach its third million—the 2000 census put our population at 2,926,000. But Iowa Farm Bureau statistics show that we are converting about 20,000 acres of prime farmland to urban use each year. Why is this happening? It is partly because government policies encourage the development of open spaces. Tax abatements exempt developers from paying levies on the growing value of their land for a specific period of time. State and local governments also provide new sewers and roads, further encouraging expansion into open spaces.

The small town of Manning, Iowa, with a population of some 1,000 people, illustrates the dilemma that advanced telecommunications presents. Manning is struggling for economic opportunity, as are many of Iowa's 950 communities—800 of which have 1,000 or fewer residents. These towns are always looking for ways to encourage business development. Manning has one of the 100-plus locally owned telephone companies in Iowa, and the town's leaders decided to install a telecommunications system that would give Manning residents high-speed Internet access. The goal was to create a "smart park"—a building connected to the Internet that allows technical workers, particularly engineers, to work in small-town Iowa while communicating with employers all over the world.

Manning's success in this venture has prompted a number of small businesses in our state to find the niche within the global economy that allows them to locate and thrive in rural communities. But as we improve our communications systems and attract more people to small communities, we create demand for more housing units and service-related businesses. There is only so much land available, and development pressures are beginning to degrade our environment.

Meanwhile, managerial and technological efficiencies allow farmers to raise yields so quickly that we have not been able to find ways to use all the corn and soybeans farmers grow. The results are escalating crop surpluses, lower farm prices, and a vicious cycle in which producers grow more to sell more. Compaction from heavy equipment threatens the long-term viability and productivity of our soil. Heavy reliance on farm chemicals places soil and water quality at risk. The U.S. Environmental Protection Agency recently identified 157 impaired waterways in our relatively small state—a finding that has jolted us to seriously consider initiatives to clean our water.

So how does a state go about addressing these conservation issues? States have had two principal responses. One is the start of smart growth policies, which encourage governments to establish parameters and incentives to ensure that public monies are used only in areas that need to be developed—so we can preserve our open spaces as best we can.

Maryland is probably the leader in this effort. Governor Parris Glendening has mounted a remarkable smart growth program that encourages development and economic expansion, but only in places that have appropriate infrastructure and environmental attributes. A Maryland video presentation on smart growth strikingly illustrates the evolving population in the state since the 1600s and the ever-diminishing open space in that state—as well as in the rest of the United States.

The other way states have approached conservation is by encouraging good stewardship of private land. That's the way we have done it in Iowa. Iowa has the highest percentage of privately owned land of any state in the country, which is another way of saying we have the least amount of publicly owned land. About 95 percent of our land is privately owned, and the 5 percent that is publicly owned includes roads and streets. We have no national parks, and we have a few state parks. An effort is under way to gain some kind of federal protection for the Loess Hills area in western Iowa, but for the most part we are private landowners. That is the ethic of our state.

Because private land rights are so dominant in Iowa, we are responding with conservation initiatives designed for private land. We have invested $11 million from the state budget in a comprehensive water quality improvement program, focused in part on encouraging volunteers and nonprofit groups to communicate person to person with farmers about the need to install buffer strips. Located between farm fields and nearby rivers and streams, these grassy strips provide a natural filter for pesticides and chemicals. With 72,000 miles of rivers and streams in our state, there are plenty of opportunities to plant buffer strips.

To protect fragile working lands and guard against sudden flooding, we are also encouraging farmers to preserve and re-create wetlands. Some waterways have recorded a 58 percent drop in suspended nutrients from this relatively simple practice. Part of this initiative includes compensating farmers for creating these wetlands, because we are asking them to take land out of production. The federal government assists us with a program that provides a relatively fair rental rate for farmers who preserve wetlands. Many farmers are reluctant to join these programs because they believe that government restrictions, government involvement, and government direction will translate into orders on how to use the land. That's why it is important to create a network of people who will talk to farmers one-on-one.

However, these two very good programs—buffer strips and wetland preservation—have not reached their full potential because we do not have a national environmental program that highlights their importance. Although many people care passionately about the environment, the average person on the street does not share this passion. I thought most Iowans would be very concerned about their natural resources. During my campaign in 1998, I fully expected the environment to be a key issue, because the state was debating the location and size of large-scale livestock confinements, whose waste creates serious pollution. This was a fairly hot topic as well as a significant one to the future of agriculture and the quality of life in communities across the state. But I was shocked to see that the environment was not one of the priority issues on Iowa voters' minds.

I believe we have not done as good a job as we need to in educating people at an early age about the connection between the environment and their quality of life. One way to remedy this is to use the power of the Internet to its full capacity. The Internet can help us create an educational system and an environmental ethic that will make farmers, business owners, and landowners ask: What is in the best interest of the land? What is in the best interest of the water? What is in the best interest of the air?

The Internet is an extraordinarily effective tool for disseminating information at relatively low cost to thousands, even millions, of people. It also is a tool for distance learning. For example, in Iowa, small rural schools cannot afford the cost of an environmental studies program. Through the Internet and the Iowa Communications Network (ICN), a teacher can reach ten, fifteen, and sometimes as many as ninety schools at a time. Our system is interactive, so young people in a school 100 or 200 miles away from the central site can ask questions. In fact, our students talked directly to scientists on the Galapagos Islands via the Internet and the ICN and had an active exchange regarding the scientists' results.

The Internet also is extraordinarily powerful in persuading people. We are just beginning to learn how to employ it effectively in political campaigns, and I think its use as a persuasive tool will be perfected over time. In Iowa, a political radio talk show posts on the Internet. The Internet also is where the younger generation reads its newspapers and does its research. We need to utilize that power. The election of Vicente Fox as Mexico's president was influenced by thousands of e-mails. I believe we will eventually vote and conduct much of the political process on the Internet. This growing use will allow more people to participate in the political process.

Over time, the Internet will become a powerful survey tool as well. The ability to gather and post information about what the public thinks on an Internet site is very powerful. If we are to create an environmental ethic, we must have the resources to do it, which means we must engage public policy experts by suggesting that ordinary folks are interested. There will be a time in the not-so-distant future when Iowans won't need to go door-to-door to convince farmers to participate in our state's clean water initiative. The Internet will provide exciting opportunities for engaging all of us in creating an environmental ethic and preserving our open spaces.

As with most powerful forces, the Internet can be used for good or for bad. I'm convinced that we will use the Internet for good. As the world's population continues to expand, we will become more concerned not just about our own little area but about the environment as a whole, because we will understand that we are interconnected. What happens in India is going to have an impact on me in Mt. Pleasant, Iowa. I believe the Internet will empower us to respond.

This book also should help us respond, by stimulating discussion and providing a blueprint for the interface between information technology and the environment. If we use the Internet to jump-start private and public land conservation as well as public education, we will be able to chart a more ecologically sound course for the planet we share.

ACKNOWLEDGMENTS

This book is the outgrowth of a remarkable meeting held at Harvard's John F. Kennedy School of Government in June 2000. At that meeting, an invited group of distinguished nonprofit leaders, private entrepreneurs, government officials, and academic researchers convened to consider the challenges of conservation in the Internet age. The meeting was distinctive because of both its interdisciplinary nature and the high level of energy that participants brought to the gathering. Demographers who typically spend long hours poring over county-by-county growth trends discussed supercomputer networks with ornithologists. Local government officials had a chance to carefully consider—and challenge—specialists in the development of information technology. Native Americans from the rural West had the opportunity to test their ideas on urban colleagues from the East Coast. And policy makers accustomed to considering the national interest dined with natural historians who have written about the interpretation of dinosaur bones.

It is to all the people who attended that meeting—especially to the authors who contributed chapters to this book—and to many more who continue to correspond face-to-face, by telephone, by e-mail, and through their writing that I owe my gratitude for making this book possible. I would also like to mention a number of individuals who have, in many ways large and small, contributed to the successful completion of this book and to the work of the Internet and Conservation Project, an effort based at the A. Alfred Taubman Center for State and Local Government at Harvard's John F. Kennedy School of Government.

First, thank you to my three loyal research assistants—Jennifer Reese, Noella Gray, and Dana Serovy—who have, in serial order over the course of the past three years, been the heart and soul of this effort. Thanks also to all of my colleagues at the Taubman Center who have shared their best ideas and good humor with me, including David Luberoff, Arnold Howitt, Charles Euchner, Gail Christopher, Dani Shefer, Mary Graham, David Weil, Kate Foster, Antonio Wendland, Archon Fung, Bob Behn, Brian Ellis, Cathleen Sarkis, Christina Marchand, David Fox, Deborah Voutselas, Guy Stuart, Jean Capizzi, Jennifer Johnson, Jonathan Richmond, Jose Gomez-Ibanez, Juliette Kayyem, Kara O'Sullivan, Karena Cronin, Kim Williams, Linda Kaboolian, Louise Kennedy, Martin West, Megan Sampson, Mingus Mapps, Rebecca Storro, Robert Putnam, Robyn Pangi, Sandra Garron, Shelley Kilday, Siobhan McLaughlin, Sophie Delano, Steve Goldsmith, Tom Polseno, Tom Sander, and Zoe Clarkwest.

My deep appreciation also goes to the faculty and staff of the other centers in the Kennedy School and throughout Harvard who have repeatedly allowed me, in my capacity as a fellow at the Taubman Center, to join in their classes and discussions. In particular, thanks to Henry Lee, John Holdren, Bill Clark, Lewis Branscomb, Deborah Hurley, Spiro Pollalis, Jeffrey Huang, David Foster, John O'Keefe, Bob Cook, Alice Ingerson, Calestous Juma, Douglas Causey, Clayton Christensen, Jaime Hoyte, Richard Forman, E. O. Wilson, and Philip Lovejoy.

Perhaps my best opportunity for learning from colleagues at Harvard came when Dr. Charles H. W. Foster, also known as Hank, gave me the wonderful opportunity to coteach a course module—Innovation and Entrepreneurship for Conservation and the Environment—with him at the Kennedy School in the fall 2000 semester. To Hank Foster, and to all the students and class participants who gave me detailed insight into the process of innovation, I offer my genuine appreciation and deepest respect.

I was lucky enough to have the professional support, encouragement, and friendship of many, many individuals outside of Harvard on this project as well. Bob Bryan, Larry Morris, Ken Hoffman, Kathy Blanchard, Sally Milliken, Tom Horn, Jessica Brown, Brent Mitchell, Anne-Seymour St. John, and the rest of the staff, as well as the board of directors, at the Quebec-Labrador Foundation have offered critical support and encouragement along the way. Thanks also to my many colleagues on the board and staff of the Massachusetts Audubon Society, including but certainly not limited to Laura Johnson, Jerry Bertrand, Gary Clayton, Andy Kendall, Banks Poor, John Mitchell, Laurie Bennett, Adrian Ayson, Lee Spelke, Franz Colloredo-Mansfeld, Merloyd Luddington, Appy Chandler, Tom Litwin, Teri Henderson, and John Fuller. Special mention also to the board and staff of the University of Kansas Natural History Museum and Biodiversity Research Center—particularly to my friend Leonard Krishtalka—and to my colleagues at the Lincoln Institute for Land Policy—most notably Armando Carbonell—who have challenged me to expand my ideas. I am further inspired by the board and staff of the New England Forestry Foundation, who have clearly demonstrated to me that old records in conservation are meant to be broken. Thanks also to colleagues on the National Science Foundation's Long Term Ecological Research 20-year review committee.

Many kinds of assistance contributed to the preparation of the June 2000 conference and of this book. The editors who have guided the preparation of the book, including Sandra Hackman in Massachusetts and Heather Boyer at Island Press, have done a wonderful job and have shown enormous goodwill and patience. Further critical support was offered by Island Press staffers James Nuzum, Chace Caven, Cecilia González, and Alphonse MacDonald. The enterprise has been kept afloat with support from many corners, including key financial and in-kind support from a number of donors. I am most grateful to the Virginia Wellington Cabot Foundation, the Ronald and Gladys Harriman Foundation, William Haney, Susan Whitehead, Bert and Joan Berkley, Linda Mason and Roger Brown, Blackboard.com, the Bullock Family Fund of the Boston Foundation, Centra Software, the Forest History Society, Forrester Research, and I-Group/Hotbank NE for their faith in and support for this project. The effort also benefited from the expert advice of an outstanding advisory board, including several individuals mentioned above, as well as Paul Judge, Chip Collins, and Andrew McLeod, whose experience in journalism, business, and public finance, respectively, was invaluable. Thanks also to Neil Karbank, Scott Brown,

John McAlpin, Ned Sullivan, Massimo Loda, Charles Hirschhorn, Tim Kaine, David Miles, Cindy Link, Peter Stein, Brian Bosworth, Deanna Douglas, Spencer Phillips, Larry Anderson, Wes Ward, Char Miller, Ian Bowles, Jay Espy, Peter Howell, Nora Mitchell, Jean Hocker, Hillary Pennington, Jim Price, and their families for all of their contributions of thought-provoking artwork, ideas, and encouragement.

The most heartfelt thanks I have saved for last. My love to my brother, Tom, my sister, Jean, and the entire Levitt family; to Bert, Joan, and all of the Berkley family; and most especially to my love, my conscience, and my closest advisor, Jane, and our three beautiful children, Will, Daniel, and Laura.

Let me close by dedicating this book to Alan Altshuler, director of the Taubman Center and senior member of the Harvard faculty, who was gracious enough to invite me to bring my unconventional ideas to this remarkable place. Thank you, Alan, for serving as a mentor, a friend, an intellectual resource, and a source of invaluable perspective.

James N. Levitt
Cambridge, Massachusetts
April 2002

CONSERVATION IN
THE INTERNET AGE

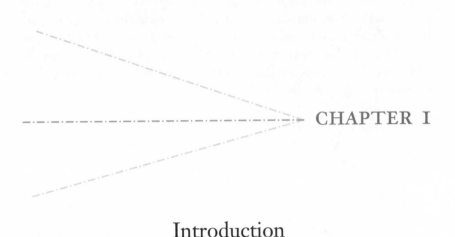

CHAPTER I

Introduction

James N. Levitt

Since the founding of the Republic, generation after generation of Americans has fallen in love with magnificent and, in several instances, seemingly magical communications and transportation networks that have brought us closer together and helped to transform the way we live, work, trade, learn, and play. New networks have played critical roles in the vast, largely beneficial changes we have seen in North America since the late 1700s, when the words *communication* and *transportation* were virtually synonymous. If you wanted to get a complex message to your spouse or a head of state in 1776, it had to be physically carried, as fast as a horse or a courier on foot could travel. To expand our ability to travel and communicate—a capacity that did not substantially change from Julius Caesar's day to George Washington's—has been a central element of the American dream.[1]

Our passion for new pathways was unambiguously expressed in the early 1800s as reports from the successful Lewis and Clark expedition filtered through the land—one Washington, D.C., resident reported, on Captain Lewis's celebrated return to the capital city: "Never did a similar event excite more joy."[2] Pre– and post–Civil War entrepreneurs who masterminded the construction of transcontinental railroad and telegraph lines understood the importance of their efforts to the nation. The oratorical expressions of gratitude made when the Pacific railroad was completed were nearly religious in fervor. John Taylor, speaking in Salt Lake City in 1869, characterized the achievement as "so stupendous that we can scarcely find words to express our sentiments or give vent to our admiration."[3] In the mid–twentieth century, Americans' love affair with automobiles and the open road was widely acknowledged, as we were encouraged on nationally syndicated television to "see the U-S-A in your Chev-ro-let." And, in our own era, we have witnessed "irrational exuberance" regarding the economic promise of the Internet— a network that promises us anywhere, anytime access to almost anything.[4]

But fervor subsides, and in the afterglow Americans have repeatedly come to understand that new networks not only expand economic and social frontiers, but also present the United States with new challenges, including the potential for dramatic environmental disruptions. Coming behind Lewis and Clark, with the enthusiastic approval of Thomas Jefferson, the American Fur Company scoured the West for beaver pelts until, within decades, the business became uneconomical, due to the overaggressive harvesting of beavers. The railroad and telegraph networks built after the Civil War similarly helped to accelerate the disruption of vast areas of prairie habitat; with the encouragement of the U.S. Army, buffalo hunters, many of whom literally shot their rifles out of the windows of moving trains, nearly wiped out seemingly limitless populations of bison on the Great Plains. And the automobile and petroleum industries, central to the post–World War II ascendancy of American industry, are now closely associated with myriad environmental challenges, including the widely recognized threat of global warming.

Yet, remarkably, even as they have enabled both economic growth and environmental disruption, new networks and the entrepreneurs who created them have also set the context for landmark conservation initiatives. These efforts to responsibly steward land and biodiversity have left permanent marks on our historical record, on the landscape, and on the national atlas. It was the American Fur Company's steamboat *Yellow Stone* that carried George Catlin up the Missouri River to make his wonderful paintings of Native Americans and wildlife in 1832.[5] Railroad entrepreneurs Jay Cooke and Frederick Billings were key players in the effort to create the world's first national park at Yosemite in 1872.[6] Laurance Rockefeller, John D. Rockefeller's grandson and an accomplished venture capitalist and aviation investor in his own right, is a major figure in the history of the American conservation movement in the twentieth century. In the 1950s and 1960s, Rockefeller was essential to the conception of the Land and Water Conservation Fund, a federal program that helps states and localities acquire conservation land, in part with revenues derived from the Highway Trust Fund and offshore oil leases.[7]

As we enter the twenty-first century, Americans are beginning to see, as earlier generations came to see, that the current crop of new networks exemplified by the Internet and FedEx-style express delivery networks, in addition to reducing the friction of distance and generally contributing to our economic well-being, may also be associated with disruptive environmental impacts. An emerging body of empirical research, as well as anecdotal reports, suggests that new communications and transportation networks may be significant enablers of what demographers call selective deconcentration—also known as *sprawl*—in rural, suburban, and urban communities across the nation. The same demographic shifts are associated with an accelerated loss of open space at local, state, and national levels, as well as with the degradation of wildlife habitat in regions ranging from Cape Cod to the Greater Yellowstone ecosystem.

As in earlier eras, however, there is also good news. Conservationists are finding new and powerful applications for the Internet and its relatives that have the potential to change the game in conservation science, education, advocacy, constituency and membership development, fund-raising, and administration, as well as the management of natural resources. Indeed, entrepreneurial individuals and organizations throughout the United States and in far-flung locations around the globe are undertaking on-the-ground conservation projects of unprecedented scale and scope with the assistance of new technologies and the active support of captains of the new economy.

The critical need for innovative, constructive conservation strategies continues to mount in the face of myriad conservation and environmental challenges, ranging from local loss of wildlife habitat to global climate change. This book examines, in a cross-disciplinary fashion, the emerging threats and opportunities to the conservation of land and biodiversity associated with pervasive new communications and transportation networks. It considers the potentially constructive and disruptive conservation impacts of the Internet in four parts.

Precedents and Prospects

The first part offers a broad view of the impacts of emerging communications and transportation networks on society and the natural landscape.

In Chapter 2, James Levitt examines the historical precedents to the network-enabled land use and conservation challenges that we now face. In the fine chapter that follows, William Mitchell, dean of MIT's School of Architecture and Planning, shows how the fragmentation and recombination of the built environment spawned by technological change offer both exciting social opportunities and significant challenges.

The Potential Impact of New Networks on Land Use and Biodiversity

The second part examines emerging empirical and anecdotal evidence that the Internet age has sparked dramatic changes in demographics, land use patterns, and biological systems. In Chapter 4, Kenneth Johnson, professor of sociology and demography at Loyola University, analyzes the rural rebound of the 1990s, a decade that saw a net positive migration of Americans from metropolitan counties to nonmetropolitan areas rich in natural and recreational amenities. Ralph Grossi, president of the American Farmland Trust, shows in Chapter 5 that America consumed open space at an unprecedented rate during the 1990s in metropolitan, metropolitan-adjacent, and nonmetropolitan counties alike. In Chapter 6, John Pitkin, president of Analysis and Forecasting, Inc., and James Levitt examine the relatively high level of Internet penetration in households in two fast-growing nonmetropolitan counties in Central Oregon, with a particular focus on usage in households that came to the area at least in part to enjoy its natural and recreational amenities. The second part concludes with a chapter by Andrew Hansen and Jay Rotella, both associate professors of ecology at Montana State University. Hansen and Rotella show how an influx of affluent private landholders in the Greater Yellowstone region appears to be disturbing the habitat of many resident wildlife species.

Harnessing the Power of New Networks to Achieve Conservation Objectives

In the third part, leading conservationists show how they are, even in the context of network-related environmental disruption, using the Internet to dramatically improve their effectiveness. In Chapter 8, Leonard Krishtalka, director of the Natural History Museum and Biodiversity Research Center at the University of Kansas, presents Species Analyst, a Web-based project that allows researchers from around the world to pool and analyze an unprecedented wealth of data on humanity's biological heritage. John

Fitzpatrick, director of the Cornell Laboratory of Ornithology, and Frank Gill, senior vice president for science at the National Audubon Society, in Chapter 9 look at the remarkable progress in citizen science and interactive education made possible by BirdSource, an imaginative Web-based venture. In Chapter 10, Jacob Scherr, senior attorney for the Natural Resources Defense Council, tells the story of the successful campaign—fueled by the Internet and e-mail—to stop industrial development of Laguna San Ignacio in Baja California, Mexico, breeding grounds for the endangered gray whale. Finally, in Chapter 11, William Roper, executive director of the Orton Institute, and Brian Muller, assistant professor of planning and design at the University of Colorado, show how pioneering interactive computer simulation and visualization tools can enhance community planning.

The Internet Age as Context for Conservation Innovation

In the final part of the printed book, the authors look ahead to opportunities for landmark conservation initiatives in the Internet age. In Chapter 12, Bob Durand, Massachusetts secretary of environmental affairs, and Sharon McGregor, Massachusetts assistant secretary for environmental affairs, show how their state is using geographic information systems (GIS) tools fashioned for use by local towns and planning boards, coupled with an innovative Community Preservation Act, to protect and preserve the state's landscape and natural resources. Joel Hirschhorn, director of natural resources policy studies for the National Governors Association, examines in Chapter 13 the growing interest of local, state, and national agencies in conserving land—not only to protect fragile natural resources but also to stimulate sustainable economic development. In Chapter 14, Peter Stein, general partner of Lyme Timber, which recently helped to forge the largest conservation land-easement deal in American history, and James Levitt consider the role of network entrepreneurs as key innovators in the history and future of the American conservation movement. Finally, Levitt concludes with an overview of opportunities for landmark conservation initiatives in coming decades, informed by the thinking of conservation leaders from the public, private, nonprofit, and academic sectors.

Web Site: A Sampler of Conservation in the Internet Age

Several of the essays in this volume focus on projects that were designed to capitalize on digital tools for conservation. To the extent that these projects have achieved their goals and are becoming models for other conservationists, they are effectively changing the game in a broad cross section of professional fields. Such projects include the Species Analyst, developed at the University of Kansas; the BirdSource project, jointly developed by the National Audubon Society and the Cornell Laboratory of Ornithology; the campaign to save Laguna San Ignacio spearheaded by the Natural Resources Defense Council; and the Commonwealth of Massachusetts's efforts to use Web-based tools to understand threats to biodiversity and promote community preservation. In other chapters, including those written by Kenneth Johnson, Ralph Grossi, Andrew Hansen, and Jay Rotella, the authors employ GIS technologies to contribute to our understanding of changing demographic and land use patterns in North America.

To give readers greater insight into the look and feel of these new information technology–based tools and projects, Island Press has created a Web site, available at www.islandpress.org/internetage, with links to the organizations and applications discussed in this volume as well as to other relevant conservation-related materials available on the World Wide Web. Many of these Web site addresses are also listed in the following chapters.[8] Our hope as authors is that the essays and examples offered here may inspire others to advance the state of the art to levels that we are not yet able to imagine at this dawn of the new century.

NOTES

1. Stephen E. Ambrose, in his book on the building of the transcontinental railway, *Nothing Like It in the World* (New York: Simon and Schuster, 2000, p. 25), quotes Thomas Curtis Clarke, who opened his 1889 essay "The Building of a Railway" with the following: "The world of to-day differs from that of Napoleon more than his world differed from that of Julius Caesar; and this change has chiefly been made by railways."

2. Stephen E. Ambrose, *Undaunted Courage* (New York: Simon and Schuster, 1996), 419.

3. Ambrose, *Nothing Like It in the World*, 368.

4. "Irrational exuberance" is a phrase introduced by Federal Reserve chairman Alan Greenspan in a speech delivered on December 5, 1996. Robert J. Schiller, a Yale economist who had several days earlier made a presentation to Greenspan on his views regarding stock market overvaluation, has written a book on the subject also titled *Irrational Exuberance* (Princeton, N.J.: Princeton

University Press, 2000). Both Greenspan's speech and Schiller's book are discussed in Louis Uchitelle's article "He Didn't Say It. But He Knew It," which appeared in the *New York Times,* Money and Business section, on Sunday, April 30, 2000.

5. Donald Jackson, *Voyages of the Steamboat Yellow Stone* (Norman: University of Oklahoma Press, 1985), 31.

6. Richard West Sellars, *Preserving Nature in the National Parks: A History* (New Haven, Conn.: Yale University Press, 1997), 9.

7. Robin Winks, *Laurance S. Rockefeller: Catalyst for Conservation* (Washington, D.C.: Island Press, 1997), 134.

8. Web site addresses such as www.islandpress.org are referred to technically as universal resource locators, or URLs.

PART I

PRECEDENTS AND PROSPECTS

Today, in the early twenty-first century, the Internet and other advanced communications and transportation systems are associated with wide-ranging, dynamic, and volatile shifts in North American economic systems, demographic patterns, and land use configurations. Our interrelationships with the natural and built environments are undergoing both dramatic and subtle changes at many levels.

The history of the United States and its international neighbors includes compelling precedents that can help put in perspective the current process of change related to the spread of new communications and transportation networks. During the 1860s and 1870s, for example, the completion of transcontinental railroads and telegraph networks was associated with very significant changes in North American demographics, land use patterns, and environmental conditions. Similar cycles of change have occurred in more recent years, associated with the spread of interstate highways, passenger airline networks, and analog communications systems. Each of these new networks played a part in enabling people to settle in new areas while remaining linked to families, associates, and markets far from their homes. Some of these changes were predicted by contemporary observers, whereas others were largely unexpected.

The following two chapters explore these historic precedents and consider how they may inform our prospects for the future. First, James Levitt examines the constructive and disruptive impacts of historic and present-day transportation and communications networks. Next, William Mitchell examines the fragmentation and reconfiguration that are often associated with new network growth. Both essays shed light on the ways new networks may influence our styles of life and work in the years ahead.

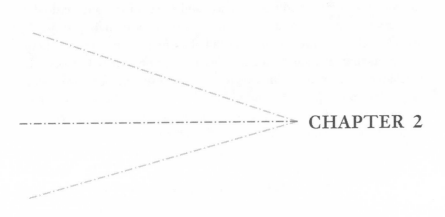

CHAPTER 2

Networks and Nature in the American Experience

James N. Levitt

The story of changing land use patterns, environmental disruption, and conservation across the United States and North America offers a rich diversity of themes; it is a complex story with many threads. It is a tale of conquest and of ongoing contests for political, economic, and social power among the many cultures present on the continent throughout history and in our own day.[1] It is a story that, since the time Columbus first set foot on Hispaniola, has been characterized by a dramatically growing population, spreading itself across the landscape in ever-evolving demographic configurations. Consider, for example, that the U.S. population grew from 3.9 million individuals in 1790 to some 280 million in 2000, and that this number is expected to expand to some 400 million by 2050—a projected 100-fold growth in the space of 260 years.[2]

The story is also one of an amazing surge in affluence shared by many of the continent's citizens. Subsistence farmers barely hanging on to their homesteads in Kansas in the 1850s would be awed to walk into a modern supermarket sited on what was once their land, finding apples from New

Zealand, water imported from France, and sweatshirts manufactured in China wheeled out the door by thousands of locals living in nearby homes, each with indoor plumbing and plentiful heat. It is a story of rapidly advancing technology of numerous kinds, from Cyrus McCormick's mechanical reaper to Willis Carrier's air conditioner.[3] It is also a story of the ever more powerful and ubiquitous communications and transportation networks that have enabled the spread of people and structures across the land. These networks are the subject of this essay and of this book.

The Significance of Networks in American Culture

Throughout history, Americans have attributed great significance to new transportation and communications networks and to the individuals who discover, build, and advocate for them. In the earliest days of the Republic, Thomas Jefferson, himself a network entrepreneur who sought to build roads, engineer canals, and launch ambitious expeditions to the West, actively sought information about the life of Christopher Columbus, an early icon of the adventurous American spirit.[4] Stories about Columbus's life and precedent-shattering explorations so impressed Jefferson and his compatriot George Washington that they named the District of Columbia in his honor.[5]

Enthusiasm for network pioneers, and for building new links to the West, continued unabated in the nineteenth century. Thomas Hart Benton, a powerful U.S. senator from Missouri, championed westward expansion and the idea of a transcontinental railroad throughout most of his long career. As an advocate of westward growth, he was greeted by attendees at the 1849 Pacific Railroad Convention in St. Louis with "rapturous applause." His oration was stirring:

> We live in extraordinary times. . . . Mechanical genius has again triumphed over the obstacles of nature . . . the conveyance being invented which annihilates both time and space. Let us rise to the grandeur of the occasion. Let us complete the grand design of Columbus by putting Europe and Asia into communication, and that to our advantage, through the heart of our country. Let us give to his ships, converted into cars, a continued course unknown to all former times. Let us make the iron road, and make it from sea to sea . . . adorned with its crowning honor, a colossal statue of Columbus . . . pointing with outstretched arm to the western horizon, and saying to the flying passengers "There is the East, there is India."[6]

U.S. President Bill Clinton, in a December 1999 speech on economic growth, revisited Benton's theme, emphasizing the importance of pathfinders and network entrepreneurs from both the private and the public sectors to our national identity:

> We know that our nation has always prospered when government has invested in giving people the opportunity to make the most of their vision and their dreams, from financing the Louisiana Purchase and the Lewis and Clark expedition to the interstate highway system and the space program. . . . The American people have always been a bold and innovative bunch. We are always drawn to uncharted lands over the next horizon. . . . That's what we do. Today, thanks to wise investments made by government and the private sector over many years, the American people have before them the unexplored continent of cyberspace and the prospect of discovering what is in the black holes in outer space.[7]

Given our deep and historic ties to explorers, pioneers, and network entrepreneurs, it is no wonder that when a new set of actors again reduces barriers of time and space on our continent and throughout the world, we are ready and willing to lionize them and their achievements.

Networks and Economic Growth

Our enthusiasm for new transportation and communications networks is grounded in economic common sense. We have long known that new pathways that reduce the cost and friction of doing business with the rest of the world are critical to the nation's wealth and prosperity. In 1803, Jefferson emphasized to Captain Meriwether Lewis the central importance of establishing new trade and communications routes as he launched Lewis and William Clark on their famous expedition: "The object of your mission is to explore the Missouri River . . . as, by its course and communication with the water of the Pacific ocean, may offer the most direct and practicable water communication across the continent, for the purpose of commerce."[8]

Commentators in our own day similarly have recognized the considerable impact of new networks. Michael Mandel, in the cover article on the "Internet Age" for the seventieth anniversary issue of *Business Week*, proclaimed that "the power to navigate the world at the click of a mouse is a force that is transforming our lives like none before."[9]

An important benefit of new networks stems from what economists call *network effects* (or *network externalities*). Network effects are particularly well suited to democratic societies, as their benefits can spread so widely. On a publicly accessible network, new members gain access to each existing member, and existing members extend their reach to each new member; economic activity expands as the size of the entire pie gets bigger and as common resources can be pooled.

The citizens of Santa Cruz, California, expressed their genuine enthusiasm for such an effect when, in the 1870s, a new rail connection opened to their town. The local newspaper trumpeted: "At last our enterprising young city is in full connection with the rest of mankind."[10] George Gilder expressed similar enthusiasm in the early days of the commercial Internet, referring to a dictum attributed to entrepreneur Bob Metcalfe: "Metcalfe's law of the telecosm . . . [shows] the magic of interconnections: connect any number, 'n,' of machines . . . and you get 'n' squared potential value. Think of phones without networks or cars without roads. Conversely, imagine the benefits of linking up tens of millions of computers and sense the exponential power of the telecosm."[11]

Indeed, we have observed dramatic economic benefits associated with new networks repeatedly in our history. With the opening of the Erie Canal in the early nineteenth century, the price of transporting a sack of grain from the Great Lakes to the Hudson River plummeted, encouraging the migration of farmers to the Midwest and helping catapult New York City past Philadelphia as the continent's busiest port. Similarly, new telegraph networks were key to managing transcontinental rail systems, themselves engines of western growth following the American Civil War.[12] Sam Walton relied heavily on sites with highway intersections and in-house computer networks in building Wal-Mart, the largest private employer in the United States in 2000.[13] And, despite the recent decline in the market value of dot-com companies, a new generation of technologies, exemplified by the Internet, wireless telephone systems, and express delivery networks, continues to transform daily life in cities, suburbs, rural communities, and even the most remote wilderness areas in the United States and in the world, from Arkansas to Antarctica.

Overoptimistic Social Forecasts

Unfortunately, although we are good at recognizing the power of new networks to help reshape economies, Americans have a history of excessive optimism regarding the ability of such networks to solve a wide variety of social problems. In the 1880s, for instance, prominent commentators predicted that by allowing the leaders of the world's great nations to communicate instantaneously, the global telegraph network could help usher in an era of lasting peace. Henry Field spoke of the transoceanic cable as "a living, fleshy bond between the severed portions of the human family, along which pulses of human tenderness will run backward and forward forever. By such strong ties does it tend to bind the human race in unity, peace and concord."[14] Field offered that flawed insight prior to the outbreaks of World War I and World War II.

In a similar vein, Nicholas Negroponte, head of the MIT Media Lab, told an audience in Brussels in 1997 that "the Internet would do no less than bring world peace by breaking down national borders." Twenty years from now, he said, children who are used to finding out about other countries through the click of a mouse "are not going to know what nationalism is."[15] In light of subsequent nationalistic atrocities committed in Kosovo and the tragic destruction of the World Trade Center in New York, Negroponte's prognostications seem no more likely than Henry Field's.

Our tendency to make overoptimistic social forecasts extends to more mundane matters. In the late 1970s, Alvin Toffler confidently predicted that with the advent of advanced telecommunications techniques, "paperless offices" would become common in corporate America. His comment caused real concern among paper manufacturers, but they need not have worried. Although the use of paper has shifted—we now send fewer first-class letters and use more paper for printers and packaging—world paper consumption tripled between the 1960s and the 1990s, and forecasts show global consumption growing another 50 percent by 2010.[16]

In the early and mid-1990s, sources ranging from the U.S. Department of Energy to AT&T excitedly predicted that telecommuting and e-commerce would make significant dents in per capita rates of automobile and truck travel as well as in associated energy consumption.[17] Analysts such as Patricia Mokhtarian at the University of California at Davis point to a quite different outcome. Mokhtarian's long-term study of California telecommuters shows that while some telecommuters may marginally reduce their own annual mileage logged in automobiles, any gains are likely to be swamped by long-term nationwide growth in annual per-person vehicle miles traveled. As Mokhtarian notes, "Historically, transportation and communication have been complements to each other, both increasing concurrently, rather than substitutes for one another. And we have no reason to expect that relationship to change."[18]

Myopia Regarding the Environmental Effects of Networks

Just as we have a history of making overly optimistic social forecasts regarding new network technologies, we have failed to recognize and mitigate potentially disruptive environmental impacts associated with emerging communications and transportation systems. Two brief histories serve to illustrate.

In his landmark 1864 work, *Man and Nature*, George Perkins Marsh, then Abraham Lincoln's ambassador to Italy (see Figure 2.1), warned of many ways in which the "destructiveness of man" can disturb natural systems. Having served as Vermont's railroad commissioner, and being a vocal critic of railroad corporations, Marsh was quite familiar with the changes associated with new transportation grids spreading across the land. In *Man and Nature*, he noted that:

> the modern increased facilities of transportation have brought distant markets within the reach of the professional hunter, and thereby given a new impulse to his destructive propensities. Not only do all Great Britain and Ireland contribute to the supply of game for the British capital, but the canvas-back duck of the Potomac, and even the prairie-hen from the basin of the Mississippi, may be found at the stalls of the London poulterer. . . . Reproduction, in cultivated countries, cannot keep pace with this excessive destruction, and there is no doubt that all the wild birds which are chased for their flesh or their plumage are diminishing with a rapidity which justifies the fear that the last of them will soon follow the dodo and the wingless awk.[19]

Despite the literary success of Marsh's tome, Congress took little notice or action over the next decade to prevent such "excessive destruction." It took more than a decade after its creation in 1872 for Yellowstone National Park to be designated to serve as a sanctuary for the bison, which were also nearly extirpated by amateurs and professionals who made extensive use of western railroads to facilitate their hunt.[20] And not until 1903, nearly forty years after Marsh published his warning, did Theodore Roosevelt, with the encouragement of the rapidly growing Audubon movement, establish the first federal preserve for seriously endangered migratory birds on Pelican Island, Florida.

Figure 2.1: Photograph of George Perkins Marsh by Civil War–era photographer Matthew Brady. (Courtesy of the U.S. Library of Congress, Brady-Handy Collection.)

Marsh's observations highlight the negative flip side of seemingly magical network effects: when a new network offers substantial numbers of people fresh access to a relatively untapped natural resource, and those people can exploit that resource at little or no incremental cost, they can quickly degrade or destroy it. To use a phrase popularized by ecologist Garrett Hardin in 1968, low-cost access provided by new networks can enable a "tragedy of the commons."[21] In North America, birds, bison, forests, and soil, as well as clean water and fresh air, have all suffered at various points from such relatively unfettered access and unrestricted use.

A second case of shortsightedness regarding the environmental impact of new networks is related to President Eisenhower's bold proposal, during his first administration, to build a national system of interstate highways.[22] The term *environmental impact* is used advisedly; it did not come into common parlance until newly created state and federal environmental protection agencies began requiring developers to submit environmental impact statements in the 1960s and 1970s.[23]

The U.S. Senate's Subcommittee on Public Roads met in February and March 1954 to hear testimony regarding Eisenhower's initiative, which his advisor Francis du Pont had shaped into the proposed Federal-Aid Highway Act.[24] During Senate hearings that winter, one enthusiastic booster after another testified regarding the cost, engineering challenges, and potential benefits of the interstates and the related forest highway system. Senator Wayne Morse, of Oregon, argued that "the economic changes in our country in the last forty years have made interstate commerce now very much dependent upon a Federal highway system." Major General Paul Yount, chief of transportation for the U.S. Army, held that the proposed system was important "to serve the national defense." Officials responsible for managing the nation's natural resources, such as Edward Cliff, assistant chief of the U.S. Forest Service, emphasized that "the need for the improvement of the forest highway system is urgent," to provide a link from forest to mill as well as public access for recreation. In related testimony, Leo Bodine, executive vice president of the National Lumber Manufacturers Association, lobbied forcefully for expanded federal funding for "more timber access roads, [which] must be provided to open up old-growth timber stands and to salvage blowdown and insect-infested timber on national forests."[25]

The proposed interstate highway system was indeed a grand scheme replete with opportunities for the nation. But records of the Senate hearings reflect very little substantial concern with the system's possible environmental and conservation impacts. The strongest recorded statement appears in a letter from the Pennsylvania Association of Township Supervisors endorsing a local "Don't Be a Litterbug" campaign.[26]

Not until the following year, in testimony before the U.S. House of Representatives, did Howard Zahniser (see Figure 2.2), executive director of the Wilderness Society, sound a serious note of caution, urging Congress to adopt a policy whereby "no highways shall be constructed or planned . . . in or across any national park or national monument, any national wildlife refuge, or any duly designated wilderness, wild, primitive, or roadless area within the national forests."[27] During his testimony,

Figure 2.2: Howard Zahniser, executive director of the Wilderness Society (1945–64), warned of the potentially disruptive consequences of interstate highways to wilderness in testimony before Congress in 1955. (Courtesy of the Wilderness Socitey.)

Zahniser took pains to assure members of Congress that he was not opposed to road development; he simply wanted to ensure that certain natural areas would be preserved intact.

In retrospect, we can see the prescience of Zahniser's testimony. As he warned, high-speed roadways built in the last half of the twentieth century have been associated with serious disruptive impacts on wilderness areas and critical wildlife habitat—including the disruption of spotted owl habitat in the old-growth forests that Leo Bodine wanted to access for logging.[28] The new roadways are also associated with myriad other environmental impacts, including urban, suburban, and rural sprawl; an increase in air pollution in many parts of the country; and the generation of carbon dioxide and other gases associated with global climate change—an issue not on the radar screens of Congress in the 1950s despite predictions of a "greenhouse effect" by Alexander Graham Bell much earlier in the century.[29]

Zahniser was not the first to point out the potential threats of new highways, nor was he the last. The group that founded the Wilderness Society, including Bob Marshall, Aldo Leopold, and Benton MacKaye, had been trying to underscore the issue for decades. In the first issue of the *Living Wilderness*, printed in 1935, they argued as follows in the essay "A Summons to Save the Wilderness":

> Ten years of warfare in Congress saved the National Parks System from water power and irrigation, but left the primitive decimated elsewhere. What little is left is passing before a popular craze and an administrative fashion. The craze is to build all the highways everywhere while billions may yet be borrowed from the unlucky future. The fashion is to barber and manicure wild America as smartly as the modern girl. Our duty is clear.[30]

Despite the foresight and eloquence of Zahniser and his colleagues, neither the public nor Congress was noticeably moved, and President Eisenhower signed the Federal-Aid Highway Act of 1956 into law without any significant provisions regarding conservation or environmental protection. Only with the publication of Rachel Carson's *Silent Spring* in 1962 did widespread public concern for wildlife and wilderness re-emerge. The Wilderness Act did not become law until September 1964, four months after Zahniser's death, and not until six years later, with the 1970 Clean Air Act, did the United States institute the first national program to manage air quality.

The traditional myopia of Americans, as well as of observers from Europe and elsewhere, regarding the potentially disruptive environmental impacts of new networks has extended into the Internet era. A number of popular books published in the late 1990s by prominent Internet enthusiasts that predicted bright social futures in the twenty-first century generally ignored or gave short shrift to the potentially disruptive environmental impacts of the current generation of new networks.[31] Environmentalists are prone to similar oversights: several recent high-level analyses of sprawl have focused on the already well-explored connection between the growing use of cars and highways and the loss of open space, rather than on the potential impact of newer networks.[32]

However, as unbridled enthusiasm for the Internet and related networks has substantially subsided, our myopia regarding the impact of the Internet and related networks on land use and biodiversity is due for correction. The essays in this volume argue for more farsighted attention to the strong hypothesis that the Internet, express delivery systems, and associated networks may play an important role in the mix of technological development, growing affluence, and demographic change that can prove disruptive to our landscape and biodiversity. It appears that the new networks affect the shaping of settlement patterns and landscape usage patterns in at least two ways:

1. Most directly, the Internet, express delivery systems, and related networks serve as *key enablers* of deconcentrating settlement and development patterns in North America.[33] By reducing the friction of distance, the new networks give individuals, households, and organizations greater mobility and locational flexibility, increasing people's ability to travel, live, and work where they like in urban, suburban, and nonmetropolitan areas, including those with relatively high levels of natural amenities. The new networks offer this growing mobility and locational flexibility in combination with existing infrastructure; they work cumulatively with railroad, highway, airport, telephone, and media networks already in place.

2. Indirectly, the new networks stimulate greater mobility and locational flexibility by *contributing to general economic expansion and growing levels of personal affluence*. With more affluence, people have a greater ability to travel, to build bigger residences and workplaces, and to move to preferred locations that may have previously been unaffordable.

Americans have good reason to be concerned that new communications and transportation networks may unleash unexpected and disruptive forces on the landscape. However, these changes also bring considerable reason for hope. Indeed, over the course of American history, similar conditions have provided the context for landmark conservation initiatives that have made a lasting mark in the field of conservation—and on the national atlas. This reiterated pattern includes several components:

- New communications and transportation systems rapidly proliferate across the land.
- Enabled in part by the new networks, demographic and land use patterns change in significant ways.
- The demographic and land use changes become closely associated with dramatic environmental disruptions.
- In the context of network proliferation, changes in demographic and land use patterns, and the associated environmental disruptions, we have achieved significant advances in conservation practice as well as landmark conservation innovations.

Examples of such innovations—described below in some detail—include the creation of western Indian reservations following the Louisiana Purchase in 1803, the establishment of the world's first national park in Yellowstone in 1872, the dramatic expansion of the national forest system around the turn of the twentieth century, and the conception of the Land and Water Conservation Fund in the 1950s and 1960s. While such historical precedents cannot, of course, guarantee that similar initiatives will occur again, they illustrate the type of achievement that may be possible in coming decades.

Thomas Jefferson, Lewis and Clark, and Western Indian Reservations

In June 1803, Thomas Jefferson wrote a letter of instruction that sent Captains Meriwether Lewis and William Clark on their epic voyage of discovery up the Missouri River and on to the Pacific Ocean. In launching the explorers, Jefferson fulfilled an ambition he had harbored for most of his adult life and that his father, Peter, had harbored before him.

In the 1740s, Peter Jefferson, a talented surveyor and land speculator, and several of his close associates founded a "Company of Adventurers called the Loyal Company." The company successfully applied for a patent from the Crown for some 800,000 acres in what is now eastern Kentucky—territory that the company successfully explored and mapped in 1749. Proceeding to consider more ambitious projects, the company prepared plans to explore the land along the little-known Missouri River far to the west, hoping to find a passable route to the Pacific Ocean. Unfortunately for Peter Jefferson and his associates, the plans were abandoned when the French and Indian War broke out.[34] Peter died in 1754, leaving his surveying tools, his land, and very likely his fascination with the West to his oldest son, Thomas.

Thomas Jefferson (see Figure 2.3) went on to develop his own skills as a surveyor, cartographer, and naturalist, and to promote prior to 1800 at least three other unsuccessful expeditionary schemes aimed at finding a transcontinental route to the Pacific.[35] After reading, as president, reports of Canadian overland expeditions to the Pacific, Jefferson finally began to pull together the elements that would become the celebrated Corps of Discovery. As an experienced network entrepreneur who had also worked to establish a canal connecting the Potomac with the Ohio River, Jefferson was well aware that the success of Lewis and Clark's mission depended on more than their ability to navigate rivers and cross

Figure 2.3: Thomas Jefferson, network entrepreneur. (Courtesy of the U.S. Library of Congress, Prints and Photographs Division.)

mountains.[36] To win funding from Congress, the expedition would have to be cast as an effort that was in the national interest. In his appeal to Congress some six months before the Louisiana Purchase was consummated, Jefferson emphasized that the effort to establish an interior route to the Pacific would be a boon to both commerce and geographical knowledge—knowledge that Congress understood could eventually be of value to fur trappers as well as military strategists. Once the news spread in July 1803 that the United States had purchased the Louisiana Territory from Napoleon, the expedition's national importance became apparent. Americans could realistically imagine the rapid expansion of an "empire for liberty" across the continent.[37]

The president also knew that the explorers would have to maintain friendly relations with native Indians west of the Mississippi. As a token of his interest in "Peace and Friendship," Jefferson cast a medal showing a white man's hand shaking an Indian's hand on one side and Jefferson's image on the other. Lewis and Clark were to hand these medals to native chiefs to underscore that Americans wanted mutually beneficial interaction (see Figure 2.4).

Beyond keeping peace with Indians already in the West, Jefferson imagined he might be able to reserve land west of the Mississippi to resettle eastern Indians displaced by pioneers then settling the Ohio Valley and elsewhere. He aimed to "live in perpetual peace with the Indians, to cultivate an affectionate attachment from them, by everything just and liberal we can do for them within the bounds of reason, and by giving them effectual protection against wrongs from our own people."[38] As part of this plan, Jefferson hoped to stem the surge of white migrants into the Louisiana Territory until the bulk of the land east of the Mississippi had been fully settled by European Americans.

As we now know, stemming the flow of European American settlers west of the Mississippi was impossible. Even before the Louisiana Purchase, with a land grant from the Spanish government, Daniel Boone and his followers established a village, now called Boonesville, in central Missouri in 1799. In May 1804, some ten days after Lewis and Clark had begun their journey, the largely American-born residents of Boone's Settlement flocked to greet the explorers and sell them corn and butter.[39]

In the first decades of the nineteenth century, many more Americans followed the paths blazed by Daniel Boone, Lewis and Clark, and others. By 1821, Missouri had enough residents to be admitted to the Union. The new settlers spreading throughout the Mississippi and lower Missouri Valleys in search of good land brought massive ecological change, quickly converting vast prairies and forests to agrarian purposes. And with Jefferson's enthusiastic encouragement, a German immigrant named John Jacob Astor amassed one of the first grand fortunes in the United States.[40]

Astor's fortune came from the highly efficient trapping of beaver to supply the fur trade. He built the American Fur Company and associated firms into an enterprise that managed a continental network stretching from the Pacific Coast through St. Louis to New York, with associated commercial outposts as far away as Asia and Europe. The company was so effective at scouring the countryside for pelts that by the early 1830s the beaver population had declined sharply, marking an early and dramatic transcontinental environmental disruption. In the process, the livelihoods of Indians and white trappers alike were profoundly undermined. Astor shrewdly got out of the fur business in 1834, reinvesting

Figure 2.4: Jefferson had the Peace and Friendship Medal cast for Lewis and Clark to distribute to Indian leaders during their expedition. (Reproduced by permission of the Smithsonian Institution, NNC, Douglas Mudd.)

proceeds into his already substantial holdings in New York City real estate. Astor Place, near Washington Square in Manhattan, still bears the family name.

Although Jefferson was unable to keep white settlers on the east bank of the Mississippi, his policy of moving eastern Indians to the West was eventually largely accomplished. The resulting establishment of Indian reservations in the western United States has made an indelible imprint on American land use patterns. Today, more than 97 percent of the 56 million acres under the jurisdiction of the U.S. Bureau of Indian Affairs is west of the Mississippi.[41]

These reservations have exerted an important impact on the natural world that Jefferson himself studied so intently. Native lands from North Dakota to Arizona to Alaska encompass key portions of ecosystems that, to this day, provide significant habitat for a rich diversity of species, from bison to desert tortoises, caribou, and salmon.[42] In a recent example illustrating the role of Indian land in the American mosaic of land stewardship, the Nez Perce nation, in partnership with the Trust for Public Land, recently regained control of a 10,000-acre tract that had been out of their control for 120 years. The Nez Perce plan to keep the land uninhabited in perpetuity, so their people and others can appreciate its visual and biological riches.[43]

Railroads and the Creation of the World's First National Park

By the mid-1800s, enthusiasm for building a new set of transportation and communications networks—railroads and telegraph networks—had reached a fever pitch in North America. As noted earlier in this chapter, by the 1840s Thomas Hart Benton and his contemporaries sought to establish a variety of railroad routes across the continent. The exact location of the lead railroad route, which would necessarily need to include rights-of-way for telegraph lines required to coordinate train movement, was one of several hotly debated issues that divided northern and southern interests prior to the Civil War.

The debate and the potential routes were well known to Abraham Lincoln, who had entered politics at age twenty-three, unsuccessfully running for a seat in the Illinois House of Representatives. As a candidate for the legislature, Lincoln gave campaign speeches supporting a plan to build a railroad from the Illinois River to Springfield, saying "no other improvement . . . can equal in utility the railroad."[44] During the 1840s and 1850s, in addition to his political activities, Lincoln built a busy law practice, making his mark by brilliantly defending railroad interests in tax and liability cases.

As U.S. president, Lincoln was a firm friend of the railroads and steamship lines that promised to bind the nation together from east to west, even as the Confederacy was trying to tear it apart from north to south. Consistent with this set of interests, Lincoln apparently was sympathetic to legislation sponsored by Republican senator John Conness, of California, to grant federal land to that state to create a reservation at Yosemite. The legislation had the backing of prominent Californians, including Jessie Benton Fremont (Thomas Hart Benton's daughter and the wife of John C. Fremont, 1856 Republican candidate for president and a key California landowner), Frederick Billings (the Fremonts' lawyer), Frederick Law Olmstead (who had found his way to California on assignment to manage the Fremonts' Mariposa estate), and steamship line executive Israel Ward Raymond.[45] The Californians, skillful at promoting Yosemite to easterners with the help of Carleton Watkins's photographs (the *New York Times* described them in the 1860s as "indescribably unique and beautiful"[46]), apparently saw in Yosemite an edenic wonder worthy of national attention, and a potentially powerful stimulus to investment, in-migration, and a tourist trade that would use steamships and eventually railroads to travel to the West.

This group's efforts were complemented by those of the Union Pacific Railroad, which provided free transportation to Yosemite in 1863 for the writer Fitz Hugh Ludlow and the painter Albert Bierstadt.[47] Ludlow's words and Bierstadt's images further burnished Yosemite's reputation as a place at least the equal of the Swiss Alps. Ludlow, for example, in a June 1864 article for the *Atlantic Monthly*, characterized Yosemite Falls as "the grandest mountain-waterfall in the known world."[48] Lincoln signed the Yosemite bill on June 30, 1864—just three days before he approved the Pacific Railroad Act, which granted additional land to the Union Pacific and Central Pacific so they could follow through on their grand plans to build a steam-powered transcontinental connection.

When the Union Pacific and Central Pacific railroads joined at Promontory Point, Utah, in 1869, the West changed forever. Within only two decades, as Frederick Jackson Turner famously noted, the "frontier" was no longer a reality. New residents poured into the region on steel wheels, boosting population density to the point that the entire region was soon considered "settled."[49]

Along with the demographic change came dramatic environmental disruption. Sodbusters converted shortgrass and tallgrass prairies to wheatfields, huge cattle drives and feedlots became common across the Great Plains, and the bison herds upon which Plains Indians depended were almost entirely eliminated, in part by military design. General Philip Sheridan, directing the U.S. Army in conflicts with the Indians, commanded: "Let them kill, skin and sell until the buffalo is exterminated, as it is the only way to bring lasting peace and allow civilization to advance."[50] The railroads were important aids in extirpating the buffalo, serving as moving shooting platforms, jumping-off points for hunting expeditions, and a means of transporting skins back to eastern markets (see Figure 2.5).

In the context of such sweeping changes, a proposal emerged to create the world's first national park in the Yellowstone region. The initiative positioned Yellowstone as a national park rather than a state reservation in part because its proposed boundaries fell within territories that had not yet achieved statehood. Much like the Yosemite proposal, the Yellowstone initiative was strongly backed by railroad interests that exhibited a combination of national interest, enlightened self-interest, and, in the case of several key executives such as Frederick Billings, a deep affection for nature.[51]

Figure 2.5: The bison population was almost entirely wiped out in the 1870s by professional and amateur hunters. This photo shows an 1876 buffalo hide yard in Dodge City, Kansas, that claimed to include some 40,000 buffalo hides. (Courtesy of the National Archives and Records Administration, Still Picture Branch.)

Billings, who had worked with the Fremonts in California, was by the early 1870s back in the East, working closely with Northern Pacific Railroad chief Jay Cooke to finance construction of the line from Minnesota to the Pacific Northwest. Cooke, known to President Grant as one of the principal dealers in Union bonds during the Civil War, wanted to keep the Yellowstone area—which was not far south of the Northern Pacific's land grant corridor—free of "squatters and claimants" who could gain control of the area's natural amenities.[52] Cooke helped pay for writer and entrepreneur Nathaniel Langford and painter Thomas Moran to accompany, respectively, the Hayden and Washburn expeditions to Yellowstone. Inspired by these visits, Langford's oratory and prose, along with Moran's magnificent painting *Grand Canyon of the Yellowstone* (see Figure 2.6), were key tools used in lobbying Congress to create the national park.

On March 1, 1872, President Grant signed legislation creating Yellowstone National Park. This first-of-its-kind reservation has, over the past 130 years, served as a landmark conservation innovation. The creation of the park, in retrospect, rates highly on criteria Alan Altshuler and Robert

Behn consider in their work *Innovation in American Government:* the park's creation was novel, effective, significant, and highly replicable, both in the United States and around the world.[53] Today, the U.S. National Park Service system encompasses more than 83 million acres within 384 areas and serves as a model for conservation professionals in nations as diverse as Suriname and China.

Gifford Pinchot and National Forests

Gifford Pinchot, the first chief of the U.S. Forest Service and a central figure in emerging national conservation policy in the early twentieth century, was, like Thomas Jefferson, significantly influenced by the dreams and aspirations of his father. Pinchot's father, James, came from a family that had made money timbering in Pennsylvania and was himself a prominent and successful wall-coverings merchant in New York.

James Pinchot was deeply concerned about the fate of the nation's forests. He was an early and long-standing supporter of the American Forestry Association, now known as American Forests, founded in 1875. James and Mary Eno Pinchot named their oldest son after Hudson River School painter Sanford Gifford, whose work *Hunter Mountain Twilight* depicted a landscape decimated by years of overaggressive forestry. The elder Pinchots purchased the painting and passed it on to young Gifford, apparently to remind him of his namesake. *Hunter Mountain Twilight* hung above the mantle in Gifford Pinchot's home for most of his adult life. As Pinchot scholar Char Miller points out, the painting's presence underscored that "Gifford [Pinchot's] job was to put the trees back. That's what his family geared him for."[54]

The Pinchot family, in the late nineteenth and early twentieth century, was fully engaged in the burgeoning New York economy. The city, like others in North America and Europe, was growing both upward and outward—upward with the assistance of electric-powered elevators that made early skyscrapers practical; outward first with horse-drawn trolleys, which made street life chaotic (the Brooklyn Dodgers got their name from the borough's denizens who artfully dodged the every-which-way traffic), and then with electric-powered subway trains, which ferried a labor force to and from work each day on a scale unprecedented in American history. The newly available electric web also powered a telephone network that gave stock traders such as Edward Harriman, man-

Figure 2.6: Thomas Moran's painting of the Grand Canyon of the Yellowstone was an important part of the 1872 lobbying effort to create the world's first national park at Yellowstone. The painting was so well received that it was purchased by the federal government and displayed in the Senate lobby. (Reproduced by permission of the Smithsonian American Art Museum, Gift of George D. Pratt.)

agers of mammoth retail stores such as Macy's, and the city's burgeoning population the ability to communicate instantaneously by voice from one end of the metropolis to the other. Contemporary observers marveled at the communications device: in 1879, when Rutherford B. Hayes had the first telephone installed in the White House, he deemed it "one of the greatest events since creation."[55]

The growth of American urban centers, enabled in part by electric power grids, telephone networks, water and sewer districts, and transit systems, was reflected in the national census. By 1900, the share of the U.S. population living in rural settings had declined to about 60 percent—down markedly from the 95 percent recorded during the first national census, in 1790.[56] An exodus from farms to cities was clearly under way, complementing massive immigration to American urban centers from Europe and elsewhere.

In the midst of rapid population growth, New Yorkers witnessed the construction of such grand edifices as the Flatiron Building (see Figure 2.7), built in 1902 just blocks from the Fifth Avenue Hotel, where Gifford

Pinchot grew up (his grandfather Amos Eno owned the famous hotel, which hosted such Gilded Age elite as Mark Twain and Ulysses S. Grant). But the urban story was not entirely a happy one. Just miles away from Fifth Avenue, Jacob Riis recorded his impressions, in vivid prose and photographic detail, in his book *How the Other Half Lives*, showing terribly crowded tenements and abhorrent public health conditions (see Figure 2.8).[57] Riis's work was an inspiration to the social reformers who associated with James Pinchot and his children—men and women who wanted to change the sometimes squalid urban conditions for the better.

Demand for wood around the turn of the century seemed boundless, fueled by farmers and ranchers, who erected ever-longer rows of barbed-wire fences; builders, who expanded the housing stock across the nation; and telephone, electric power, and railroad companies, who demanded steadily expanding volumes of new and replacement poles and ties. Because effective wood preservatives had yet to be widely used, much of the nation's wooden infrastructure had to be replaced periodically. The health of North American forests fell to a low point, with the volume of timber cut greatly exceeding that of timber growth. In addition to fears of a "timber famine," policy makers and commentators inspired by Marsh's *Man and Nature* warned that unless forest decimation was halted, the nation's mightiest and most commercially important rivers, including the Hudson, might dry up and no longer be navigable.[58] For example, nationally recognized botanist Charles Sprague Sargent wrote in 1883:

> Judging by the effect already produced upon the water supply in our rivers by forest destruction in the State, we believe that there is no exaggeration in stating that the future prosperity and even the commercial existence of this State are bound up with these forests and that, unless we are prepared to abandon the natural advantages which made New York what it is, we must stop, and stop at once, any further destruction of the Adirondack forests.[59]

It was in this cultural context that several landmark conservation innovations were born. In Massachusetts, Charles Eliot, son of a president of Harvard and protégé of Frederick Law Olmstead, spearheaded the creation of the first regional land trust, now known as the Trustees of Reservations (TTOR). In 1891 Eliot succeeded in convincing the Massachusetts legislature to charter a nongovernmental organization that could preserve remaining "bits of scenery which possess uncommon beauty and more than the usual refreshing power." Eliot argued that cities such as Boston urgently needed rural

Figure 2.7: The Flatiron Building, designed by Daniel Burnham, was completed in 1902 at the intersection of Fifth Avenue and Broadway in New York. The city's oldest remaining skyscraper, it is 285 feet tall and originally had its own electric generator to power the elevators and supply heat. (Courtesy of the U.S. Library of Congress, Detroit Publishing Company Collection.)

retreats larger than fifty acres and accessible to both rich and poor, where citizens could enjoy "the subtle influence which the skies and seas, clouds and shadows, woods and fields, and all that mingling of natural and human which we call landscape sheds upon human life. . . . It is an influence which has a most peculiar value as antidote to the poisonous struggling and excitement of city life."[60] Eliot's innovation was quickly copied in this country and abroad. Besides setting the precedent for the modern U.S. land trust movement, TTOR helped inspire, for example, the creation of Britain's National Trust—a fact cited by Queen Elizabeth in her congratulatory note to TTOR during centennial celebrations in 1991.

Figure 2.8: Jacob Riis's photographs of New York, such as this one taken in 1890, "Children Sleeping in Mulberry Street," helped raise the awareness of affluent New Yorkers, including Theodore Roosevelt, regarding the plight of poor people in their town. (From Jacob Riis, *How the Other Half Lives*, 1890)

During the same era that TTOR was being created, various political forces, including the American Forestry Association, lobbied intensively for forest protection. In 1891, during Benjamin Harrison's administration, the first 13 million acres of national forest reserves were set aside.[61] President Cleveland added 5 million acres to the reserve system shortly thereafter. In New York State—largely in response to the testimony of commercial interests concerned with the potential loss of navigable rivers, hydropower, and water to fight fires—an 1894 constitutional convention succeeded in designating as "forever wild" state land within the boundaries of the Adirondack forest preserve. As noted by historian Paul Schneider: "The legal structure that has resulted in the largest remnant of forest wilderness east of the Mississippi River was largely created as a cushion for industrial prosperity."[62]

During his tenure as governor of New York and as president of the United States, Theodore Roosevelt, another son of a prominent New York merchant and social reformer, called on Gifford Pinchot to be his principal conservation advisor. Together they engineered a vast expansion of public conservation programs. While the two men were unsuccessful in loosening "forever wild" policies in the Adirondacks, they were nonetheless masterful in building broad public support for the careful management of natural resources on a nationwide basis. Throughout their conservation campaign, they consistently enlisted the support of network entrepreneurs and their associates.

For example, the team convinced *National Geographic* publisher Gilbert H. Grosvenor, Alexander Graham Bell's son-in-law, who almost always shied away from "pieces of a controversial nature," to print a 1905 Roosevelt speech, "Forests Vital to Our Future," as well as a 1908 piece penned by Pinchot (see Box 2.1).[63] And at the first-ever Conference of Governors convened in Washington in 1908, which Pinchot financed partly from his own purse, steel magnate Andrew Carnegie addressed the assembled dignitaries in support of the cause: "In conclusion, Mr. President and Governors of our States, it seems to me our duty is: first, conservation of forests, for no forests, not long navigable rivers; no rivers, not cheap transportation."[64]

The impact of Pinchot and his boss on the conservation of land and habitat for preserving biodiversity was, of course, immense. By the time Roosevelt left office in 1909, the size of the national forest system had tripled to more than 150 million acres. Pinchot, as the first chief of the U.S. Forest Service in the Department of Agriculture under Roosevelt and then briefly under William Howard Taft, launched a government agency legendary for its professionalism and devotion to task. Even after Pinchot and Roosevelt departed from national office, conservation goals that they had advocated were realized: for example, in 1911, the federal government passed the Weeks Act, paving the way for the creation of national forests for the first time in New England and the South.

Gifford Pinchot went on to become a founder of the Progressive Party and a two-term governor of Pennsylvania, where he continued to fight for planned use of natural resources as well as the expansion of road and electric power systems. He passed away in 1946 after a long and distinguished career. The example that Pinchot and Theodore Roosevelt set in build-

Box 2.1

Gifford Pinchot was quite skilled at getting support from a variety of sources and sectors. For example, he convinced the usually reticent Gilbert H. Grosvenor, the publisher of *National Geographic*, to publish Pinchot's policy piece in the magazine. In this piece, Pinchot argued:

> Not only could [Americans] not build transcontinental railroad lines without millions upon millions of railroad ties cut from the forests of the country, but they could not mine the iron and coal, except as the forest gave them means of timbering their mines, transporting the ore, and disposing of the finished product. The whole civilization which they built up was conditioned on iron, coal and timber.

Source: Gifford Pinchot addressing the National Geographic Society, January 31, 1908, reprinted in *National Geographic*, May 1908.

ing a coalition of support for conservation in the midst of dramatic technological, demographic, and environmental change continues to be highly relevant in our own time.

Laurance Rockefeller and the Land and Water Conservation Fund

By the end of World War II, American prosperity had become a driving force in the world economy. As noted previously, President Eisenhower intended to expand the country's economic base and improve national defense by building a coast-to-coast system of interstate highways. Legislation enabling the interstates passed in 1956. Soon afterward, in 1958, Eisenhower appointed a commission to address the nation's growing appetite for outdoor leisure. The commission considered recreational opportunities not only in places that could be reached by public transit systems but also at sites accessible principally by "private motor car."[65]

After reviewing a list of possible chairs, the Eisenhower administration asked Laurance Spelman Rockefeller to head its Outdoor Recreation Resources Review Commission (ORRRC).

Box 2.2

Laurance Rockefeller showed considerable skill in explaining the link between conservation, business sense, and national prosperity. In a 1965 speech addressing the 70th Congress of American Industry, sponsored by the National Association of Manufacturers, he noted:

> The preservation and enhancement of the environment has become an important factor of modern business. Business can take this development in stride in the same way it has, over the years, taken in stride other steps which seemed at the time to be broad social rather than economic obligations. . . . This will turn out in the end to be just plain good business. [The] awakening of an appetite for natural beauty is simply an extension into the outdoors of the desirability of good working and living surroundings.

Source: Robin Winks, *Laurance S. Rockefeller: Catalyst for Conservation* (Washington, D.C.: Island Press, 1997), 57.

Laurance Rockefeller, grandson of John D. Rockefeller and an heir to the family's Standard Oil fortune, had by that time already pursued a remarkable career as a member of the first generation of American venture capitalists, an early investor in the passenger air transportation sector, a nature tourism entrepreneur, and a philanthropist with wide-ranging interests.[66] Laurance's interest in conservation was nourished by his father, John D. Rockefeller Jr., himself a passionate conservationist, and by his own wife, Mary French Rockefeller, granddaughter of Frederick Billings, the railroad entrepreneur present at the creation of Yosemite and Yellowstone. Laurance Rockefeller proved to be an exceptional choice to head the commission, putting enormous energy and focus into the project. He was indefatigable in his work for the ORRRC, patiently explaining to fellow entrepreneurs and to policy makers that to provide for the nation's outdoor recreational needs is to serve the national interest, the well-being of its citizens, and the nation's collective, long-term economic prosperity (see Box 2.2).

The ORRRC and its immediate successors did their work during the administrations of Presidents Eisenhower, Kennedy, and Johnson, when new networks were once again enabling dramatic changes on the

American landscape. The baby boom generation, with federal financing provided under the GI bill and other programs, fueled the burgeoning move to metropolitan areas, but with a twist: the 1970 census was the first to show that a majority of metropolitan residents were living in suburban communities rather than central cities.[67] Interstates and improved state and local roads allowed workers living in dispersed housing sites to commute to offices located in central cities or at more suburban locations across town.[68]

In the midst of what became known as the "go-go" economy, Americans with rising disposable incomes responded favorably to Dinah Shore, who, on new, widely watched national television networks, urged them to "see the U-S-A in your Chev-ro-let." Between 1960 and 1970, the number of registered vehicles in the United States jumped from 74 million to 111 million, and annual vehicle miles traveled shot from 719 million to 1.1 billion. Americans were equally disposed to follow a suggestion made by ESSO (Standard Oil of New Jersey) to "Put a Tiger in Your Tank," boosting motor vehicle fuel consumption from 58 billion to 92 billion gallons per year—an increase of 58 percent in ten years.[69]

Amid these changes, a powerful wave of public concern arose that focused on starkly evident environmental disruptions. Rachel Carson set off an explosion of interest in environmental quality with the 1962 publication of her book *Silent Spring*, which documented the role of the pesticide DDT, as applied to urban, suburban, and rural landscapes, in decimating bird populations. National attention focused on Cleveland's burning Cuyahoga River. Protests against highway expansion and oil-drilling operations erupted from New York to New Orleans to Santa Barbara. Canadian singer Joni Mitchell warned that North Americans had "paved paradise and put up a parking lot."

Following President John F. Kennedy's assassination in 1963—and in the context of mounting racial tensions, national concern over the conflict in Vietnam, and growing public alarm regarding the environment—Lyndon Johnson moved quickly to institute his vision of the Great Society. In part because of the profound interest of his wife, Lady Bird, in conserving natural beauty, Johnson in September 1964 signed both the Wilderness Act (the cause championed by Howard Zahniser in 1955 at the House hearings on the interstate highway system) and the act authorizing the Land and Water Conservation Fund (LWCF), championed by Laurance Rockefeller and the ORRRC. The LWCF was designed to direct funds to federal, state, and local agencies to acquire open space for recreational purposes. Given his family's oil industry background, it is notable that Rockefeller supported early plans to finance the LWCF from the Highway Trust Fund (itself funded from taxes on gasoline for motor vehicles), as well as later plans to tap revenues from federal offshore oil leases.[70]

The LWCF has had a pervasive and significant impact: all 50 states have participated in the program, which, in its first thirty years, has helped to finance more than 37,000 state and local projects.[71] The LWCF, in partnership with towns, counties, states, and nonprofit land trusts and conservation organizations across the nation, has proved to be one of the key sources for funding conservation land acquisitions and easements in the late twentieth century and into the twenty-first century.

The LWCF was only one of several landmark conservation initiatives realized in the 1960s and early 1970s, many of which were not directly related to Rockefeller's work. For example, *Silent Spring* crystallized many of the ideas that led to the creation of the Environmental Protection Agency in 1970.[72] However, as the conservation community seeks to rally support behind national initiatives that respond to the current challenges of open space consumption and habitat loss, the reinvigoration of Rockefeller's innovation is high on the list of priorities. As Laurance Rockefeller and Lady Bird Johnson pointed out in a *Washington Post* editorial in 2000, great challenges lie ahead: "We are not done, nor will we ever be. While protecting our natural resources is often a quiet, steady exercise, sometimes great moments of opportunity arise. We are at such a moment now."[73]

The Conservation Challenges of the Twenty-First Century

During the same decade that interstate highway construction was reaching full throttle and the Land and Water Conservation Fund was launched with money from the Highway Trust Fund, a new generation of communications and transportation networks was being conceived. Like that for the interstate highways, the rationale for building the Internet was in part military. The U.S. Department of Defense Advanced Research Projects Agency (DARPA) funded the first precursor to the Internet in the late 1960s. Scientists at Bolt Baranek & Newman (BBN) in Cambridge, Massachusetts, were awarded a contract to build a system that could connect researchers working at disparate locations. The BBN researchers employed packet switching technology to allow decentralized but interconnected computers to continue operating even if one of the nodes failed—a function critical to military and political leadership in the event of a conventional or nuclear attack on command facilities. The system, initially called ARPAnet, allowed researchers at UCLA, the University of California Santa Barbara, Stanford Research Institute, and the University of Utah to communicate by computer as early as fall 1969.[74] Growing from the humble ARPAnet beginnings, the public Internet mushroomed over subsequent decades and had attracted some 236 million active users around the globe by July 2001.[75]

During the same era that the Internet's precursor was being hatched, Fred Smith, an undergraduate at Yale, wrote a senior paper outlining his idea that companies with significant investments in information infrastructures would be well served by a rapid-delivery network that could provide expensive spare parts for computers on short notice. Although Smith's professor graded the paper a "C," after graduation and service as a Marine in Vietnam, Smith built upon his idea, launching a new business called Federal Express. The company, also known as FedEx, commenced operations in April 1973.[76] In 2001, FedEx had a physical presence in almost every major city of the world and ranked 112 on the Fortune 500 with revenues of $19.6 billion.

In effect, less than forty years since they were conceived, these novel and still-evolving communications and transportation networks reach almost every center of commerce and exchange on earth. Given that the initial intent of network designers was to serve widely dispersed locations, it is not surprising that during the last decade of the twentieth century the networks appear to have enabled increasingly widely dispersed demographic, economic, and land use patterns in North America. As detailed respectively in Chapters 4, 5, and 7 of this book, in the context of a general growth in U.S. population during the 1990s we have witnessed: (1) a positive net in-migration to rural areas, particularly those nonmetropolitan areas with high levels of natural amenities; (2) an accelerated rate of U.S. open space consumption; and (3) new settlement patterns in environmentally sensitive locations such as the Greater Yellowstone ecosystem that appear to be closely associated with habitat disruption and severe pressures on wildlife populations. If, as forecast, twenty-first century communications and transportation networks become even faster, cheaper, and more ubiquitous than they were at the close of the twentieth century, the accompanying social changes may yield increasingly powerful cumulative impacts on open space and biodiversity.

It is consistent with the American experience in earlier eras that, in the context of the present surge in technological, demographic, and land use change, there is also heightened public concern regarding land use and the environment. For example, in the 2000 elections, voters approved 84 percent of local and state proposals to acquire open space, committing nearly $8 billion.[77] The total funds committed for such use from 1998 to 2000 exceeded $20 billion.

Also consistent with historical precedent, land and biodiversity conservation initiatives have enjoyed the support of an impressive array of entrepreneurs and their families associated with the new communications and transportation networks. Ted Turner, the legendary cable entrepreneur, is also reportedly the largest private landowner in the United States. Turner and his family are committed to managing their land to preserve open space and wildlife habitat and strongly support environmental and land trust initiatives in the United States and abroad.

Similarly, Gordon Moore, a founder of Intel, is a remarkably generous supporter of biodiversity initiatives, having dedicated a share of his Intel stock worth several billion dollars to a new foundation that will finance conservation, educational, and scientific initiatives.[78] Also, the foundations set up by William Hewlett and David Packard, now advised by the next generation of Hewletts and Packards, maintain long-standing commitments to conserving landscapes and habitats.

The open question is whether our generation will realize a set of landmark conservation innovations that may someday equal the novelty, significance, effectiveness, and replicability of the first national park, the expansive American national forest system, the world's first regional land trust, or the Land and Water Conservation Fund. Recent news from the field is encouraging.

Conservationists have recently launched a broad spectrum of inventive efforts engaging the public, private, nonprofit, and academic sectors, and more such efforts appear imminent. In Florida, the state and federal governments have recently committed some $7.8 billion to complete the Everglades Restoration Project, the largest such effort in world history, which is designed to restore a reliable flow of water to that peerless "river of grass."[79] In Maine in 2000, the New England Forestry Foundation and the Pingree Forest Partnership completed the largest conservation easement deal in the history of the American land trust movement, safeguarding from unmanaged development some 760,000 acres of privately owned and managed forestry land.[80] And at sites across the continent, from the Pacific Northwest to the Gulf of Maine, conservationists are considering new, highly ambitious efforts, even amid volatile economic and political conditions.

In their attempts to achieve landmark initiatives, conservationists will have at their disposal a remarkable set of scientific, educational, advocacy, and administrative tools enabled by the Internet and Internet-related technologies. Natural scientists across the globe can now share information and ideas in ways that were simply not feasible as recently as 1990. Educators can engage legions of citizen scientists in the purposeful observation of migrating birds and changing water quality, thereby building scientific databases of considerable value as well as powerful communities of interest. And advocates for habitat conservation can rely on electronic communication to mount global advocacy campaigns of surprising immediacy and breadth.

Even with all of these remarkable tools, however, the fundamental tasks facing the conservation community are quite similar to the ones faced by innovators in earlier eras. Public managers and policy makers, the leaders of nongovernmental organizations, private entrepreneurs, and distinguished academics must demonstrate in a bipartisan, cross-sectoral fashion that their ideas make good sense and good policy. Historical precedent indicates that among the initiatives successful in the long term will be those that are broadly understood as being in the national and international interest; enhancing the security and sustainability of highly valued natural resources and amenities; improving the prospects for our future prosperity; and widening our freedom to pursue happiness in concert with the natural world, with which we are fundamentally linked.

NOTES

1. See Patricia Nelson Limerick, *The Legacy of Conquest* (New York: Norton, 1987).
2. Inter-university Consortium for Political and Social Research (ICPSR), "United States Historical Census Data Browser," available at http://fisher.lib.virginia.edu/census. The total U.S. population in 1790 is recorded as 3,893,874, for sixteen states, including the thirteen original colonies and Maine, Vermont,

and Kentucky. Also see estimate for 2050 from Population Estimates Program, Population Division, U.S. Census Bureau, Washington, D.C. "Annual Projections of the Total Resident Population as of July 1 (NP-T1)," available at www.census.gov/population/projections/nation/summary/np-t1.pdf. The middle series forecast issued by the U.S. Census Bureau indicates that there will be 403,687,000 Americans in the year 2050.

3. Cyrus McCormick, known as the "father of modern agriculture," invented the horse-drawn mechanical reaper in 1831. A brief description of the invention is available at http://web.mit.edu/invent/www/inventorsI-Q/mccormick.html. Willis Haviland Carrier, a twenty-five-year-old engineer, invented the modern air conditioner in 1901. A brief description of the invention is available at http://web.mit.edu/invent/www/inventorsA-H/carrier.html.

4. Thomas Jefferson, Thomas Jefferson Papers Series 1, General Corres-pondence, 1651–1827, Library of Congress, available at http://memory.loc.gov. See Thomas Jefferson's letter to David Humphreys, April 11, 1791.

5. Thomas Jefferson, Thomas Jefferson Papers Series 1, General Correpondence, 1651–1827, Library of Congress, available at http://memory.loc.gov. See Jefferson's "Draft of an Act for Establishing the Government for the Territory of Columbia," July 1790.

6. William Meigs, *The Life of Thomas Hart Benton* (New York: Lippincott, 1904), 421–22; Charles van Ravenssway, *St. Louis: An Informal History of the City and Its People, 1764–1865* (St. Louis: Missouri Historical Society, 1991), 393. Note that Theodore Roosevelt, himself an ardent advocate of building new water and irri-gation networks, strongly admired Benton. In the biography of Benton that he wrote in 1887 (*Thomas H. Benton*, [Cambridge, Mass.: Riverside Press of Houghton Mifflin], 35), Roosevelt offered the following: "Benton was deeply imbued with the masterful, overbearing spirit of the West,—a spirit whose mani-festations were not always agreeable, but the possession of which is certainly a most healthy sign of the virile strength of a young community. He thoroughly appre-ciated that he was helping to shape the future of a country, whose wonderful devel-opment is the most important feature in the history of the nineteenth century."

7. Bill Clinton, remarks on economic growth, December 3, 1999, available in October 2000 at www.whitehouse.gov/WH/EOP/OSTP/html/0014_5.html.

8. Thomas Jefferson, instructions to Captain Meriwether Lewis, June 20, 1803, in Thomas Jefferson, *Thomas Jefferson: Writings*, ed. Merrill D. Peterson (New York: Library of America, 1984).

9. Michael J. Mandel, "The Internet Age," *Business Week*, October 4, 1999.

10. See information posted at http://americanhistory.si.edu/csr/locomove/locojh.htm.

11. George Gilder, "Metcalfe's Law and Legacy," *Forbes ASAP*, September 13, 1993, available at www.gildertech.com/public/telecosm_series/metcalf.html.

12. Alfred Chandler Jr., *The Visible Hand* (Cambridge: Belknap Press of Harvard University Press, 1977).

13. See, for example, "Discounting Dynamo: Sam Walton," *Time 100*, c. 2000, available at www.time.com/time/time100/builder/profile/walton.html.

14. Quoted in Tom Standage's remarkable book on the impact of the telegraph on nineteenth-century society, *The Victorian Internet* (New York: Walker, 1998), 104.

15. Quoted in Standage, *The Victorian Internet*, 207. The quote was first reported by CNN Interactive on November 27, 1997.

16. OECD Environment, "Sectoral Studies: Rethinking Paper Consumption," in *Sustainable Consumption*, available at www.oecd.org/env/consumption/scp24c.htm; World Resources Institute, "No End to Paperwork," in *Global Trends*, last modified September 18, 2000, available at www.wri.org/trends/paperwk.html.

17. Frances Cairncross, *The Death of Distance: How the Communications Revolution Will Change Our Lives* (Cambridge: Harvard Business School Press, 1997), 235.

18. Patricia Mokhtarian, "Now That Travel Can Be Virtual, Will Congestion Virtually Disappear?" *Scientific American*, October 1997, available at www.sciam.com/1097issue/1097mokhtarian.html.

19. George Perkins Marsh, *Man and Nature* (1864; reprint, Cambridge: Belknap Press of Harvard University Press, 1965), 87.

20. Richard West Sellars, *Preserving Nature in the National Parks: A History* (New Haven, Conn.: Yale University Press, 1997), 24; William Cronin, *Nature's Metropolis* (New York: Norton, 1991), 214–18.

21. Garrett Hardin, "The Tragedy of the Commons," *Science* 162 (1968): 1243–48.

22. Tom Lewis, *Divided Highways* (New York: Penguin, 1997), 52–53. The idea of a national superhighway system was seriously considered by the Franklin Roosevelt administration in the 1930s; it was not undertaken, however, because Roosevelt's principal advisor on roads, Thomas MacDonald, deemed it "economically impractical."

23. Alvin L. Alm, "NEPA: Past, Present, and Future," *EPA Journal* (January/February 1988), available at www.epa.gov/history/topics/nepa/01.htm. The 1970 National Environmental Policy Act required that "environmental impact statements" (EIS) be prepared for "major federal actions having a significant effect on the environment." The newly created President's Council on Environmental Quality had to come up with regulations regarding the preparation of EIS documents.

24. Tom Lewis, *Divided Highways*, 100.

25. U.S. Senate Committee on Public Works, *Federal-Aid Highway Act of 1954: Hearings before a Subcommittee of the Committee on Public Works* (Washington, D.C.: U.S. Government Printing Office, 1954), 109, 133–34, 159, 276.

26. U.S. Senate Committee on Public Works, *Federal-Aid Highway Act of 1954*, 338.

27. U.S. House of Representatives Committee on Public Works, *National Highway*

Program: Hearings before the Committee on Public Works (Washington, D.C.: U.S. Government Printing Office, 1955), 710, 711.

28. Richard T. T. Forman, "Estimate of the Area Affected Ecologically by the Road System in the United States," *Conservation Biology* 14 (2000): 31–35. See also William Cromie, "Roads Scholar Visits Most Remote Spots," *Harvard University Gazette*, June 14, 2001, 1. Available at www.news.harvard.edu/gazette/2001/06.14/01-roadsscholar.html.

29. F. Kaid Benfield, Matthew D. Raimi, and Donald D. T. Chen, *Once There Were Greenfields: How Urban Sprawl Is Undermining America's Environment, Economy and Social Fabric* (New York: Natural Resources Defense Council, 1999). For example, see chap. 2, "Paving Paradise: Sprawl and the Environment," 29. Regarding the increase in air pollution, see p. 56.

Regarding Bell, see Edwin S. Grosvenor and Morgan Wesson, *Alexander Graham Bell: The Life and Times of the Man Who Invented the Telephone* (New York: Abrams, 1997), 275. Grosvenor (Bell's great-grandson) and Wesson explain: "The few scientists who thought about [air pollution in 1917] were convinced that dirtier air would mean that the climate would cool as the sun's warming rays were blocked. Bell reasoned differently. 'While we would lose some of the sun's heat,' he wrote, 'we would gain some of the earth's heat which is normally radiated into space. . . . I am inclined to think we would have some sort of greenhouse effect. . . . The net result is that the green-house becomes a hot-house.'"

30. The Wilderness Society, "A Summons to Save the Wilderness," *Living Wilderness* 1 (September 1935), cited in Paul Shiver Sutter, "Driven Wild: The Intellectual and Cultural Origins of Wilderness" (Ph.D. thesis, University of Kansas, 1997, available as UMI microform 9827540 from University of Michigan, Ann Arbor).

31. For instance, see Cairncross, *The Death of Distance*. Cairncross's broad-ranging account gave little attention to potential environmental impacts associated with new communications technologies save for several sentences reiterating the message of an AT&T study that showed how telecommuting might reduce automobile fuel consumption (235).

In a second example, Michael Dertouzos, head of the MIT Laboratory for Computer Science, wrote *What Will Be: How the New World of Information Will Change Our Lives* (New York: HarperCollins, 1997), which featured a foreword by Microsoft's Bill Gates. While covering daily life, pleasure, health, learning, business and organizations, and a number of other topics, Dertouzos paid little attention to environmental issues. However, he did point out, accurately, that advanced sensing devices, which he referred to as "electronic noses," could be a great boon in the detection of toxic and explosive materials (73).

32. For example, Benfield, Raimi, and Chen's book *Once There Were Greenfields*,

published in 1999 by the Natural Resources Defense Council and the Surface Transportation Policy Project, provides an excellent overview of the literature associating sprawl with cars, highways, and general surface transportation issues. However, the volume does not consider in any depth how newer networks might be an influence in the mix of factors influencing deconcentration.

Similarly, a section of the Sierra Club's Web site (www.sierraclub.org/sprawl/), made available in summer 2001, offers extensive literature on the impact of highways and automobiles on sprawl; the analysis does not go on to consider in depth how other, newer communications and transportation networks may influence settlement patterns.

33. Both the Internet and express delivery networks such as FedEx are exemplary cases of emerging global communications and transportation networks; associated wired and wireless communications networks, ranging from cellular phone networks to self-organizing sensor networks, as well as increasingly sophisticated physical transportation and logistics networks, including sophisticated supply-chain systems, are included here by reference as "associated networks."

The author has explored most extensively the potential impact of new networks on the landscape and biodiversity of North America.

34. Guy Meriwether Benson, William Irwin, and Heather Moore, "Loyal Company Grant, July 12, 1749," chap. 16 in *Exploring the West from Monticello: A Perspective in Maps from Columbus to Lewis and Clark* (1995), Alderman Library, University of Virginia, available at www.lib.virginia.edu/exhibits/lewis_clark/ch3-16.html.

35. Benson, Irwin, and Moore, "To the Western Ocean: Planning the Lewis and Clark Expedition," chap. 4 in *Exploring the West from Monticello* (1995), Alderman Library, University of Virginia, available at www.lib.virginia.edu/exhibits/lewis_clark/ch4.html.

36. Thomas Jefferson, *Thomas Jefferson: Writings*, ed. Merrill D. Peterson (New York: Library of America, 1984), 786–89. Jefferson wrote to George Washington on April 16, 1784, with elaborate details regarding a possible canal connecting the Potomac and the Ohio, asking Washington if he might consider leading such an effort.

37. Stephen E. Ambrose, *Undaunted Courage* (New York: Simon and Schuster, 1996), 78, 79, 101.

38. Jefferson, *Thomas Jefferson: Writings*, 1118, 1138. Jefferson spelled out his ideas for dealing with Indians in a letter he wrote to William Henry Harrison on February 27, 1803, before the announcement of the Louisiana Purchase, and in a second letter he wrote to John C. Breckenridge from Monticello on August 12, 1803, after the Louisiana Purchase had been announced.

39. Ambrose, *Undaunted Courage*, 144.

40. Thomas Jefferson, letter to John Jacob Astor, May 24, 1812, quoted in Gary

Wills et al., *Thomas Jefferson: Genius of Liberty* (Washington, D.C.: Viking Studio with Library of Congress, 2000).

41. Data on lands under Bureau of Indian Affairs jurisdiction, by state, available in July 2001 at www.doi.gov/bia/realty/state.html.

42. G. Frank Williss, *Do Things Right the First Time: The National Park Service and the Alaska National Interest Lands Conservation Act of 1980* (National Park Service, September 1985), available at www.cr.nps.gov/history/online_books/williss/adhi.htm.

43. Trust for Public Land, "Case Statement for Tribal Lands Initiative," available at www.tpl.org/tier3_cdl.cfm?content_item_id=1183&folder_id=217.

44. Stephen E. Ambrose, *Nothing Like It in the World* (New York: Simon and Schuster, 2000), 27.

45. Robin Winks, *Frederick Billings* (New York: Oxford University Press, 1991), 282.

46. Douglas R. Nickel, *Carleton Watkins: The Art of Perception* (San Francisco: San Francisco Museum of Modern Art, 1999), 10.

47. Joshua Scott Johns, "All Aboard: The Role of the Railroads in Protecting, Promoting, and Selling Yellowstone and Yosemite National Parks" (master's thesis, University of Virginia, 1996), available at http://xroads.virginia.edu/~MA96/RAILROAD/home.html. See also Alfred Runte, *Trains of Discovery: Western Railroads and the National Parks* (Boulder, Colo.: Rinehart, 1990).

48. Fitz Hugh Ludlow, "Seven Weeks in the Great Yo-Semite," *Atlantic Monthly,* June 1864, 739–54.

49. Frederick Jackson Turner, "The Problem of the West," *Atlantic Monthly,* September 1896, available at www.theatlantic.com/issues/95sep/ets/turn.htm.

50. General Philip Sheridan, quoted on the Web site of the Museum of Westward Expansion, St. Louis, Missouri, available at www.nps.gov/jeff/mus-hunters.htm. The Museum of Westward Expansion itself is located on the same site as the landmark Gateway Arch, alongside the Mississippi River in St. Louis. The site is the former home of the museum assembled by William Clark of the Lewis and Clark expedition during the period in the early 1800s when he served as Indian Agent for the U.S. government and dealt extensively with the Indian population west of the Mississippi.

51. Winks, *Frederick Billings,* 286. Winks describes how Billings, who saw no conflict between commerce and conservation, felt that "God would not forgive those who destroyed his greatest creations."

52. Sellars, *Preserving Nature in the National Parks,* 9. Sellars quotes a Cooke letter on the subject written in October 1871.

53. Alan Altshuler and Robert Behn, eds., *Innovation in American Government: Challenges, Opportunities and Dilemmas* (Washington, D.C.: Brookings

Institution Press, 1997). The four criteria are considered, for example, in an essay by Altshuler and Marc Zegans, "Innovation and Public Management," on pp. 74 and 75 of the book. Note that the criteria are those used in selecting award-winning innovations in the annual Innovations in American Government competition sponsored by the Ford Foundation and Harvard's Kennedy School of Government.

54. Kirk Johnson, "Hunter Mountain Paintings Spurred Recovery of Land," *New York Times*, June 7, 2001, available at www.pinchot.org/gt/news/nytimes_huntermtn.htm.

55. Public Broadcasting Service (PBS), *The American Experience*, transcript of the episode "The Telephone," narrated by David McCullough, p. 7, available at www.pbs.org/wgbh/amex/telephone/filmmore/transcript/index.html.

56. James N. Levitt, "Innovating on the Land: Conservation on the Working Landscape in American History" (paper prepared for Private Lands, Public Benefits: A Policy Summit on Working Lands Conservation, National Governors Association, March 16, 2001). See this paper for further details on the conservation innovations of Pinchot, Charles Eliot, and others, including Arbor Day founder Sterling Morton, a Nebraska Democrat who served in President Cleveland's second administration as secretary of agriculture and as president of the American Forestry Association.

57. Jacob Riis, *How the Other Half Lives* (New York: Scribner's, 1890), available at www.yale.edu/amstud/inforev/riis/title.html.

58. Quote from Theodore Roosevelt, "Seventh Annual Message to Congress," December 3, 1907, available at www.pbs.org/weta/thewest/resources/archives/eight/trconserv.htm. For background on Roosevelt's conservation policies, see Gifford Pinchot, *Breaking New Ground* (1947; reprint, Washington, D.C.: Island Press, 1998).

59. Paul Jamieson, *The Adirondack Reader* (Lake George, N.Y.: Adirondack Mountain Club, 1982), 2.

60. Gordon Abbott, *Saving Special Places* (Ipswich, Mass.: Ipswich Press, 1993), 6–10.

61. American Forests, "Timeline of American Forests," available at www.americanforests.org/about_us/history_timeline.php.

62. Paul Schneider, *The Adirondacks: A History of America's First Wilderness* (New York: Holt, 1997), 227.

63. C.D.B. Bryan, *The National Geographic Society: 100 Years of Adventure and Discovery* (New York: Abrams, 1987), 405.

64. *Proceedings of a Conference of Governors: In the White House, May 13–15, 1908* (Washington, D.C.: U.S. Government Printing Office, 1908), 24.

65. Robin Winks, *Laurance S. Rockefeller: Catalyst for Conservation* (Washington, D.C.: Island Press, 1997), 124.

66. Winks, *Laurance S. Rockefeller*, 121.

67. Alan Altshuler et al., eds., National Research Council, *Governance and Opportunity in Metropolitan America* (Washington, D.C.: National Academy Press, 1999), 25.

68. Judy Davis, for Parsons Brinckerhoff Quade & Douglas, "Consequences of the Development of the Interstate Highway System for Transit," Transportation Research Board, National Research Council, Transit Cooperative Research Program, *Research Results Digest* 21 (August 1997): 3, available at http://nationalacademies.org/trb/publications/tcrp/tcrp_rrd_21.pdf. Davis reports that one of the two main positions in transportation literature regarding the effects of the interstate highway system on transit is that "the interstate highway system facilitated the suburbanization of households and jobs, creating origins and destinations that were difficult for conventional transit to serve."

69. Infoplease.com, "Motor Vehicle Fuel Consumption and Travel in the U.S., 1960–1999," available at www.infoplease.com/ipa/A0004727.html.

70. Winks, *Laurance S. Rockefeller*, 136, 138.

71. For additional information on the Land and Water Conservation Fund, see www.ncrc.nps.gov/programs/lwcf.

72. Jack Lewis, "The Birth of EPA," *EPA Journal* (November 1985), available at www.epa.gov/history/topics/epa/15c.htm.

73. Lady Bird Johnson and Laurance Rockefeller, "A New Conservation Century," *Washington Post*, September 14, 2000, available in October 2001 at www.ilparks.org/update-cara9-15-00.htm.

74. R. T. Griffiths, "History of the Internet: Internet for Historians," (Leiden, Netherlands: Universiteit Leiden, 1999), available at www.let.leidenuniv.nl/history/ivh/frame_theorie.html.

75. Michael Pastore, "The World's Online Populations," *Cyberatlas: The Big Picture*, available at http://cyberatlas.internet.com/big_picture/geographics/article/0,,5911_151151,00.html.

76. Todd Lappin, "The Airline of the Internet," *Wired*, December 1996, available at www.wired.com/wired/archive/4.12/ffedex_pr.html. Fred Smith, the founder of Federal Express, wrote a now-famous term paper as a Yale undergraduate that discussed the need for a rapid-delivery service that would make spare parts for high-value computers available on short notice. The idea eventually was a key part of the concept Smith used to launch Federal Express in the early 1970s.

77. Land Trust Alliance, "84% of Referenda Passed: More than $7.4 Billion Committed to Open Space Protection," *Public Policy* (December 1, 2000), available in October 2001 at www.lta.org/publicpolicy/referenda2000.htm.

78. "The 400 Richest People in America, 2001 Edition," *Forbes*, September 2001, 142.

79. Florida Department of Environmental Protection, external affairs communication, "Governor Bush Announces Full-Funding Commitment to Everglades Restoration," January 18, 2000, available in October 2001 at www.dep.state.fl.us/secretary/comm/2000/00-009.htm.

80. See www.neforestry.org/ for additional information on the New England Forestry Foundation and the Pingree Forest Partnership.

CHAPTER 3

The Internet, New Urban Patterns, and Conservation

William J. Mitchell

Networks and Urban Form

From earliest times, cities have been organized around distribution and communication networks—street and road systems, water supply and sewer networks, electrical grids, and so on. In particular contexts, the structures of these networks, and their relationships to patterns of land use, have been fundamental determinants of the overall *efficiency* with which the urban system operates, the *equity* with which resources and opportunities are distributed within it.[1] So it is with the most recent arrival—the Internet.

As new technologies have emerged, and as new types of networks have become possible, these new types of networks have frequently been superimposed upon existing urban fabric. This has led, in general, to a restructuring of activity patterns, changes in efficiency and equity, and eventual adaptation of the spatial distribution of activities in response. The Los Angeles region, for example, developed initially as a network of small towns linked by an extensive rail system. Then the rail system was replaced by an automobile transportation system, infill development took place among the nodes of the former network, and a new spatial pattern emerged. The costs and benefits of this new system were very different, and it radically transformed the distribution of access to resources and opportunities within the region. Inhabitants who were affluent enough to own automobiles gained freedom of mobility and access to the many employment, residential, retail, and recreational resources of the region. Those who were not gained few of the advantages, but could not avoid sharing the downsides of congestion and smog.

When we trace these processes over time, it becomes clear that there is a complex, chicken-and-egg relationship between construction of networks and emerging patterns of land development. Existing patterns of land use generate demands for network service, while construction of new networks (particularly at the urban fringes) typically stimulates development. There is an ongoing, mutually recursive process in which network construction responds to land use and land use responds to network construction.

The latest type of network infrastructure to be superimposed on cities is that of digital telecommunications. The nodes are servers (and, increasingly server farms), routers and switches of various kinds, and end user devices such as personal computers and cell phones. The links consist of fiber-optic channels, repurposed copper (cable modem systems operating over cable television networks, DSL operating over telephone wires, and simple telephone dial-up), terrestrial wireless links, and satellite wireless links. It is all held together by standard telecommunications protocols such as TCP/IP—much as railroad networks are held together by the standardization of gauges. This new infrastructure, like its predecessors, is transforming urban patterns. The effects upon resource use, efficiency, equity, and opportunities for conservation will be profound.

Selective Loosening of Spatial and Temporal Linkages

The most immediate effect of any new network is the selective loosening of spatial and temporal linkages among urban activities, and eventually, spatial and temporal reorganization of those activities in response. The process may be visualized as one of fragmentation and recombination. Consider, for example, the introduction of a piped water supply system into a village that had been organized around a central well. Dwellings need no longer be clustered within water-carrying distance of the well, and can now distribute themselves along the water supply lines. Subject to limits on the total supply, the village can spread out and become larger. Bathing need no longer take place at the central, public location; houses can get private bathrooms. (In other words, the central, public bathing space fragments, decentralizes, and recombines with private, domestic space.) And the well no longer serves as a central focus and site for exchange of news and gossip; new meeting places, new patterns of communication emerge, and the rhythms of community life change.

As the water supply network loosens certain spatial and temporal linkages among activities, latent demands for other types of linkages begin to take over and determine facility locations. Thus dwellings, freed from the constraint of proximity to the well, may be located for view or pleasant microclimate, for proximity to sites of employment, and so on. And private bathrooms become popular as people take advantage of the new locational freedom to establish convenient proximity between bathing and other domestic activities. The result of selective loosening is not simply generalized decentralization, but the emergence of a new logic of clustering.

In the case of digital telecommunications networks, the first-order effect is to loosen spatial and temporal linkages that had been created by requirements for face-to-face exchange of tokens or information. Instead of handing over cash, you can now make payments electronically, at a distance. Instead of meeting with a bank teller, across a counter, during banking hours, you can use an ATM or an electronic home banking system at any time and in any one of many widely scattered locations. Instead of visiting your local bookstore to browse through the shelves, you can browse through the Amazon.com online catalog. Instead of going to a record store to purchase a CD, you can download a file from Napster (or one of its successors) at any location with a network connection.

The resulting effects of fragmentation and recombination are already observable in many contexts.[2] For example, local bank and branch bank buildings were once prominent features of Main Streets everywhere. But many of these have now been demolished, or converted into coffee shops, restaurants, and retail stores. The space for banking transactions has fragmented and has been redistributed to become a large, geographically dispersed collection of ATM and electronic banking sites. The ATMs have combined with supermarkets, airports, student unions, office building lobbies, gambling casinos—anywhere people may need cash. And electronic home banking has combined with domestic space and office space.

Workspace has also begun to fragment and recombine. Laptop computers and cell phones now provide location-independent access to files, coworkers, and other essential resources, so loosening dependence on central office space. Furthermore, most electronically supported information work (unlike factory work) does not generate pollution, noise, and heavy service traffic, and it is not incompatible with residential and recreational activities, so it does not need to be segregated in specially zoned parts of the city. Consequently, such work is increasingly performed not only in the office, but at home, at customer locations, in hotel rooms, in airline lounges and airplane seats, in the backseats of taxis, at café tables, and so on. The long-maintained distinction between workspace and non-workspace is beginning to break down; almost any type of space may now function, at least partially and temporarily, as workspace.

Similar effects may be observed across a wide range of building types. Increasingly many homes are being wired, they are incorporating space for electronically supported activities, and they are being located to take advantage of reduced need for accessibility to central workplaces—for example, in relatively remote recreational and scenic areas. Many workplaces are being reconfigured and relocated in response to more flexible work hours and patterns, and to reduced emphasis on private office spaces and cubicles, which are far less necessary when your files are online, your desktop is on a screen, and your phone is in your pocket. Retailers are being forced to consider the roles of local stores facing intense competition from centralized regional and national distribution facilities that are allied with online ordering. University campuses, libraries, medical centers, and other types of service facilities are beginning to exploit possibilities for remote and asynchronous delivery, and are evolving new forms. As a result, virtual campuses, online libraries, telemedicine centers, and other new forms are emerging.

These processes of fragmentation and recombination are still in their early stages, they require a great deal more empirical study before we can characterize them with confidence, and it is by no means clear precisely where they will lead us. But there seems little doubt that they will, one way or another, be crucial in formation of the cities of the twenty-first century—just as railroad networks and associated land development processes were in the nineteenth century, and freeway networks in the twentieth.

Interactions Among Networks

A second common effect of new networks is to change the roles of existing networks and transform the demands made upon them. This may occur through substitution. When a piped water supply is introduced, for example, the street network no longer has to serve pedestrians carrying water containers. The function of water distribution is performed in a more efficient way by the new network, and this traffic simply disappears. But the effect may also be one of functional complementarity. Thus, when the piped water supply system allows a village to distribute itself over a wider area, the street system must also be extended, and it carries a larger volume of traffic in a new spatial and temporal pattern. Typically, effects of complementarity far outweigh those of substitution.

In the small, rural town where I grew up, outhouses and sewage removal trucks featured prominently. Then a sewer network was built, the outhouses were abandoned and replaced by toilet fixtures in bathrooms—a very welcome recombination on cold, dark nights. This reduced a component of road traffic by getting the smelly trucks off the street. But, at the same time, all those flushing toilets significantly increased demand on the town's water supply network.

In the early, heady days of digital telecommunications, there were frequent suggestions that telecommuting would simply substitute, on a large scale, for travel to the workplace. In other words, it was expected to reduce overall traffic volumes and conserve energy. There were rosy predictions of contented telecommuters, at work in their rural electronic cottages, far from the trouble and noise of the big city—a new version of the recurrent American agrarian dream. The reality has turned out to be far more complex.[3] Telecommuting serves to bring the disabled, and those with pressing childcare and elderly care responsibilities, into the workforce. It creates new opportunities for those who, for whatever reason, choose to or are forced to reside in isolated areas. It allows consultants, and others who work with widely distributed clients, to maintain their working contacts without constant travel. It facilitates mobilization of workers and continual transformation of workplaces, in flexible ways, in response to changing circumstances and demands. But there is little evidence, so far, of much direct substitution. Instead, employers and employees seek optimal mixes, which vary from context to context, of high-quality but expensive face-to-face interaction and lower-quality, but convenient and inexpensive, remote electronic interaction.

In the case of electronic home banking, it is clear that remote, electronic transactions do substitute for trips to the bank. Similarly, downloading a file from Napster substitutes for physical transportation of a CD from vendor to purchaser. This results specifically because, in these instances, the transfers have become completely immaterial; they no longer involve transportation networks at all. When you purchase a book from Amazon.com, though, the overall effect is more complex. You save yourself a trip to and from the bookstore, but you generate a trip by a delivery van to your door. It is the same with online groceries and fast food. It turns out that online retailing generates new patterns of distribution centers, and restructured but not necessarily reduced transportation demand.

In the case of social interaction, there have been optimistic suggestions that online virtual communities may partially or completely substitute for place-based communities. Conversely, more pessimistic commentators have painted grim pictures of alienated geeks, sitting in darkened rooms in their underwear, communicating solely through e-mail and instant messaging. Once again, though, there is little evidence that such direct substitution predominates. E-mail contacts frequently generate desire for travel to face-to-face meetings, and face-to-face meetings motivate e-mail follow-up. Low-cost, convenient, asynchronous e-mail is commonly (and rationally) used to coordinate scarce, high-cost opportunities for meetings among busy people.

Generally, the lesson is that it is very misleading to consider the functions and effects of any one urban network in isolation. And it is little better to consider the interactions among networks solely in terms of potential substitutions—of telecommunication for transportation, of electric-powered rail transportation for gasoline-powered automobile transportation, and so on. It is more useful to think of any building (or, at a larger scale, any neighborhood or town) as a node in multiple networks. Whenever any one of these networks becomes more or less efficient or costly, and whenever a new network is introduced, functions and demands may be redistributed—often in complex ways—across *all* the networks.

Enlarged Functional Footprints

A third effect of networks is to rescale the functional footprints of buildings and settlements. That is, networks can increase the areas over which a given set of buildings and settlements maintain significant functional interactions with other buildings and settlements and with natural systems, and the scales of the resource markets in which that set of buildings and settlements may participate. When a building depends upon rainwater catchment from its roof it establishes only a local functional footprint, for example, but when it taps into a piped water supply network it begins to interact with catchment areas and reservoirs that may be hundreds of miles away. When a city extends a road system into its hinterland, it begins to interact with distant agricultural areas. When power supply grids are inter-connected over wide areas, cities may acquire electricity from multiple, far-flung sources.

Digital telecommunication networks enlarge functional footprints dramatically, since they offer global interconnection, since cost does not increase greatly with distance (in increasingly many contexts, at least), and since they can operate at very high speed. Provided there is sufficient channel capacity, it matters little whether a server is a mile away or on the other side of the world. If you require the services of a telephone call center, for example to make an airline reservation, it matters more that it is in the right time zone than that it is nearby.

When digital telecommunication networks take over functions from networks of more limited speed and range, important scale effects may develop. Markets, for example, were once local affairs limited by the capacity of road networks to bring buyers and sellers to central points. Now, by contrast, online auctions can be conducted over vast geographic areas, and electronically supported financial markets are global, operating around the clock. Professional and intellectual communities of interest were once limited by the range and speed of mail systems, but can now operate effectively at a global scale by means of digital telecommunications.

Intelligent Management of Networks

In summary, we can expect the overlay of new networks upon cities to produce three basic types of spatial effects. First, because spatial and temporal linkages among activities are loosened by network interconnections, there will be fragmentation and recombination of established building types and urban patterns. Second, because the functions of networks are interdependent and overlapping, there will be redistribution of functions among the various distribution and communication networks and restruc-turing of demands upon them. And third, because networks establish functional interactions at a distance, there will be enlargement of the functional footprints of buildings and settlements. These effects, acting in combination, clearly have the potential to transform urban spatial patterns and functional organizations. But it is the crucial addition of a fourth effect—one that is peculiar to telecommunications networks—that provides cause for optimism that the overall result could benefit conservation. Telecommunications networks can be employed to manage and coordinate the operations of *other* types of networks, and so make them far more powerful and efficient. In effect, digital telecommunications networks add nervous systems to architectural and urban organisms, enabling them to operate more intelligently.

The critical, symbiotic interactions of railroad and telegraph systems, of telephone networks, two-way radio systems and taxi services, and of the pedal-powered shortwave radio combined with light aircraft in the Australia's Royal Flying Doctor Service, all gave us a preview of this. Today, these systems have digital descendants. The operations of transportation systems are managed by increasingly sophisticated computer and telecommunication systems that coordinate pickups and deliveries, optimize use of resources, track deliveries, and so on. Electronic sensors keep track of current traffic volumes, and make this information available both to drivers and to automated management systems. In cities like Singapore, electronic sensors located at street intersections and other strategic locations interact with devices in automobiles to implement a sophisticated electronic road pricing system. And GPS-driven, computerized navigation systems are increasingly common in vehicles of all kinds—including automobiles navigating city streets.

There is an important potential interaction between intelligent road networks and intelligent vehicles. Imagine an electronic road pricing system that dynamically varies prices according to current traffic conditions, current pollution levels, sizes of vehicles, and so on. Combine this with automobile navigation systems that can respond intelligently to current pricing—by computing the cheapest route from current location to a specified destination, by computing the fastest route subject to given cost constraints, and so on. This creates a new type of market for road capacity, and new opportunities for intelligent and efficient management of this scarce resource through pricing.

Water supply might be managed in a similar way. Price water dynamically, and equip buildings with intelligent, networked devices—Internet protocol enabled garden irrigation systems, washing machines, and so on. Then program these devices to perform their functions in ways that take account of water costs. The result, if this is structured in the right way, is an effective tool for demand management driven by economic incentives to conserve.

Electrical supply systems offer some particularly attractive opportunities for smart, efficient management through networks and distributed intelligence.[4] Consider an electrical supply system with multiple power sources, including buildings that can sometimes "run the meter backwards" through operation of photovoltaics, fuel cells, and the like. Incorporate dynamic pricing that encourages power conservation at peak demand periods. And deploy smart appliances programmed to take current power prices into account as they perform their functions.

At building scale, the various service networks and their associated appliances—heating and ventilating, electrical, lighting, elevator, water supply and waste removal, and safety systems—all offer opportunities for more intelligent and efficient operation. Buildings of the near future will integrate many more sensors to monitor interior and exterior conditions to which they must respond, many more intelligent devices, and much more sophisticated control strategies.[5]

Opportunities and Strategies for Conservation

What specific opportunities for conservation do these new conditions create? How can we take advantage of digital telecommunications to create more intelligent and efficient, less wasteful buildings, neighborhoods, and urban and regional patterns? Let me conclude by suggesting several attractive possibilities.

The first possibility is to exploit the enormous potential of inexpensive, distributed, decentralized intelligence (the grounding principle of the Internet itself) to create more responsive and efficient road and vehicle systems, electrical supply systems, and building service systems.[6] We need to think of digital telecommunication and computation networks not as stand-alone systems, but as the artificial nervous systems that coordinate and control other networks, and that turn buildings and cities into increasingly intelligent and efficient organisms. We can make them work smarter, not harder.

The second possibility is to repurpose and revitalize valued, older urban fabric by rewiring it. For example, the SoMa area of San Francisco, and the

SoHo area of New York, were decaying warehouse and industrial areas not so long ago, but the introduction of advanced telecommunications infrastructure enabled them to function as high-tech, live/work neighborhoods. Since this type of infrastructure is physically unobtrusive (unlike transportation infrastructure, for example) it does not threaten scale or historic architectural character. Furthermore, the type of economic activity that this encourages does not necessarily require new, large structures, and can often be accommodated gracefully within existing building stock. Indeed, the existing architectural character is often a strong part of the attraction. So digital telecommunications infrastructure can become a new and powerful component in strategies for architectural preservation and historic neighborhood conservation.

The third possibility is to pursue efficient, network-enabled patterns of new development that conserve land and natural resources. One interesting approach is to reinvent the traditional neighborhood (as has been eloquently advocated by the New Urbanists), but in a technologically savvy, forward-looking rather than nostalgic way.[7] The basic units could be wired, live/work dwellings. By sustaining a 24-hour population (rather than a commuter population that leaves the neighborhood half-empty much of the time), these dwellings could effectively support local services—cafés and restaurants, business centers, health clubs and recreational facilities, cleaners, day care and elderly care, and the like. It should be possible to create compact, interesting, multiuse neighborhoods that blend satisfying, pedestrian-scaled local character with electronic connection to a wider world.

The inverse of this third possibility is unmanaged, scattered development enabled by widely available telecommunications, distributed small-scale electric power generation, and micro–water treatment technologies. This may bring some welcome economic vitality to rural backwaters, but it also creates a particular threat to the landscapes of smaller communities that do not have the planning resources to deal with it positively. So there is an urgent need to develop straightforward, easy-to-apply guidelines and prototypical patterns for rural communities facing the new challenges of the Internet age—particularly the fact that being within the reach of traditional infrastructure networks is no longer essential for certain types of new development.

These possibilities do not exhaust the list, but they should suffice to suggest a general principle. The key role of the Internet in conservation is a powerful but indirect one; it provides us with an opportunity to rethink some long-standing, fundamental problems of urban form and function, and to develop some new solutions that may effectively advance the cause of intelligently responsible resource use.

NOTES

1. Stephen Graham and Simon Marvin, *Splintering Urbanism: Network Infrastructures, Technological Mobilities and the Urban Condition* (New York: Routledge, 2001).
2. William J. Mitchell, *City of Bits: Space, Place, and the Infobahn* (Cambridge: MIT Press, 1995). William J. Mitchell, *E-topia: "Urban Life, Jim—But Not As We Know It"* (Cambridge: MIT Press, 1999). Thomas A. Horan, *Digital Places: Building Our City of Bits* (Washington, D.C.: Urban Land Institute, 2000). Joel Kotkin, *The New Geography: How the Digital Revolution Is Reshaping the American Landscape* (New York: Random House, 2000).
3. Jack M. Nilles, *Managing Telework: Strategies for Managing the Virtual Workforce* (New York: John Wiley and Sons, 1998).
4. Steve Silberman, "The Intelligent Web," *Wired* (July 2001), pp. 114–27.
5. National Research Council, *Embedded Everywhere: A Research Agenda for Networked Systems of Embedded Computers* (Washington, D.C.: National Academy Press, 2001).
6. Institute for Civil Infrastructure Systems (ICIS), *Bringing Information Technology to Infrastructure*, workshop materials (New York: New York University, Institute for Civil Infrastructure Systems, July 2001).
7. Andreas Duany, Elizabeth Plater-Zyberg, and Jeff Speck, *Suburban Nation: The Rise of Sprawl and the Decline of the American Dream* (San Francisco: North Point Press, 2000). Peter Calthorpe and William Fulton, *The Regional City: Planning for the End of Sprawl* (Washington, D.C.: Island Press, 2001).

PART II

THE POTENTIAL IMPACT
OF NEW NETWORKS ON LAND
USE AND BIODIVERSITY

In the first part of this book, the author considered how new communications and transportation networks, exemplified by the Internet and global express delivery networks, are likely to be critical enablers in the reshaping of how and where we live, work, shop, play, learn, and interact with nature. If, in Craig McCaw's memorable phrasing, "the real potential of the information age is that people can live where they like" (*Business Week*, September 29, 1998), then, in the early days of the new networks' emergence, careful observers should be able to find initial anecdotal and empirical evidence of shifts in demographic and land use trends.

The chapters in this second part point to such trends in the 1990s, the first decade in which wide-scale commercial deployment of the Internet and related networks occurred. Just as social scientists are still debating the role that interstate highways played in the suburbanization of North American cities in the 1960s and 1970s—more than forty years after construction of the interstate highway system began in earnest—we are only just beginning to gain a thoughtful understanding of the role that the Internet and advanced logistics systems are playing in contemporary demographic and land use changes.

It is evident, however, that important changes are under way. The chapters here examine such changes, as well as associated disruptions to the environment and biodiversity habitat. In Chapter 4, Kenneth Johnson details the rural rebound of the 1990s, in which there was a net positive migration of Americans from metropolitan to nonmetropolitan counties, for only the second time in the century. Ralph Grossi discusses in Chapter 5 the accelerating consumption of open space in the nation during the 1990s, as evidenced by the National Resources Inventory provided by the U.S. Department of Agriculture. In Chapter 6, James Levitt and John Pitkin offer information showing that, at least in one survey, people moving to nonmetropolitan places such as Bend, Oregon, in part to take advantage of their natural amenities, are significantly more likely than others to use the Internet at home as part of their daily lives. Finally, in Chapter 7, Andrew Hansen and Jay Rotella look at how, in areas that offer magnificent natural amenities such as the Greater Yellowstone ecosystem, in-migration and open space consumption pose considerable threats to biodiversity. The authors hope that these investigations will be followed by many more that strive to understand historic changes under way on the North American landscape during the twenty-first century.

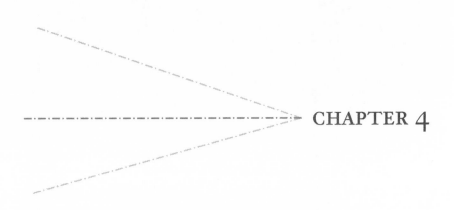

CHAPTER 4

The Rural Rebound of the
1990s and Beyond

Kenneth M. Johnson

During most of the twentieth century, rural America experienced widespread and protracted outmigration.[1] The magnitude of this loss varied from decade to decade, but the pattern was quite consistent: rural areas grew only when an excess of births over deaths offset the number of people who moved away from these communities. Many rural communities suffered population decline because outmigration was so substantial and persistent.

This historical pattern came to an unexpected end during the 1970s, when nonmetropolitan areas experienced a remarkable demographic turnaround. Fueled primarily by net in-migration, population gains in nonmetropolitan areas actually exceeded those in metropolitan areas for the first time in at least 150 years.

This turnaround generated considerable academic interest, but attention waned with the reemergence of widespread outmigration and population decline in rural America during the 1980s. In fact, the downturn of the 1980s led some to conclude that the turnaround of the 1970s was

Table 4.1: Population Change, Net Migration, and Natural Increase by Adjacency and Metropolitan Status, 1970–2000

	No. of cases	Initial population (thousands)	Population change			Net migration			Natural increase		
			Absolute change (thousands)	Percent change	Percent growing	Absolute change (thousands)	Percent change	Percent growing	Absolute change (thousands)	Percent change	Percent growing
1970–1980											
All non-metro	2,276	43,345	5,805	13.4	80.0	3,159	7.3	66.9	2,630	6.1	88.1
Nonadjacent	1,274	19,772	2,477	12.5	72.1	1,223	6.2	60.5	1,249	6.3	85.9
Adjacent	1,002	23,573	3,328	14.1	89.1	1,936	8.2	75.1	1,381	6.3	90.8
Metropolitan	834	158,884	17,146	10.8	88.6	5,948	3.7	3.4	11,198	7.0	97.8
Total	**3,110**	**202,229**	**22,937**	**11.3**	**82.0**	**9,107**	**4.5**	**68.7**	**13,830**	**6.8**	**90.7**
1980–1990											
All non-metro	2,305	49,578	1,320	2.7	45.1	−1370	−2.8	27.3	2,690	5.4	89.6
Nonadjacent	1,298	22,612	134	0.6	36.4	−1175	−5.2	20.7	1,309	5.8	87.0
Adjacent	1,007	26,966	1,186	4.4	56.3	−194	−0.7	35.8	1,382	5.1	92.9
Metropolitan	836	176,965	20,848	11.8	81.0	6575	3.7	57.7	14,271	8.1	97.7
Total	**3,141**	**226,543**	**22,168**	**9.8**	**54.7**	**5206**	**2.3**	**35.4**	**16,962**	**7.5**	**91.8**
1990–2000											
All non-metro	2,303	50,824	5,249	10.3	73.9	3,509	6.9	68.3	1,740	3.4	71.4
Nonadjacent	1,297	22,671	1,848	8.2	64.4	1,079	4.8	59.8	769	3.4	64.1
Adjacent	1,006	28,154	3,400	12.1	86.1	2,430	8.6	79.2	971	3.4	80.9
Metropolitan	837	197,963	27,383	13.8	90.1	12,044	6.1	77.5	15,338	7.7	95.2
Total	**3,140**	**248,787**	**32,631**	**13.1**	**78.2**	**15,553**	**6.3**	**70.7**	**17,078**	**6.9**	**77.8**

Notes: 1993 metropolitan status used for all periods.
Natural increase 1990–1999 from FSCPE. Natural increase projected to 4/2000 from FSCPE.
Sources: Census 2000 PL-94 data 1970-1990 Census and Federal-State Cooperative Population Estimates Program (FSCPE) .

merely a short-term deviation from historical trends. Since 1990, however, most nonmetropolitan areas have again witnessed in-migration. This trend—combined with modest natural increase (more births than deaths)—has produced another large rural population gain.[2] Such a deconcentration of the U.S. population is likely to continue as overall population growth and technological and organizational innovations diminish the importance of distance.

Migration has played a particularly profound role in the remarkable growth of recreational counties over the past three decades. This growth partly reflects the substantial number of older Americans in the migration stream, but recent research suggests that younger age-groups are also well represented. The aging of baby boomers—together with economic gains—is likely to foster continuing migration to recreational areas. The resulting growth in recreational regions and other regions with abundant natural amenities will exert a significant impact on these environmentally sensitive areas.

The Changing Demographics of Rural Areas

Nearly 74 percent of counties classified as nonmetropolitan in 1993 gained population between 1990 and 2000—compared with 45 percent in the 1980s (see Table 4.1).[3] The nonmetropolitan population stood at 56.1 million in April 2000—a gain of nearly 5.3 million since April 1990. This contrasted with an increase of fewer than 1.3 million in these areas during the 1980s.

The nonmetropolitan population grew at a slower pace (10.3 percent) than the metropolitan population (13.8 percent) between 1990 and 2000, but the gap was much narrower than during the 1980s. Gains were prevalent in the Mountain West, the Upper Great Lakes, the Ozarks, parts of the South, and rural areas of the Northeast. Widespread losses occurred in the Great Plains, the western Corn Belt, and the Mississippi Delta (see Figure 4.1).

Migration gains accounted for 67 percent of the nonmetropolitan population increase between April 1990 and April 2000. These areas recorded a net inflow of some 3,509,000 people, compared with a net outflow of 1,370,000 during the 1980s. In addition, the net migration gain between 1990 and 2000 (6.9 percent) was greater than in metropolitan areas (6.1 percent). This is a sharp contrast to the 1980s, when nonmetropolitan

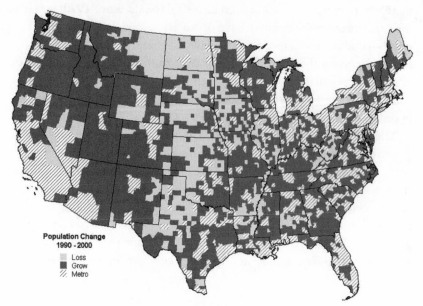

Figure 4.1: Although most nonmetropolitan counties experienced population growth between 1990 and 2000, the Great Plains, the western Corn Belt, and the Mississippi Delta lost population. (Map by Kenneth M. Johnson, Loyola University–Chicago, data from 1990 and 2000 U.S. Census.)

areas saw a net outflow of 2.8 percent, while metropolitan areas recorded a net in-migration of 3.7 percent. Migration gains were widely distributed geographically, though least prevalent in the Great Plains, West Texas, and the Mississippi Delta (see Figure 4.2).

Natural increase accounted for 33 percent of the nonmetropolitan population growth between 1990 and 2000, with births exceeding deaths by 1,740,000. However, natural increase in nonmetropolitan areas during this period was considerably lower than during the 1980s, while in metropolitan areas the rate of natural increase diminished marginally. This pattern of nonmetropolitan growth is similar to that during the turnaround decade of the 1970s, though smaller in magnitude.

The 1970s and 1990s represent a significant departure from historical demographic trends.[4] Through most of the past century, nonmetropolitan population growth was fueled entirely by natural increase, with migration losses diminishing these gains (see Figure 4.3).[5] In the 1990s and 1970s,

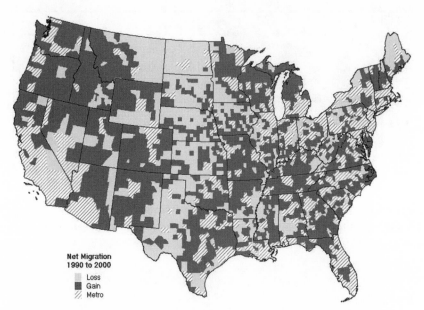

Figure 4.2: Migration gains accounted for most of the nonmetropolitan population growth between 1990 and 2000. (Map by Kenneth M. Johnson, Loyola University–Chicago, data from 1990 and 2000 U.S. Census and Federal-State Cooperative Population Estimates Program.)

in contrast, both net migration gains and natural increase fueled growth. Even the minimal migration losses and modest natural increase of the 1980s are a weak echo of the massive outmigration and substantial natural increases of the 1940s and 1950s.

The Factors Fueling Rural Growth

Part of the growth in nonmetropolitan areas is spillover from nearby metropolitan areas. More than 86 percent of nonmetropolitan counties adjacent to urban areas gained population in the 1990s, while 79 percent saw net in-migration (review data from Table 4.1). In fact, migration gains in adjacent nonmetropolitan counties (8.6 percent) significantly exceeded gains in metropolitan areas (6.1 percent). But even nonadjacent counties recorded net in-migration of 4.8 percent between 1990 and 2000, compared with a net migration loss (–5.2 percent) in

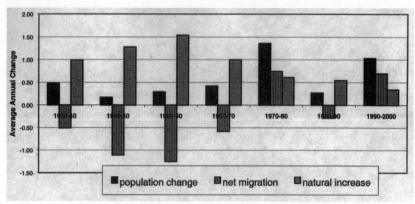

Figure 4.3: As in the 1970s, nonmetropolitan population growth in the 1990s was fueled by both net in-migration and natural increase (more births than deaths). (Data from U.S. Census 1930–2000.)

the 1980s.

Recreation centers and nonmetropolitan destinations for retirement-age migrants—largely in the Sunbelt, coastal regions, parts of the West, and the Upper Great Lakes—were among the fastest-growing counties during the 1990s (and also prominent growth nodes during the 1970s and 1980s).[6] In fact, all 190 nonmetropolitan retirement-destination counties gained population, and 99 percent saw net in-migration (see Table 4.2). Population and migration gains were also common in the 285 recreational counties.[7] Counties where much of the land is federally owned—which are concentrated in the West—similarly saw widespread growth in the 1990s, fueled by people attracted to their scenic and recreational amenities.

Counties where a large proportion of the workforce commutes to jobs in other counties, and those with economies dominated by service-sector jobs, also grew rapidly. Nonmetropolitan population gains were widespread—though more modest—in manufacturing- and government-dependent counties, with gains in the latter two types of counties more evenly balanced between natural increase and net migration.

Counties dependent on farming and mining were the least likely to gain population during the 1990s: only 49 percent of farming-dependent counties gained population. Natural decrease was also common in farming-dependent counties. Population gains were only slightly more

Table 4.2: Population Change, Net Migration, and Natural Increase in Nonmetropolitan Counties by Selected Variables, 1990–2000

County Type	No.	Population change		Net migration		Natural increase	
		Percent change	Percent growing	Percent change	Percent growing	Percent change	Percent growing
Retirement	190	28.4	100	25.9	99	2.5	59
Federal lands	269	22.3	90	16.4	83	5.9	83
Recreational	285	19.3	92	15.9	90	3.5	71
Manufacturing	506	9.5	87	6.1	76	3.4	86
Commuting	381	15.2	92	12.0	88	3.2	80
Government	243	11.5	85	5.2	74	6.3	77
Service	323	14.6	81	11.7	78	2.9	71
Nonspecialized	484	10.9	84	8.4	80	2.5	73
Transfer	381	8.5	75	6.5	69	1.9	60
Poverty	535	9.1	77	4.4	63	4.7	80
Mining	146	2.3	54	-1.5	44	3.8	81
Farming	556	6.6	49	3.9	49	2.7	53
Total non-metropolitan	2,303	10.3	74	6.9	68	3.4	71

Notes: 1993 metropolitan definition; 14 previously metro counties excluded from type analysis.

Percent change is aggregate change for all cases in category.

Recreational counties defined by Beale and Johnson (1998).

All other types defined as in Cook and Mizer (1994).

Counties are classified into one economic type (farming, mining, manufacturing, government, service and nonspecialized).

Other types are not mutually exclusive.

widespread in mining counties, with these areas experiencing net out-migration. However, even among these counties, population declines and migration losses moderated in the 1990s. Counties with persistent poverty also saw low growth rates during the 1990s, with natural increase—as in the case of mining and farming counties—accounting for most population gains.

Economic trends contributed to renewed rural growth. The recessions of the early 1980s had a more severe impact and lasted longer in nonmetropolitan areas, while the farm crisis of that decade also hurt many agricultural counties badly, resulting in widespread outmigration. However, rural employment began to recover in the late 1980s and continued to do so during the 1990s. In addition, the economic recession of 1990–92 had a greater impact on urban areas, undercutting the economic attraction of cities, particularly for rural young people.

Concern about such urban problems as crime, pollution, and poor-quality schools may also have attracted urban residents to rural areas and discouraged rural residents from moving to cities. Recent survey data suggest that many residents of the nation's largest cities would rather live in smaller places, whereas a substantial majority of rural residents are happy where they are.[8] Although these findings are consistent with earlier surveys, the diminished friction of distance—together with a healthy economy—has probably allowed more households to act upon their preferences.

The growing integration of rural communities into the national and international economy also contributes to nonmetropolitan growth. Recent improvements in transportation and communications infrastructure facilitate interaction between urban and rural areas, thereby diminishing the effect of distance. Location decisions for both firms and families now encompass a wider geographic sphere, and the result is that many now enjoy the social and environmental advantages associated with rural living while retaining easy access to metropolitan areas. For example, midwestern parts suppliers tend to cluster along interstate highways within a few hours' drive of auto assembly plants, where land is cheaper, wages are lower, and unions less common.[9] Such rural manufacturing plants contribute to widespread nonmetropolitan population growth.

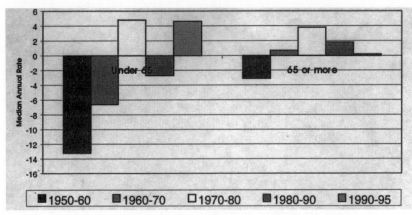

Figure 4.4: Data suggest that working-age people comprise a majority of the migrants into nonmetropolitan areas, which has significant implications for future growth and development in these areas. (Data from U.S. Census 1950–1990 and U.S. Census, Federal-State Cooperative Population Estimates Program, 1995.)

The Influence of Age on Rural Growth

Net migration to nonmetropolitan areas has always been age selective.[10] Young adults historically have left nonmetropolitan areas in substantial numbers, while the net flow of individuals at other ages has been less consistent. In some periods—notably the 1970s—rural areas saw a net influx of individuals at all ages except young adult, while in other periods such areas experienced a net migration loss of people of virtually every age.

Johnson and Fuguitt, who have examined these age-specific migration patterns in some detail, report one puzzling finding with significant implications.[11] Between 1990 and 1995, the net influx of people under age sixty-five to nonmetropolitan areas has been much higher than historical trends (see Figure 4.4), while the influx of adults sixty-five and older to such areas has been considerably less than expected. According to Fuguitt *et al.*, this pattern is consistent across regional groupings and county socioeconomic types.[12]

To be sure, retirement-destination counties have received a significant influx of older migrants, but many other areas have not. This is surprising given the historical propensity of older Americans to move to

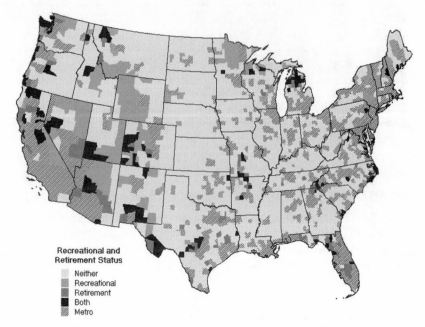

Figure 4.5: The population growth in recreational counties is likely to remain substantial. (From Beale and Johnson [1998] and Cook and Mizer [1994].)

and remain in nonmetropolitan areas. If further data substantiate this startling finding, it suggests that working-age people account for a considerable majority of the migration gain fueling the rural rebound. If the migration profile of nonmetropolitan areas is shifting, it holds significant implications for future growth and development in such areas, as the earning power, expertise, and expectations of working-age individuals and families differ from those of seniors.

A Focus on Recreational Counties

Recent population surveys show that annual migration gains in nonmetropolitan areas peaked in 1994–95 and have diminished each year since.[13] This slowdown in the late 1990s closely resembles the migration patterns during the 1970s. However, although little is known about why net migration to nonmetropolitan areas has slowed, the growth of recreational counties in particular promises to remain significant.

The nation's 285 counties with significant concentrations of recreational activity represent 16 percent of the year 2000 nonmetropolitan population and 12 percent of all counties. Recreational counties are widely distributed, but several regions contain significant concentrations (see Figure 4.5). In the Upper Great Lakes and the Northeast, many such counties are in traditionally summer-oriented lake regions, though most such areas now also encompass winter sports. In the West, many recreational counties contain popular national parks or the numerous ski resorts developed over the past generation. Recreational counties also encompass the southern Appalachians as well as the Ozarks and other nonmetropolitan regions of the Atlantic and Pacific coasts. The Great Plains, the Corn Belt, and the lower Mid-South (Louisiana, Mississippi, and Alabama) have a dearth of recreational counties.

Recreational counties enjoyed a substantial population gain of 19.3 percent between 1990 and 2000—compared with an overall growth of 10.3 percent for all nonmetropolitan counties and 13.1 percent for the nation as a whole (see Table 4.3). Migration fueled most of this growth, with the percentage gain from net migration more than twice that in nonmetropolitan areas and the nation as a whole. Newcomers are attracted to the scenic and leisure-time amenities of recreational areas, while fewer existing residents leave, because of the economic opportunities that growth fosters. Natural increase has contributed relatively little to recent population gains in these areas.

Growth rates in the 101 counties designated as both recreational and retirement destinations are the highest of any identified group. Both types of counties tend to offer natural amenities, temperate climate, and scenic advantages that attract vacationers and seasonal residents as well as retirees.[14]

Although many recreational counties are some distance from major urban centers, 105 are adjacent to metropolitan areas that contain nearly 100 million residents, and those counties are even more likely to grow than are other recreational counties. As the nation's metropolitan population continues to deconcentrate, areas near urban centers that contain scenic and recreational amenities are likely to be particularly appealing to people seeking both permanent residences and second homes.

Growth in recreational counties is not a short-term phenomenon: data spanning three turbulent decades show their sustained appeal.

Table 4.3: Aggregate Population Change, Net Migration, and Natural Increase by Recreational and Metropolitan Status, 1970–2000

	No. of cases	Initial population (thousands)	Population change			Net migration			Natural increase		
			Absolute change (thousands)	Percent change	Percent growing	Absolute change (thousands)	Percent change	Percent growing	Absolute change (thousands)	Percent change	Percent growing
1970–1980:											
Recreation	284	5,337	1,443	27.0	95.1	1,102	20.6	90.1	341	6.4	88.7
All nonmetro	2307	43,661	5,919	13.6	79.7	3,219	7.4	66.9	2,685	6.1	88.1
Metro	836	159,641	17,326	10.9	88.6	6,002	3.8	73.3	11,324	7.1	97.8
Total	**3142**	**203,302**	**23,245**	**11.4**	**82.1**	**9,221**	**4.5**	**68.6**	**14,009**	**6.9**	**90.7**
1980–1990:											
Recreation	285	6,780	934	13.8	79.6	515	7.6	61.8	420	6.2	88.8
All nonmetro	2305	49,577	1,321	2.7	45.1	-1,370	-2.8	27.5	2,690	5.4	89.6
Metro	836	176,965	20,847	11.8	81.0	6,576	3.7	57.7	14,271	8.1	97.7
Total	**3141**	**226,542**	**22,168**	**9.8**	**54.7**	**5,206**	**2.3**	**35.5**	**16,961**	**7.5**	**91.8**
1990–2000:											
Recreation	285	7,714	1,491	19.3	89.8	1,223	15.9	83.1	268	3.5	72.3
All nonmetro	2303	50,824	5,249	10.3	73.9	3,509	6.9	68.3	1,740	3.4	71.4
Metro	837	197,963	27,383	13.8	90.1	12,044	6.1	77.5	15,338	7.7	95.2
Total	**3140**	**248,791**	**32,632**	**13.1**	**78.2**	**15,553**	**6.3**	**70.7**	**17,078**	**6.9**	**77.8**

Notes: 1993 metropolitan status used.
Data for net migration and natural increase for 1990–2000 are estimates.

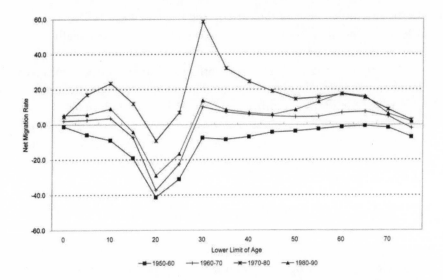

Figure 4.6: Recreational areas are experiencing population growth due to the migration of a broad age range of individuals and families. (From Johnson and Fuguitt [2000].)

During the rural population turnaround of the 1970s, gains in recreational counties were particularly rapid: 95 percent of these counties gained population, and 90 percent recorded net in-migration. During the 1980s, when most nonmetropolitan areas grew little, if at all, nearly 80 percent of recreational counties continued to gain population and more than 60 percent saw net in-migration. The population gain in recreational counties during that decade (13.8 percent) was more than five times that of all nonmetropolitan counties.

David McGranahan has documented the strong relationship between the capacity to attract and retain population and counties with high levels of natural amenities, such as a moderate climate, varied topography, and bodies of water.[15] The average 1970–96 population change in nonmetropolitan counties high on his amenity index was 120 percent, compared with 1 percent for counties low on his index.

The overall shape of the age-specific migration profile for recreational counties is consistent with that of nonmetropolitan America generally, with migration losses greatest and gains lowest among young adults. Migration gains among young adults were lowest during the

1950s and peaked during the turnaround decade of the 1970s, with the 1980s intermediate between these extremes—a pattern virtually identical to that among older adults.

Recent research suggests that recreational areas entice significant numbers of older migrants, but migration gains in such counties are also significant for adults age thirty and over and their children (see Figure 4.6). The fact that recreational counties are attracting a relatively broad cross section of the population has significant implications for planners and policy makers because demands on the environment and expectations for government services are age based.

The Impact of Growth in Recreational Regions

Recent studies suggest that rapid population change exacerbates fiscal problems in nonmetropolitan counties.[16] Such problems are likely to be especially severe in recreational counties, which face greater costs for infrastructure and personnel than do nonrecreational counties of the same size, because they must cope with the additional demands of a transient population.[17] Recent research also suggests that recreational visits to an area may represent the first link in a chain of activities that eventually leads people to migrate to the area.[18] Thus the widespread appeal of recreational and scenic areas for second homes is likely to foster even more in-migration over the next several decades as the large baby boom cohort disengages from the labor force.

Unfortunately, the scenic amenities that attract visitors and migrants often encompass fragile ecosystems. Lakes, coastal regions, and forests are likely to experience higher levels of environmental stress as the volume of human activity increases.[19] Such stresses are likely to be most pronounced at the human–nature interface. For example, many amenity areas were originally settled as small, seasonal second-home developments. However, once established, such areas have tended to grow quickly.[20] Recently, as the retirement-age population has grown and as technological and transportation innovations have made these areas more accessible, people have begun renovating and expanding modest second homes or tearing them down and replacing them with much larger year-round units. This is likely to increase the stress on water quality because septic systems designed for weekend use may not support the greater effluent produced by full-time or nearly full-time residents making frequent use of dishwashers, washing machines, and other water-hungry appliances.

Intensive development in amenity-rich areas affects riparian areas as well. As lawns and extensive landscaping replace native plants at the water's edge, runoff into lakes and streams rises because native species no longer provide filtering. Development also replaces wetlands, bogs, and fallen trees along the water's edge with docks and breakwaters. Growing use of boats and personal watercraft accelerates erosion and introduces foreign species into lakes and streams.

Population growth in amenity-rich areas also fragments forests. Such fragmentation makes it increasingly difficult to manage forests and vastly complicates the task of suppressing forest fires, because staff and equipment must be deployed to protect housing and lives along the forest edge. The use of controlled burns to manage forests is also much less feasible.

The Impact of Sprawl on Farms

Rural America was originally settled by people who subsisted—and sometimes flourished—by extracting food, fiber, and minerals from the environment. Although agriculture no longer dominates rural America as it once did, it remains an important element of the local economy and psyche in broad swaths of the country. The deconcentration of the U.S. population is exerting a significant impact on both the natural and the social environment in farming areas.

In agricultural regions, development can consume thousands of acres of prime farmland at an alarming rate. Development also fragments remaining agricultural land, making operations difficult for farmers. Rising traffic density on traditional farm-to-market roads makes moving heavy equipment from field to field difficult, and farmers must travel farther to reach dealers who service and sell parts for complex equipment and to deliver crops to wholesalers at grain elevators, dairies, and livestock yards. Development also pushes up land prices, so young farmers face enormous financial burdens getting started and older farmers have more difficulty passing their land on to the next generation. Rapid development also quickly makes farmers a minority despite their centrality to an area's character and appeal.

Rural population growth therefore jeopardizes the social as well as the natural environment in nonmetropolitan areas. Many of these areas need and welcome additional population, but rural residents are concerned about how the influx of people and businesses will influence an area. Will new people and firms alter the style and pace of life that make such regions appealing? People who move to rural areas often want to escape problems associated with urban living, but they also often expect the services typically available in urban areas.

Newcomers bring more than expectations for better services; they also bring new talents, ideas, and ways of doing things. This influx of new people and ideas is both exciting and threatening. It is exciting because it represents an infusion of human capital into communities that have lost much through the years. Newcomers bring expertise and skills that may reinvigorate existing institutions and create new ones. But such an influx also is threatening because it challenges long-established social networks and procedures. Integrating new arrivals without destroying the sense of community that makes smaller places appealing is no less daunting a task than protecting the natural environment.

The deconcentration of the U.S. population underscores the fact that urban sprawl and smart growth are particularly significant for nonmetropolitan areas. Yet much discussion of smart growth might be better characterized as abatement of suburban sprawl. Such discussion is dominated by city and suburban interests maneuvering to protect turf and access to resources. Yet, for rural communities that have coped with declining populations and resources for years, managing an influx of people and businesses represents a serious challenge that many are not fully prepared to face.

These special needs must be considered in developing smart growth policies. To manage growth in rural areas, local governments need the staff, training, legal framework, and resources to produce and enforce plans that protect the environment, public access, open space, and farmland. Any serious discussion of smart growth must recognize nonmetropolitan areas as viable partners in the policy-making process.

The Future of Rural Expansion

The slowdown in migration to nonmetropolitan areas during the late 1990s underscores the complexity of forces shaping the demographic future of nonmetropolitan areas. Deconcentration is likely to continue as the U.S. population grows and both technological and organizational innovations continue to diminish the importance of distance, but such deconcentration is likely to be selective and sensitive to temporal and cyclical factors, such as the economy.[21]

The rapid growth of the older population after 2010, the process by which the baby boom generation disengages from the labor force, and the residential decisions baby boomers make will exert a profound impact on the rate and pattern of population deconcentration. The nation has never had such a large number of affluent, well-educated, and healthy older citizens. Although most older Americans do not migrate, those who do have enormous flexibility in where they settle. Should baby boomers be attracted to the same areas that have appealed to earlier cohorts of retirees, population growth in recreational and high-amenity areas could be substantial.

Future nonmetropolitan demographic change will likely depend even more on migration because recent rural fertility patterns, together with shifts in the age structure of rural populations, have diminished the contribution natural increase can make to rural growth. The rising dependence of such areas on migration, coupled with their greater integration into the national and international economic, communications, and transportation systems, will make rural America ever more sensitive to outside forces. That the pattern of future growth in the nation's vast nonmetropolitan regions should depend on such a broad array of forces underscores how information technology has altered the world and how closely the future of the natural environment is linked to human settlement patterns.

NOTES

The research for this chapter was funded in part by a Research Joint Venture Agreement with the Urban Populations Study Group of the U.S. Forest Service and by a Cooperative Agreement with the Economic Research Service of the U.S. Department of Agriculture.

1. The terms *rural* and *nonmetropolitan* are used interchangeably here, consistent with literature on the subject. Definitions by the U.S. Census Bureau of rural and nonmetropolitan areas differ, but these distinctions are not relevant

to this chapter.

2. To avoid confusion regarding the two recent periods of rural demographic growth, the term *turnaround* is used to refer to the rural gains of the 1970s. The term *rebound* refers to the nonmetropolitan gains since 1990.

3. The 2000 population of each county comes from the PL-94 redistricting datasets released by the U.S. Census Bureau in March 2000. Natural increase between 1990 and 2000 was estimated using data from the Federal-State Cooperative Population Estimates Program (FSCPE). This FSCPE dataset reports births and deaths on an annual basis as of July 1 and covers the period from April 1, 1990, through July 1, 1999. To estimate natural increase for the nine-month period from July 1, 1999, to April 1, 2000, natural increase for the period from July 1, 1997, to July 1, 1999, was multiplied by .375 and added to natural increase from April 1, 1990, to July 1, 1999. Net migration from 1990 to 2000 was calculated as a residual of population change minus natural increase.

Counties are the unit of analysis because they have historically stable boundaries and are a basic unit for reporting fertility, mortality, and census data. Counties are delineated as metropolitan or nonmetropolitan using criteria developed by the U.S. Office of Management and Budget. The United States contains 3,142 counties or county equivalents. As of 1993, 837 counties were defined as metropolitan and 2,305 were defined as nonmetropolitan. Based on empirical and contextual analysis, 285 counties were designated as recreational. Data used to identify recreational counties came from the Census Bureau's Census of Housing and Economic Census and from the Bureau of Economic Analysis. A detailed discussion of the creation of the recreational variable is presented in Calvin L. Beale and Kenneth M. Johnson, "The Identification of Recreational Counties in Nonmetropolitan Areas of the USA," *Population Research and Policy Review* 17 (1998): 37–53.

4. Kenneth M. Johnson and Calvin L. Beale, "The Recent Revival of Widespread Population Growth in Nonmetropolitan Areas of the United States," *Rural Sociology* 59 (1994): 655–67.

5. Kenneth M. Johnson, "Recent Population Redistribution Trends in Nonmetropolitan America," *Rural Sociology* 54 (1989): 301–26.

6. Peggy J. Cook and Karen L. Mizer, *The Revised ERS County Typology: An Overview* (RDRR-89) (Washington, D.C.: Economic Research Service, U.S. Department of Agriculture, 1994).

7. Beale and Johnson, "The Identification of Recreational Counties in Nonmetropolitan Areas of the USA."

8. David L. Brown et al., "Continuities in Size of Place Preference in the

United States, 1972–1992," *Rural Sociology* 62 (1997): 408–28.

9. Thomas H. Klier and Kenneth M. Johnson, "Effect of Auto Plant Openings on Net Migration in the Auto Corridor, 1980–1997," *Economic Perspectives* 24:4 (2000): 14–29.

10. Glenn V. Fuguitt and Tim B. Heaton, "The Impact of Migration on the Nonmetropolitan Population Age Structure, 1960–1990," *Population Research and Policy Review* 14 (1995): 215–32.

11. Kenneth M. Johnson and Glenn V. Fuguitt, "Continuity and Change in Rural Migration Patterns, 1950–1995," *Rural Sociology* 65:1 (2000): 27–49.

12. Glenn V. Fuguitt et al., "Elderly Population Change in Nonmetropolitan Areas: From the Turnaround to the Rebound" (paper presented at the annual meeting of the Western Regional Science Association, Monterey, Calif., 1998).

13. Jason Schachter, "Geographic Mobility: 1999–2000," *Current Population Reports* (P20-538) (Washington, D.C.: U.S. Census Bureau, U.S. Department of Commerce, 2001).

14. Cook and Mizer, *The Revised ERS County Typology.*

15. David A. McGranahan, "Natural Amenities Drive Rural Population Change," *Agricultural Economic Report*, no. 781 (Washington, D.C.: Economic Research Service, U.S. Department of Agriculture, 1999). There is considerable overlap between the recreational counties identified here and those with high levels of natural amenities, according to McGranahan. He reports that 63 percent of the recreational counties rank high on the amenity index. Given that natural amenities are a significant factor in an area's recreational appeal, this finding is consistent with expectations.

16. Kenneth M. Johnson et al., "Local Government Fiscal Burden in Nonmetropolitan America," *Rural Sociology* 60 (1995): 381–98.

17. Beale and Johnson, "The Identification of Recreational Counties in Nonmetropolitan Areas of the USA."

18. Sue I. Stewart and Daniel J. Stynes, "Toward a Dynamic Model of Complex Tourism Choices: The Seasonal Home Location Decision," *Journal of Travel and Tourism Marketing* 3:3 (1994): 69–88.

19. David N. Wear and P. Bolstad, "Land-Use and Change in Southern Appalachian Landscapes: Spatial Analysis and Forecast Evaluation," *Ecosystems* 1 (1998): 575–94; David N. Wear, M. G. Turner, and R. J. Naiman, "Land Cover along an Urban-Rural Gradient: Implications for Water Quality," *Ecological Applications* 8:3 (1998): 619–30.

20. Volker C. Radeloff et al., "Human Demographic Trends and Landscape Level Forest Management in the Northwest Wisconsin Pine Barrens," *Forest Science*, 47:2 (2001): 229–41.

21. William H. Frey and Kenneth M. Johnson, "Concentrated Immigration, Restructuring, and the Selective Deconcentration of the U.S. Population," in *Migration into Rural Areas: Theories and Issues*, eds. Paul J. Boyle and Keith H. Halfacree (London: Wiley, 1998).

CHAPTER 5

Farmland in the Age of the Internet

"LET THEM EAT ELECTRONS"?

Ralph E. Grossi

Without doubt, the Internet is one of the great technological developments of our time, revolutionizing the way people communicate, learn, spend money, and pursue their livelihoods. As with all new developments, however, this one has benefits and drawbacks—and one of the unintended impacts of this new technology might be the destruction of vast areas of prime farmland, particularly on the outskirts of America's metropolitan areas.

No matter how many cables are stretched from town to town, and no matter how many communications satellites are sent into orbit, one factor never changes: the amount of land on Planet Earth. As we charge into the twenty-first century and the third millennium, the competition for that land is intensifying at an unprecedented rate. The symptoms of this competition are readily visible in the almost constant conflicts between private landowners and public interests over a wide range of values attached to the working landscape.

Used carefully and thoughtfully, the Internet and other modern technology could become a tool to mitigate the pressure of population growth on precious farmland. But used carelessly and thoughtlessly, these innovations could unleash even further the centrifugal forces that are spreading the residents of every metropolitan area farther across the working landscape.

What's Happening on the Land

Before looking at the positive and negative impacts of computers, e-mail, instant messaging, and Internet technology, let us first understand what is happening to America's farmland.

The management of privately owned farms, ranches, and forests has a huge impact on the environment and on the aesthetic values of open land that so many appreciate as important contributions to our quality of life. It isn't just food that is of concern. Although, historically, the farms that make up most of the working landscape have been valued first for the production of food and fiber, more recently the value of their environmental benefits has, in many regions, become as or more important. This recognition is due to many factors, including the growing base of knowledge regarding land use practices and biological diversity collected and made available by local, state, and national governments as well as by universities and nongovernmental organizations.

Nowhere is the competition for land as intense and as publicly contentious as in the geographic territory know as the "urban edge," in places that surround metropolitan centers such as San Francisco, Boston, and Washington, D.C. (see Boxes 5.1 and 5.2, later in this chapter, which address Marin County and Monterey County, California, two jurisdictions on San Francisco's urban edge). Bitter disputes are being played out in second- and third-ring suburbs all across the country as one community after another faces the significant loss of the high-quality farmland that is so much a part of its socioeconomic structure. The sprawling development that results is changing the face of America.

The most recent National Resources Inventory, the 1997 NRI, tells the story. The NRI is a comprehensive, statistically valid survey of the state of nonfederal land taken every five years by the Natural Resources Conservation Service (NRCS) of the U.S. Department of Agriculture (USDA). NRCS field staff collect data from 300,000 "primary sample units" and about 800,000 "sample points" on nonfederal lands throughout the forty-eight coterminous United States, Hawaii, Puerto Rico, and the

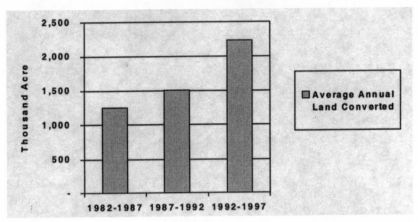

Figure 5.1: The average annual rate of land converted to developed land increased by 49 percent in the most recent five-year period measured (1992–97) versus the previous five-year period (1987–92). (Data from U.S. Department of Agriculture, Natural Resources Conservation Service, "1997 National Resources Inventory" [revised December 2000], available at www.nrcs.usda.gov/technical/NRI/1997/summary_report/body.html.)

U.S. Virgin Islands.[1] This information is supplemented with photo interpretation and remote sensing information as well as by selected visits to sample sites to characterize land use patterns. Researchers at Iowa State University's statistical laboratory use this information to develop a national portrait of soil erosion, wildlife habitat, wetlands, and conservation practices. The NRI is also the best source of data on the conversion of agricultural land to other uses.

According to the 1997 NRI, published in revised form in December 2000, developers converted 11.2 million acres of land to other uses between 1992 and 1997—an average of 2.24 million acres annually. This is a 49 percent increase over the average annual rate of 1.51 million acres reported between 1987 and 1992 and a 79% increase over the 1982–1987 rate of 1.25 million acres (see Figure 5.1). Worse, 6.2 million of those acres converted between 1992 and 1997 were productive farmland, and a high percentage—as many as 3.2 million acres—were classified as "prime farmland."

It should come as no surprise that this nation's best land is under the most pressure. Our agrarian ancestors settled on the most productive land, meaning that many of our fastest-growing cities are spreading out on the nation's best farmland. A 1990 study by the American Farmland Trust found that 82 percent of the fruit, 85 percent of the vegetables, and 79 per-cent of the dairy production in the United States is produced in metro-politan-area counties or the fastest-growing adjacent counties. A recent study by the USDA's Economic Research Service showed a similar fact: 56 percent of all agricultural production in the United States occurs in "urban influenced" counties adjacent to metropolitan statistical areas, reflecting the high correlation between land quality and population settlement. Add the fact that the land adjacent to roadways connecting farming settlements has itself been opened for urban, suburban, and exurban settlement areas, and it is no wonder that agriculture is in head-to-head competition with housing, retail, industry, transportation, and employment for the very best farmland in America.

According to the most recent NRI, Americans are fanning out across the landscape as in no other time in our history. Consider, for example, that between 1982 and 1992, we developed land at an average rate of 0.6 acre per "new" person—that is, an incremental resident in a given place. However, during the most recent reporting period—1992 to 1997—land was being converted to developed uses at an average rate of 0.9 acre per "new" per-son—a 50 percent increase.[2] These findings are consistent with the results of a recent Northern Illinois Planning Council study showing that, in recent decades, land consumption was racing ahead of population growth in one of the nation's largest metropolitan areas. The study revealed that while the population of the Greater Chicago metro area grew by 4 percent from 1970 to 1990, the urban footprint expanded by more than 50 percent.

Two significant contributing factors to the destruction of once-bucolic farming regions are the large-lot, single-family residential tract and the "ranchette." Ranchettes are five- to thirty-five-acre properties that, while requiring roads, schools, public safety, and other costly infrastructure investments like more typical suburban settlements, strive to give owners an ersatz rural feel. Despite efforts to lend such settlements a rural flavor, ranchettes are generally too small for any economically viable agriculture. Not only do ranchette and large-lot residential tract developments waste vast amounts of productive farm-land, they also often intrude upon remaining farms, setting up inevitable conflicts over farming practices and personal values. Imagine, for example, the challenges facing county commissioners who have to sort out the complaints registered by new residents in rural communi-ties regarding the unpleasant noise and smell associated with long-standing, local livestock operations. Ultimately, these conflicts force many farmers to sell out prematurely.

What impact can Internet technology have on these vast forces? Every technological development that allows people to conduct work without going to a downtown office has the potential to encourage sprawl, and the Internet is no exception. At present, despite growing traffic constraints, there are already some Manhattan workers who live in Pennsylvania and some Washington, D.C., employees who travel to their offices in the nation's capital from West Virginia. If the ability to telecommute means that at least some people will be able to live almost anywhere while delivering a work product, America's population could be spread ever more thinly across the countryside, further disrupting farm and forest operations and permanently changing the working landscape we enjoy. In fact, the impact of this option is already evident in such regions as the Rocky Mountain and Intermountain states, where a near land rush is on for rural parcels in places as disparate as the Roaring Fork Valley in Colorado, the desert country of southern Utah, and the Bitterroot Valley in Montana.

Conservation Easements in the Twenty-First Century

In the context of such rapid change in land markets—in part enabled by new communications technologies—traditional methods of protecting the working landscape, wildlife habitat, and open space, used by themselves, appear insufficient to the task. For most of the past half century, when the public decided to protect or preserve a certain landscape, a public entity purchased the land for protection or for a specific use or, alternatively, restrictively regulated the land use. But the public sector cannot possibly afford to purchase every parcel that merits protection, and the regulatory approach is limited by political and legal constraints on diminishing private property rights. Most communities simply cannot afford the litigation and compensation costs that may arise when new land use regulations are challenged in court. Consequently, even the threat of litigation often leads to reluctance to employ full regulatory powers. A critical underlying question is, how can our society fairly achieve public goals on private lands?

A mechanism used increasingly since the 1970s offers an alternate, market-based method of land preservation, and the Internet can help make it work. The mechanism is known by several names: conservation easement, conservation restriction, or purchase of development rights. A *conservation easement* is a legal contract that runs with the land in which both the easement seller (typically the landowner) and the easement buyer (typically a unit of government or a nonprofit organization) agree to certain restrictions on the development of a particular parcel of land in exchange for some type of compensation. For example, an agricultural landowner may, using a conservation easement, "sell" to a local or state government the development rights to a given parcel of land. The landholder would receive payment for relinquishing the rights to, say, build houses on the land but could continue to own and conduct farming operations on the land in the future. The county or state holding the easement would be legally empowered to keep the land open and undeveloped in perpetuity and could prevent the present or any future owner of the land from building houses or subdividing the land. In addition to the compensation that a landowner may receive for granting such an easement, there are important federal, state, and local tax incentives available that can add considerable economic benefit to the deal.

Nearly twenty states and many local governments around the country have some type of conservation easement program—for example, the Pennsylvania Agricultural Conservation Easement Purchase Program, which has played a key role in making Pennsylvania the national leader in farmland preservation; the State of California Division of Land Protection's California Farmland Conservancy Program; the Agricultural Land Preservation Program of Harford County, Maryland; and New York City's Land Acquisition and Stewardship Program.[3] Details on a wide variety of programs offered on a state-by-state basis are offered, in regularly updated online publications, by such nonprofit organizations as the American Farmland Trust, the Land Trust Alliance (LTA), and the Trust for Public Land.[4] Unfortunately, many such public programs around the nation are still woefully underfunded.

Supplementing the governmental programs are the efforts of not-for-profit organizations across North America, at the local, state, and national level, that negotiate, finance, and hold conservation easements on agricultural and forestry lands as well as critical wildlife habitats, wilderness areas, outdoor recreation venues, and historic landscapes. The activity and scope of such organizations continue to grow, as evidenced by easement deals completed in the last several years that encompass hundreds of thousands of

Box 5.1

FARMLAND PROTECTION IN
MARIN COUNTY, CALIFORNIA

Marin County's famed rainy and foggy weather provides ideal conditions for grazing. Miles of open rangeland, combined with easy access to food markets in neighboring San Francisco, made Marin Northern California's leading dairy county for much of the twentieth century.

COMPREHENSIVE PLANNING

In 1973, when the loss of farmland had become a crisis in this conservation-oriented county, the board of supervisors adopted the Marin Countywide Plan. The plan's Agriculture Element is designed to "enhance, promote and protect agricultural land uses and the agricultural industry in Marin County." It does so, in part, by dividing the county into three corridors: the City-Centered Corridor, on the eastern side of the county; the Inland Rural Corridor, in the middle of the county; and the Coastal Recreation Corridor, along the western side. Each of the corridors is intended to focus on priority land uses; in the Inland Rural and Coastal Recreation corridors, agriculture is a designated priority.

AGRICULTURAL PROTECTION ZONING

The Marin County zoning ordinance was amended in 1973 to implement the new plan and has been modified several times since. It establishes seven agricultural zoning districts. Densities for properties in each district are assigned on the county's zoning map. On 90 percent of the land within these districts, the maximum density and/or minimum lot size is one house per sixty acres. Prior to 1973, the minimum lot size in most of the county was one acre.

The effectiveness of agricultural zoning, particularly sixty-acre zoning, quickly became—and has remained—an issue of debate in Marin. Some residents, particularly county officials, believe Marin

Box 5.1 continued

County's zoning and land use regulations have sent the majority of interested home buyers elsewhere. On the other hand, some believe that the zoning has not done what it was established to do, as large-lot home sites are slowly consuming land that traditionally had been used for agricultural purposes.

Despite these opposing views, a general consensus prevailed that the new zoning required farmers to make a financial sacrifice because it limited the development potential of their land.

THE BIRTH OF MALT
(THE MARIN AGRICULTURAL LAND TRUST)

Marin found an effective way to address these problems in 1980, when the private, nonprofit Marin Agricultural Land Trust was formed to purchase conservation easements on the county's agricultural land. At the time, there were a total of 431 nonprofit, local land trusts in the United States; however, MALT was the first to focus specifically on the preservation of agricultural land. For owners of agricultural land, MALT represented the "carrot" that went with the zoning "stick." By acting as a buyer of conservation easements from private agricultural landowners, it provided the permanence that zoning lacked and pre-sented an alternative to development that farmers and ranchers could choose if they were faced with having to sell their land.

As of fall 2001, MALT holds forty-five easements on 30,330 acres of land. The average easement covers 670 acres. Purchase prices have ranged from $275 to $1,500 per acre and currently average $980 per acre. Easement prices range from 25 to 50 percent of the unrestricted mar-ket value of the property. Farmers generally consider these prices fair.

INTENSIFYING CHALLENGES

The challenges facing conservationists striving to protect agricultural land and open space in Marin County, just to the north of the San Francisco Bay, are intensifying as we enter the twenty-first century. As has been extensively reported by the press, the Bay Area continued in the 1990s to grow in terms of population, unprecedented levels of per-sonal wealth, and the availability of some of the world's most advanced communications networks and technologies. Homes in Marin County, many of which are prized by Bay Area high-technology entrepreneurs for their green vistas and proximity to open space, have risen dramatically in value in the past five years, from an average sales price of $433,999 in 1995 to an average of $841,568 in 2000.*

As reported by MALT, ongoing threats to open space include substantial development pressure ("the beauty of West Marin increasingly attracts non-agricultural buyers who are eager and able to purchase farm properties for luxury home sites and non-agricultural uses") and high land prices ("because of the proximity of metropolitan San Francisco and its population of six million people, land prices have escalated far above values based on farm incomes. Ranching families often can't afford to transfer property to the next generation. Young farmers find it nearly impossible to buy land to get started").**

If the level of connectivity provided by the Internet and related networks continues to grow in the context of rising local levels of affluence, it is quite likely that the pressures on the Marin landscape will become increasingly intense. In the context of such potential developments over the next several decades, it will take creative and persistent efforts by MALT and conservation-minded citizens to continue making substantial strides to protect the region's traditionally green and open landscape.

* Average home prices reported by West Bay Real Estate on its "Marin County Cities" Web site, available at http://westbayre.com/marin.htm.
** Information on the Marin Agricultural Land Trust and the current threats to open space and agriculture in the region is available at www.malt.org/preserve/threat.html.

acres, as well as the continuing growth of the land trust movement as documented by the LTA's National Land Trust Census (see Figures 5.2 and 5.3).[5] Note that, as the use of conservation easements has evolved over the past several decades around the country, the need for customized local programs that reflect community values has led to the establishment of a broad range of costs for conservation restrictions. The price of a conservation easement on forest or farmland may vary widely, from less than $40 per acre in Northern

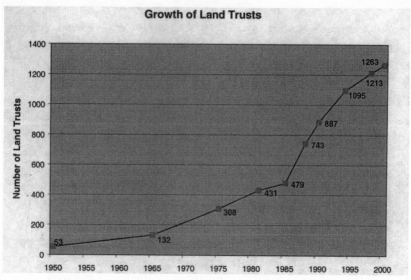

Figure 5.2: The land trust movement has gained momentum during the past fifty years, as evidenced by the increase in the number of land trusts throughout the nation. (Data from the Land Trust Alliance, www.lta.org.)

Maine to $2,000 per acre in Marin County, California, to $5,000 to $10,000 per acre in Montgomery County, Maryland, to a stunning $40,000 per acre for a small farm in a town in Massachusetts.

How Emerging Technologies Can Help

How can the Internet facilitate the operation of the market for conservation easements? First, Internet-based geographic information system (GIS) mapping tools, digital photography technologies, and scenario-testing software may enable local, state, and national decision makers to evaluate the impact of a given conservation or development initiative, helping to build consensus and political support for necessary funding. Second, new communications technologies may provide important methods by which land parcels can be advertised to prospective conservation buyers in the public and private sectors. As funds become available to purchase development rights, communities and farmers might be able to much more easily reach closure on deals of mutual benefit. We have seen that Internet-based auction markets such as eBay have made markets for a wide variety of goods

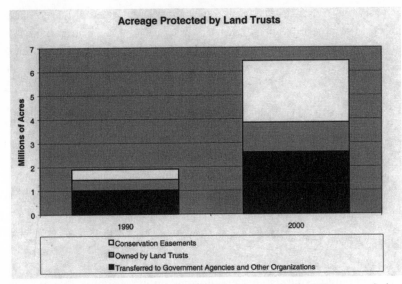

Figure 5.3: The total acreage protected by land trusts grew by 238 percent in just a decade. (Data from the Land Trust Alliance, www.lta.org.)

and services more efficient; perhaps similar tools can do the same for the land conservation market.

Using computers to solve conflicts between farmers and developers is not as far-fetched as it might sound. Meeting the food and fiber needs of future generations will prove daunting—the American Farm Bureau Federation calculates that U.S. farmers will have to double food production by 2025 to feed a growing domestic population and allow the United States to remain competitive in international markets. As noted above, the agricultural land base on which these farmers will operate is shrinking each year. To increase production on a shrinking land base, farmers clearly will have to operate on the very front line of new technology.

And so they are. According to a recent USDA survey, farm family computer ownership and Internet connectivity increased rapidly in the past several years. Between 1997 and 2001, U.S. farms with computer access rose from 38 percent to 55 percent. Correspondingly, farms with Internet access grew from 13 percent to 43 percent.[6] Advances in connectivity are being complemented throughout agriculture with a vast array of emerging technologies. For example, new genetic engineering, remote sensing, and GIS technologies are already allowing researchers at agricultural research stations and some innovative farmers to apply just the right amount of fertilizer to a crop with specific genetic characteristics at a precise field location, using techniques that are sometimes called "precision farming." Such advances promise to dramatically increase per-acre yields as well as improve agricultural input efficiency.

We cannot, however, take such gains on a widespread, sustainable basis for granted. It would be a tragic mistake to assume that as a nation we need not maintain a critical mass of prime farmland to meet future food and fiber demand. No matter how productive agriculture becomes, it will still require the highest-quality land for the most efficient production of food and fiber.

It is the latest in communications systems that hold the greatest promise of saving farmland, for these systems have the ability to reach and educate people in ways never imagined just a few years ago. The key to improved public policy is a better-informed public—a public that not only will demand improved conservation policy on every governmental level but will also know enough to blow the whistle when such policies fail to protect the land.

Today, suburban developers often have greater familiarity and access to land use information than the public. Development insiders may well learn about properties available for sale weeks, months, or years before relevant information on those properties is posted in the newspaper. Land parcels that look unchanged from decades past might actually have a new zoning designation or be slated to adjoin a future highway off-ramp or even to become an airport. This kind of information is not only worth millions of dollars to developers but also tends to negate any preservation action by a small, disorganized pro-farm or conservation group.

The Internet may play an important part in leveling the playing field. With computerized GIS and with the growing trend toward "e-government" (placing public information on the World Wide Web), small, modestly funded organizations and even individual citizens can begin to tap the kind of power that large developers and large government agencies presently have. Wired homes will be able to receive complex information about such planning factors as roads, sewers, schools, power generation, ecological carrying capacity, the ratio of jobs to residents, and much more. Using new software similar to the urban-planning game "Sim City," average citizens will be able to create and test what-if scenarios relating to

Box 5.2

MAPPING THE FUTURE OF FARMLAND:
ONE COMMUNITY'S INTERACTIVE APPROACH

The General Plan of Monterey County, California, was due for an update in 2001. In addition to getting input from county residents up front, the county decided to put technology to use. At an interactive Web site hosted by the Monterey County 21st Century Program/General Plan Update, county residents could plan their own futures for the county's growth and submit them to the County Planning Commission for consideration (see the County's planning Web site at www.co.monterey.ca.us/gpu/).

Part of the interaction involved getting residents to understand the role and meaning of the land use planning process. As the planning Web site put it, the county was at a crossroads and had to ask itself, "Is Monterey County just a place to receive the spillover from the Bay Area or is it a place that cultivates its own identity? Will Monterey County continue to prosper yet also protect the very elements which make it unique—its stunning natural environment, its valuable agricultural land and economy, and its vibrant historic cities and communities?"

Once visitors clicked in to the planning process, they were asked to balance trade-offs, just as professional planners must do. Different key considerations included economic growth, county services, agricultural conservation, and transportation. If the county really were to grow at such and such a rate, for example, it would need to prioritize which agricultural lands it wished to keep, because it would not be able to keep them all. Site visitors could access maps that showed how the county would look under different scenarios.

In the first three months that the interactive process ran, 102 visitors responded to the challenge and created an alternative plan for Monterey County. After weeding out duplicates, the County was left with seventy-one unique plans. Was it worth the effort for those who produced the site, ensuring that they captured the public's input? That evaluation is yet to be done. But by opening up the planning process, Monterey County took just one of the possible roads that makes the Internet age hold such promise for the future of land conservation.

growth, pollution, education, food production, environmental benefits, and other variables and so be better informed to make more proactive use of conservation tools, including conservation easements. In some communities, such forward-looking analyses may be facilitated by the government agencies themselves; elsewhere, it may be up to organizations such as American Farmland Trust to serve as facilitator by helping to convene policy makers, local communities, and landowners in appropriate forums. In short, technology offers great potential to dramatically increase public access and participation in the planning process.

Some communities are already testing the potential use of Internet technology in assisting the local planning process. In Monterey County, California, the board of supervisors, through its planning commission, created a Web site to increase public input as they updated the County General Plan. Similarly, Pitkin County, Colorado, provides detailed online information regarding ongoing planning task force work; the county also posts all building permit applications online.[7]

Moreover, computerized software tools can facilitate mathematical calculations to reveal all kinds of information relevant to community planners, including the visual, ecological, and financial impacts of a given development proposal. As explored in Chapters 11 and 12, software now being deployed may significantly enhance the public's ability to envision various futures for landscapes and communities.

And there's more. Computers and the Internet could help farms stay in business by increasing their profitability. At present, the majority of farmers, even those near big cities, end up selling their produce to middlepersons who often ship the food long distances and keep eighty cents of every food dollar. Some farmers find that only one buyer controls the market for a given commodity, leaving little competition and invariably a lower price. However, a growing movement is pushing for the establishment of direct-to-consumer outlets, such as urban "greenmarkets," where local farmers (that is, those within a hundred miles or so) sell directly to consumers in central-city locations. A related development known as community supported agriculture enables city dwellers to prepurchase a share of a local farm's output, guaranteeing weekly deliveries of seasonal produce and providing much-needed up-front cash for the farmer. Information technologies, including Web advertising, community agriculture listservs, targeted e-mailing, and up-to-the-minute reports on food availability, can facilitate both of these approaches, which not only provide city dwellers with fresher food at lower prices but also return 100 percent of the profits to the farmer.

Real-time information about markets is already putting farmers in a stronger negotiating position and opening new markets for their products. And, as noted, "precision farming" technologies currently being developed by researchers in the public, private, and academic sectors are making it possible for farmers to use techniques such as satellite-fed, up-to-the-minute weather information to better time planting, fertilizing, and harvesting.[8]

In sum, we are poised on the cusp of great technological change. If we use this new technology wisely, we can improve the lives of all our citizens, both on the land and in the cities. If we blindly rush to make lifestyle decisions that ignore external impacts, we truly may reach a point where a Marie Antoinette of the future has to tell the hungry poor, "Let them eat electrons."

NOTES

1. U.S. Department of Agriculture, Natural Resources Conservation Service, "1997 National Resources Inventory, National Results" (January 9, 2001), available at www.nrcs.usda.gov/technical/NRI/1997/national_results.html. See this site for additional information on NRI results, methods, and data resources.

2. Calculated by dividing the change in *developed land* reported by the 1997 NRI by the change in *population estimates* reported by the U.S. Census Bureau for different time periods. The "1997 National Resources Inventory Summary Report," released in revised form on January 9, 2001, by the Natural Resources Conservation Service of the U.S. Department of Agriculture, is available at www.nrcs.usda.gov/technical/NRI/1997/summary_report/body.html.

3. Commonwealth of Pennsylvania, Department of Agriculture, "Pennsylvania Continues to Lead Nation in Farmland Preservation," available at http://sites.state.pa.us/PA_Exec/Agriculture/bureaus/farmland_protection/press_release.htm. For information on the California Farmland Conservancy Program, see www.consrv.ca.gov/dlrp/cfcp/index.htm. For information on the Harford County, Maryland, Agricultural Land Preservation Program, see www.co.ha.md.us/ag_preservation.html. For information on the New York City Land Acquisition and Stewardship Program, which is used to protect lands associated with the city's water supply in upstate New York, see www.ci.nyc.ny.us/html/dep/html/news/easement.html.

4. For more information from the American Farmland Trust on state-by-state agricultural easement programs, see www.farmland.org/research.

5. Information on recommended land trust standards and practices, on federal tax law relevant to conservation easements, and on the location of various land trusts in North America is available from the Land Trust Alliance at www.lta.org.

6. U.S. Department of Agriculture, National Agricultural Statistics Service, "Farm Computer Usage and Ownership" (July 30, 2001), available at usda.mannlib.cornell.edu/reports/nassr/other/computer/empc0701.txt.

7. Pitkin County Government, "Building Permits," available at http://www.pitkingov.com/sitepages/pid7.php.

8. For a sample listing of Internet sites relevant to precision farming techniques, see the Missouri Precision Agriculture Center at www.fse.missouri.edu/mpac/index.htm.

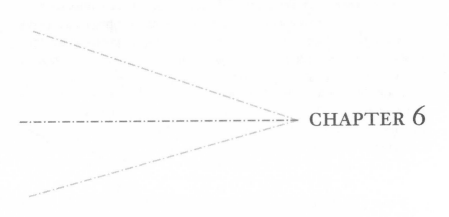

CHAPTER 6

Internet Use in a High-Growth, Amenity-Rich Region

James N. Levitt and John R. Pitkin

"The real potential of the information age is that people can live where they like." So says Craig McCaw, the entrepreneur dubbed by *Business Week* as "the prophet of telecom."[1] McCaw's record in the industry is indeed impressive. He built McCaw Cellular virtually from scratch into an industry giant that he sold to AT&T in 1994 for more than $12 billion. McCaw, along with Bill Gates, Boeing, and Saudi Prince Alwaleed bin Talal, are now betting billions on the success of Teledesic, a low-earth-orbit satellite network that plans to offer high-speed data connectivity to 95 percent of the earth's landmass by 2005.

And where do people like to live? As detailed in Chapter 4, a positive net migration of Americans to nonmetropolitan counties occurred during the 1970s and the 1990s. Recent research by the U.S. Department of Agriculture's Economic Research Service (ERS) suggests why: "People once moved to rural areas in search of fertile farmland, mineral deposits and timber; today people move to rural areas in search of a pleasant environment for residence and recreation."[2] The ERS's David McGranahan, in an article titled "Natural

Amenities Drive Rural Population Change," shows that counties with the highest levels of natural amenities—based on a scale accounting for a county's climate, topography, and water surface area—saw their population multiply by several times between 1970 and 1996. Counties with low levels of natural amenities tended to lose population.[3]

If both McCaw and McGranahan are right, we might expect to see relatively intensive use of information technology among people who move to rural counties that are rich in natural amenities. Rather than *causing* such people to move, the new networks would *enable* them to "live where they like."

The idea that new technologies are a key factor in recent nonmetropolitan growth finds support among prominent observers of the rural scene. Kenneth Johnson and Calvin Beale, for example, in their widely read article "The Rural Rebound," argue:

> Driving the revival is a potent blend of economic, social, and technological forces. Improvements in communications technology and transportation have sharply reduced the "friction of distance" that once hobbled rural areas in the competition with the great metropolitan centers for people and commerce. In practical terms, rural areas are now much less isolated than they were only a few decades ago. Satellite technology, fax machines, and the Internet are among the most familiar aids, rendering distance virtually irrelevant in the transmission of information. Other sources of change are less obvious. Decades of steady state and federal investment in roads and airports—building and widening of highways, runway paving, subsidies for equipment purchases—have also made an enormous difference.[4]

Yet empirical research detailing the role of new networks in rapidly growing nonmetropolitan and small metropolitan counties high in natural amenities is modest and inconclusive.[5] Therefore, to better understand the relationship between new communications technologies and changes in American demographic and land use patterns, we conducted a telephone survey of two rapidly growing counties in the Intermountain West: Deschutes County, Oregon, home of the city of Bend; and adjacent Crook County, seated in Prineville (see Figure 6.1).[6]

What we learned gives credence to the idea that, at least in Central Oregon, significant numbers of people are moving to the area to enjoy its natural and recreational features, and such "amenity-influenced in-migrants" appear to be relatively heavy users of new information technologies. By migrating in substantial numbers to Central Oregon, such

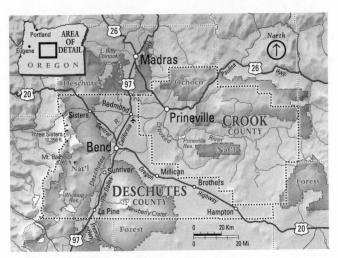

Figure 6.1: Deschutes and Crook Counties, Oregon, are amenity-rich regions, offering great scenic beauty and a wide variety of recreational opportunities. (Map by David Deis, Dreamline Cartography.)

households are helping propel dramatic changes in the region's character and economy, as employment shifts toward the service sector, and consumption of open space, fragmentation of wildlife habitat, and stress on water resources accelerate.

Ironically, the rapidly growing, relatively Internet-savvy population of Deschutes and Crook Counties presents both problems and opportunities for local conservation initiatives and organizations. We found that Internet penetration is particularly high among households that reported themselves members of conservation organizations. Although conservation groups face a stiff challenge in preserving the natural amenities of the Deschutes/Crook Counties area, those groups also have the opportunity to use the Web, e-mail, and other forms of electronic communications to seek widespread support for conservation efforts.

A Place to Pursue Dreams

Bend, Oregon, and the surrounding communities in Deschutes and Crook Counties are booming. As described glowingly in a May 2001 story in *USA Today*, the Bend area is "a magnet for vacationers with direct air service . . . in the middle of a recreational paradise. . . . Downtown Bend throbs with

vitality."[7] As one local guidebook explains in dramatic language, Central Oregon is "in the grand tradition of the West, . . . a place where people come to pursue their dreams and make them real."[8] Census statistics show that over the past several decades, the area has become highly attractive to in-migrants as well as visitors.[9]

Two keys to the area's attractiveness are its natural and recreational amenities: the region is blessed with what twenty-first-century Americans consider high levels of each. A towering range of mountains lies to the west of Deschutes and Crook Counties, a federally designated wilderness area lies to the north, and a network of rivers, reservoirs, and canyons winds through the region. In fact, David McGranahan's "natural amenity scale" ranks Deschutes County and Crook County among the top 4 percent of all counties in the continental United States.[10]

These natural amenities provide a broad array of recreational offerings. Mt. Bachelor, just east of Bend, is a major skiing destination in the Pacific Northwest. The area boasts several highly regarded golf resorts, including those at Sunriver and Black Butte Ranch. The region also offers extensive opportunities for hiking, boating, horseback riding, fishing, birdwatching, and other wildlife excursions at such sites as the Deschutes National Forest, the Mill Creek Wilderness of the Ochoco National Forest, the Newbury National Volcanic Monument, the Ochoco Lake State Park, and the Prineville Reservoir State Park.

The Central Oregon climate favors outdoor activities. Average annual rainfall in Bend, on the dry eastern side of the Cascade Mountains, is about 11 inches—less than one-quarter of that recorded in Eugene, on the Cascades' wet western side. Temperature swings in Bend are relatively moderate compared with those in much of the country: mean average temperatures range from 41 degrees F in January to 68 degrees F in the summer.[11] The relatively dry, mild climate contrasts favorably with the wetter climates of such metropolitan centers as Portland and Seattle. As explained by the local guidebook: "What's so special about Central Oregon, and why do so many people who come here to visit eventually decide to stay and make it their home? For many, the attraction is at first physical—forests of pine, sky-scraping mountains and sunny days that seem to go on forever."[12]

A Changing Region

Despite its current popularity, Central Oregon was, as recently as the mid–nineteenth century, a place that most European Americans wanted to avoid. In the 1840s and 1850s, migrants heading to Oregon's lush Willamette Valley on the wet western slope of the Cascade Mountains considered arid Central Oregon a desert to be crossed on the way to the promised land.

The first whites to establish homesteads in Central Oregon—generally along the Deschutes River and its tributaries, including the Crooked River—did so in the 1860s, after the opening of a toll road from Eugene, some eighty miles to the west.[13] The road serviced Central Oregon's emerging cattle, sheep, mining, and timber operations as well as scattered homesteads. In a move that ultimately contributed to the resolution of the infamous nineteenth-century range wars between cattle herders and sheep-herders in Central Oregon, the federal government in 1893 established the Cascades Forest Reserve. That reserve served as the predecessor of several national forest units now operating in the area, including the Deschutes National Forest, which occupies much of the Cascades' eastern slopes above the Deschutes River.[14]

The Central Oregon area grew slowly into the early 1900s. Bend incorporated as a city in 1904 with a population of about 500. James J. Hill's Great Northern Railroad reached Bend in 1911, after Hill completed a deal with Edward H. Harriman of Union Pacific to share the single track running to Bend from the north.[15] Not until the 1920s did an automobile bridge over the Crooked River allow U.S. Highway 97—still the region's principal north–south artery—to pass through the area, giving trucks access to the area's lumber mills. When a military air base in the town of Redmond, midway between Bend and Prineville, was converted to civilian use following World War II, regular passenger air service was introduced to the region. And in the mid-1950s, the Mt. Bachelor Ski Area, in the Deschutes National Forest just west of Bend, helped establish outdoor recreation as an important part of the local economy. By 1970, the combined population of Deschutes and Crook Counties had grown to 40,000 people.[16]

Population Boom Brings Economic Diversification

From 1970 to 1999, the population of Deschutes County alone more than tripled—from 30,000 to more than 100,000. Crook County's growth has been less dramatic, rising from 9,900 to about 17,700. Recent growth in the two-county area occurred in two spurts (see Figure 6.2). Population rose from 40,000 to about 75,000 between 1970 and 1980; held relatively steady between 1980 and 1985, when it totaled 78,000; and then took off to some 128,000 by 1999.[17] This double-humped growth pattern generally

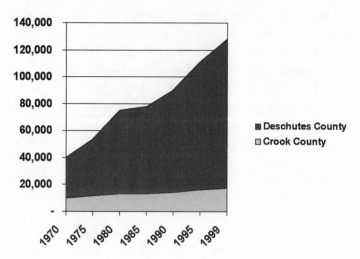

Figure 6.2: Since 1970, the population of Deschutes and Crook Counties has more than tripled. (Data from U.S. Census Bureau and Texas A&M Real Estate Center, http://recenter.tamu.edu.)

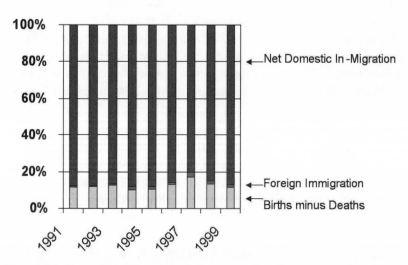

Figure 6.3: From 80 to 90 percent of recent growth in the area is net domestic in-migration. (Data from U.S. Census Bureau and Texas A&M Real Estate Center, http://recenter.tamu.edu.)

reflects the national "rural rebound" patterns described by Kenneth Johnson in Chapter 4.

By far the dominant portion of this growth—from 80 to 90 percent— has come from in-migration by people already living in the United States (see Figure 6.3). The second-biggest factor is natural increase—local births minus deaths. Foreign in-migration did not account for more than 3 percent of population growth in any year during the 1990s.

Population growth has corresponded to a dramatic change in the area's employment mix. In 1970, wage-earning jobs in the service sector accounted for only 18 percent of the Deschutes County total; by 1997, the service-sector share had risen to 31 percent. In contrast, wages from agricultural services, fishing, and forestry accounted for less than 2 percent of the 1997 total (see Figure 6.4). Income from service-sector jobs—as well as from dividends, interest, rent, and transfer payments such as Social Security—has also been growing, both in absolute numbers and as a share of the total (see Figure 6.5).

The economic transition occurring in Central Oregon is similar to that under way in a wide variety of rapidly growing nonmetropolitan areas around the country. The traditional mix of farming, ranching, mining, and other extractive industries is becoming less important.[18] Information and service-intensive industries, such as tourism, retail, and investment services, have generally continued to grow.[19] In Deschutes and Crook Counties, two companies exemplify such growth: the area is home to the headquarters of the Les Schwab Tire Company—the largest tire retailer in the Pacific Northwest—in Prineville, and to a Charles Schwab investment-services walk-in site in downtown Bend.

Amenity-influenced, technology-enabled in-migrants appear to be important participants in changing the culture, economic mix, and land use patterns of Deschutes and Crook Counties. Currently one-third of the two-county area's population and rising, such households are potential customers for new homes in such previously undeveloped and environmentally sensitive areas as La Pine Basin, south of Bend.

Along with the changes in local economics and demographics, use of air and surface transportation networks in Central Oregon has become more intensive in recent decades. Between 1988 and 1998, use of state roads in the two-county area rose by some 236 million vehicle miles traveled (VMTs), at a compound annual growth rate of 4.4 percent (see Figure 6.6). Use of the local airport has climbed even more sharply. The annual num-

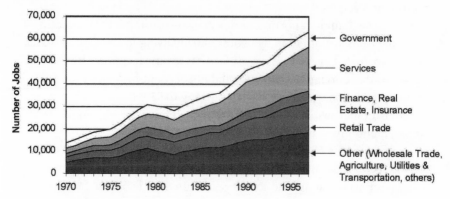

Figure 6.4: Recent growth in wage-earning jobs is strongest in the service and retail trade sectors. (Data from Bureau of Economic Analysis, *Regional Economic Information System 1969–1996* [CD-ROM] [Washington, D.C.: U.S. Department of Commerce, 1999].)

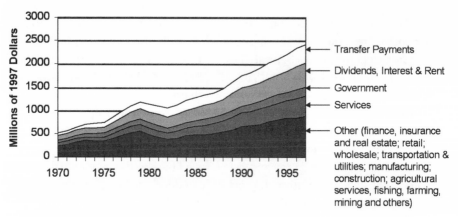

Figure 6.5: Most recent income growth (expressed in millions of 1997 dollars) comes from transfer payments, investments, and service jobs. (Data from Bureau of Economic Analysis, *Regional Economic Information System 1969–1996* [CD-ROM] [Washington, D.C.: U.S. Department of Commerce, 1999].)

ber of passengers boarding planes at the Redmond, Oregon, airport increased from about 26,000 in 1986 to some 127,000 in 1998, at a compound annual growth rate of 14 percent (see Figure 6.7). Both inbound and outbound travel between Redmond and such locations as Portland, Seattle, and San Francisco—all major air hubs within a two-hour flight of Central Oregon—contribute to the rise in transportation network utilization.

As noted in Chapter 1, the growth of surface and air network use concurrent with a growing penetration of electronic communications networks such as the Internet in area households is consistent with historical precedent. Telecommunications industry executives have long understood, to paraphrase a comment made by a seasoned telephone executive, that "the more you talk, the more you travel; and the more you travel, the more you talk."[20]

Associated Environmental Challenges

By the mid-1990s, local and state officials had identified several environmental problems related to Central Oregon's rapidly growing population and development of open space. For example, a state-funded study regarding a thirty-square-mile area in La Pine Basin, in southern Deschutes County, reported that "existing development (4,071 lots) and continued development of vacant lots (4,901 lots less than 1.5 acres in size)" could result in the following regional problems:

- Groundwater pollution that could affect drinking water and public health
- Additional loss of wildlife habitat, including mule deer migration corridors and riparian and wetland habitat
- Increased threat of forest fire
- More air pollution from traffic on unpaved roads[21]

The state-funded Deschutes Ground Water Work Group grappled with similar issues in its report:

Deschutes County is the fastest growing county in Oregon. People from throughout the West are drawn here by the area's many recreational and economic opportunities, and high quality of life. But underlying the region's natural splendor and robust economy is the need for water. Unfortunately, the

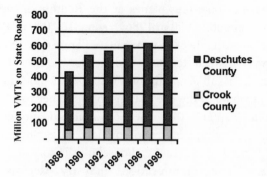

Figure 6.6: Vehicle traffic on Deschutes and Crook Counties state roads continues to increase. (Data from Oregon Department of Transportation, www.odot.state.or.us.)

region's surface water has been fully appropriated—eliminating it as a source of water to sustain growth in the region.

The authors called on the Oregon Water Resources Department, development advocates, and conservationists to identify "ways to supply water necessary to accommodate growth while protecting instream values represented by scenic waterway flows and instream water rights."[22]

These challenges are underscored by the recent U.S. Interior Department designation of the area's famed steelhead salmon as "threatened" under the Federal Endangered Species Act. To improve conditions for the steelhead, Congress appropriated $2 million a year from 2002 to 2006 for the Deschutes Resources Conservancy (DRC) to conduct a wide array of water and land conservation efforts.[23]

One recent effort entailed restoring the Camp Polk Meadow, near the Deschutes County town of Sisters. The DRC worked with the nonprofit Deschutes Basin Land Trust, which received an $800,000 donation from Portland General Electric, Oregon's largest electric utility, to acquire the 385-acre property. The press release announcing the effort highlighted the project's importance to conservation efforts:

> "Camp Polk Meadow is one of the crown jewels of historic steelhead habitat in the basin," said Land Trust Executive Director Brad Chalfant. "The property was on the verge of being subdivided, and now it will be protected

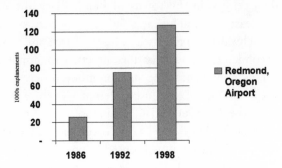

Figure 6.7: Regional demand on the air transportation network has also climbed sharply in recent years. (Data from City of Redmond, Oregon, www.redmond.or.us/stats.htm.)

because of PGE's donation." . . . Biologists and conservation organizations say that the meadow has the upper basin's best potential spawning ground for summer steelhead, a threatened species.[24]

A Telephone Survey of the Deschutes and Crook Counties Area

To better understand the role of new communications in the changes occurring in Deschutes and Crook Counties, we conducted a telephone survey of area residents in early 2000. We surveyed households whose members had resided in the two-county area for ten years or more—representing some 50 percent of all residents—as well as more recent in-migrants.[25] We specifically aimed to record Internet use by households whose members were attracted to the area by its natural and recreational amenities and by households that include members of conservation organizations.

During the period we conducted our survey, consumers and businesses who wanted to log on to the Internet from Deschutes and Crook Counties had several choices. National Internet providers, such as America Online, offered local numbers in Bend for dial-up access. Local providers, including BendNet, EmpireNet, and Coinet, also offered service at speeds up to 56K per second at prices comparable to those offered by the national providers. Several local sites, including a café in downtown Bend, offered walk-in Internet access. Bend Cable Communications offered higher-speed cable modem service in the towns of Bend, Black Butte, Sisters, and Redmond. At least one local business had developed custom, satellite-based services: Les Schwab contracted with Hughes Electronics to build a VSAT system that connected its headquarters directly with hundreds of its retail outlets for transaction processing and communications.[26]

Based in Cambridge, Massachusetts, Jennifer Reese, a research assistant affiliated with the Internet and Conservation Project at the A. Alfred Taubman Center for State and Local Government at Harvard's John F. Kennedy School of Government, used a computerized questionnaire between March 7 and May 4, 2000, to complete 347 phone interviews with Deschutes and Crook County heads of households.[27] Typically conducted in twelve minutes or less, the interviews included queries regarding the household's location and length of residence in the area, its use of computers and the Internet, the family's housing and land use, its reasons for moving to the area, its water and sewer connections, its affiliations with conservation organizations, and its income.[28] Because of the interviewer's meticulous efforts, the number of invalid or inconsistent item responses was negligible.[29]

We found that during spring 2000, 60 percent of households in Deschutes and Crook Counties used the Internet at home. Of that group, 90 percent gained access to the Internet through lower-bandwidth dial-up connections; the remaining 10 percent relied on higher-bandwidth cable modems or satellite connections. This 60 percent household penetration rate was well above the national average. Nielsen/Net Ratings reported that 52 percent of U.S. homes had access to the Internet in July 2000; Media Metrix reported that 51 percent had such access as of April 2000.[30]

Our survey results also showed that Internet households in the Deschutes/Crook Counties area, like Internet households throughout the United States, differed on average from non-Internet households on such important factors as income, household size, and age. Not surprisingly, households with higher incomes were more likely to have Internet access; on average, a 50 percent rise in income corresponded to a 10 percent rise in household Internet penetration rate. Internet penetration peaked among families with three or four members, falling off in smaller or larger households.

Interestingly, Internet use in the Bend area peaked in households where the average age of the household head is forty-nine—dropping off by about a third in families with younger or older household heads. The relatively lower use among younger households may be partly due to the fact that no four-year college or university was located in the region (Central Oregon Community College offered a two-year course in the area, and Oregon State University had, at the time the survey was conducted, announced plans to open a Cascades campus in Bend in the near future). The area probably lacked a concentration of young adults enrolled in undergraduate and graduate programs—a group that national surveys show includes relatively intensive users of the Internet and related technologies.

Our survey also showed that some 62 percent of all recent in-migrant households had access to the Internet at home. This is only 5 percent more than households of longer-term residents—a modest margin, not large enough to be characterized as statistically significant. Furthermore, households with postal zip codes within the city of Bend (60 percent) and those with zip codes for outlying towns and rural communities showed no difference in Internet access.

However, we did find a key difference in the reasons cited for moving to the area between recent in-migrant households with Internet access and those without Internet access. The interviewer asked recent in-migrant respondents to answer, in their own words, "Why did you first move to Deschutes [or Crook] County?" The interviewer coded the response into one of eight categories: new job, transferred, to be near family or friends, to be near recreational activities/locations, to be near the area's natural amenities or the environment, to have a lower cost of living, to retire or plan for retirement, or to pursue a business opportunity.[31] The interviewer also asked about the respondent's secondary reason for moving to the area: "People can have more than one reason for moving to a place. Did any of the following reasons also affect your decision?" The interviewer asked the interviewee to respond to each of the eight coded reasons listed above.

Some 24 percent of recent in-migrant respondents (about 12 percent of the sample population) mentioned the region's natural or recreational amenities as a primary reason for moving to the area. About 43 percent of such households (21 percent of the sample population) mentioned natural or recreational amenities as a secondary reason for moving to the area. Thus, some 67 percent of recently migrated households—about one-third of the sample total population—mentioned natural or recreational amenities as a primary or secondary reason for moving to the Deschutes/Crook Counties area.

Of these "amenity-influenced recent in-migrant households," some 68 percent reported household Internet access—substantially higher than the

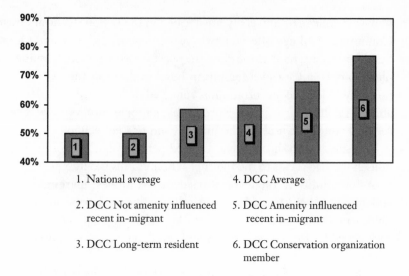

1. National average

2. DCC Not amenity influenced recent in-migrant

3. DCC Long-term resident

4. DCC Average

5. DCC Amenity inflluenced recent in-migrant

6. DCC Conservation organization member

Figure 6.8: Household Internet penetration is high among Deschutes/Crook Counties (DCC) amenity-influenced in-migrants and conservation organization members. (Data from telephone survey directed by authors.)

national rate and significantly higher than the average rate for all Deschutes/Crook Counties area respondents. The segment of recent in-migrant households that did not mention recreational or natural amenities as a factor in their move—one-third of the recent in-migrants, or one-sixth of the sample population—was much less likely to have Internet access: only about 50 percent of that population reported having Internet access at home. The latter rate of Internet access is typical of national household Internet penetration rates at that time (see Figure 6.8).

Regression modeling shows that the 18 percent difference in household Internet penetration among the two groups of recent in-migrants is not entirely attributable to differences in income, the age of the head of household, or household size.[32] Even after controlling for those factors, a statistically significant 12 percent spread between the two groups remains. In brief, it appears that households that moved to the Deschutes/Crook Counties area in the 1990s at least in part to be near nature or recreational opportunities are more likely than other in-migrants to use the Internet at home as one element of a lifestyle that enables them to "live where they like." The amenity-influenced in-migrant households are also more likely than other in-migrant households to use faxes at home (32 percent versus 24 percent) and cell phones in their cars (65 percent versus 46 percent).

Like amenity-influenced in-migrants, households that identified one of their members as affiliated with at least one conservation group were also significantly more likely to have Internet access at home. Among the survey population in the two-county area—including both long-term residents and more recent in-migrants—one household in five maintains at least one membership in a conservation group. Of these, 77 percent use the Internet at home, compared with a penetration of 55 percent among nonmember households. Differences in age, household size, and income account for less than a fifth of this measured difference. The variation in Internet use between households that are members of at least one conservation group and those that are not is equivalent to a 150 percent increase in household income, or an increase in household size from one to four persons.[33]

So What?

Our research effort was modest and exploratory rather than comprehensive and definitive: we focused on a single geographic location at a single point in time. Yet the results lend credibility to the idea that the Internet is becoming an integral part of the everyday lives of people who are influenced by the availability of natural and recreational amenities in deciding where to live.

We do not argue that the Internet somehow *causes* people to move to amenity-rich areas, just as we are not aware of definitive evidence that new interstate highways caused middle-class residents of U.S. cities to move to the suburbs in the 1960s and 1970s.[34] Rather, we hypothesize that the Internet, fax machines, cell phones, and related technologies that "reduce the friction of distance" *enable* people to move to physically and recreationally attractive regions.

Paradoxically, while in-migrants attracted to an area's natural amenities put pressure on the very resources they appear to value, technology-enabled in-migrants may also become important participants in the conservation response to development pressures. Because amenity-influenced in-migrants apparently are more likely than others to affiliate with conservation groups, they could provide such organizations with membership income, additional contributions to capital campaigns, renewed enthusiasm, and political muscle. And because such in-migrants are relatively likely to have access to the Internet at home, conservation groups may find it easier to reach and enlist them in supporting advocacy campaigns and appeals for financial support.

Indeed, local conservationists offer anecdotal reports that birding and other natural history–related activities in the Deschutes watershed—much of it relying in part on electronic communications—are attracting growing levels of interest, particularly among recent in-migrants. For example, the Deschutes Basin Land Trust is making increasing use of electronic communications to inform its growing list of members about local conservation efforts—in one case, some twenty enthusiastic birders who rely heavily on e-mail for coordination are systematically monitoring and recording bird breeding and migration activity at the Deschutes Basin Land Trust's Camp Polk Meadows property.[35] Many of these individuals actively participate in the statewide Oregon Birders Online listserv, which posts up-to-date information on bird migration activity throughout the state.[36] A young couple who obtained graduate training in ornithology in Montana before migrating to Bend in 1999 took over the local Wild Birds Unlimited franchise, which posts information and a "shop online" capability on the Web that caters to local birders.[37] And the growing number of local birders have made important contributions to a remarkably detailed Oregon Breeding Bird Atlas, which became available on CD-ROM in early 2002.[38]

New Directions for Research

Our research findings open new avenues for investigation. We hope that researchers interested in the interrelationship of nature and networks—the impact of new communications and transportation networks on land use and natural systems—will focus growing attention on this field of inquiry and pursue a larger set of questions:

- Will initial indications that household Internet penetration tends to be relatively high in faster-growing nonmetropolitan areas in Oregon stand up over years and decades?
- What will be the longer-term settlement patterns of amenity-influenced in-migrants? Now residing in a high-amenity region, will they stay, or are they likely to continue to relocate? What are the complementary consumption patterns of second-home owners, who were not included in this survey?

- Over the coming decade, what will primary- and secondary-home settlement patterns look like, particularly as higher-bandwidth, always-on Internet service (via cable, fixed wireless, and satellite technologies) becomes more pervasively available at affordable prices?
- Is the Deschutes/Crook Counties region an anomaly, or are similar patterns of Internet use, particularly among amenity-influenced recent in-migrants, occurring in other amenity-rich areas in the United States and beyond?
- What is the systematic impact of new settlement patterns on biodiversity, critical habitat, and the health of indigenous species in the Deschutes/Crook Counties area and in other areas rich in natural and recreational amenities in the United States and abroad?

If the Internet and its successors do indeed exert an enabling influence on settlement patterns over the coming fifty years comparable to that of interstate highways over the past half century, then the time to start thinking about how to deal with such forces is now. The recently arrived and soon-to-come residents of high-amenity regions will play critical roles in shaping the character of such places for generations. A key question for conservationists is how they can employ in-migrants' talents, interests, affluence, and love for the land—as well as their embrace of information technologies—to sustain the very resources that attracted them to beautiful yet vulnerable areas.

NOTES

1. Peter Elstrom, "Craig McCaw: The Prophet of Telecom," *Business Week*, September 29, 1998.
2. U.S. Department of Agriculture, Economic Research Service, "Natural Amenities and Rural Growth," available at www.ers.usda.gov/topics/view.asp?T=104024.
3. David A. McGranahan, Food and Rural Economics Division, Economic Research Service U.S. Department of Agriculture, "Natural Amenities Drive Rural Population Change," *Agricultural Economic Report*, no. 781 (1999).
4. Kenneth M. Johnson and Calvin L. Beale, "The Rural Rebound," *Wilson Quarterly* 12 (1998): 16–27, available at http://wwics.si.edu/outreach/wq/wqselect/rural.htm.

5. For example, in a number of publications over the past several decades, William Beyers has pointed out the importance of producer services and information technology in changing rural employment trends. Beyers has also said that the primary locational attraction of "lone eagles and high fliers" (independent, locationally flexible, technology-enabled workers) to rural areas is the "high quality of life." See, for example, William Beyers and David P. Lindahl, "Lone Eagles and High Fliers in Rural Producer Services," *Rural Development Perspectives* 11:3 (1996): 2–10, available at www.ers.usda.gov/Publications/RDP/rdp696/rdp696a.pdf; and William Beyers, "Trends in Producer Services Growth in the Rural Heartland," in *Economic Forces Shaping the Rural Heartland* (Kansas City, Mo.: Federal Reserve Bank of Kansas City, April 1996), available at www.kc.frb.org/PUBLICAT/heartlnd/hrtbeyer.pdf.

Priscilla Salant and her colleagues concluded that, while the few "lone eagles" did not comprise a major force behind in-migration to nonmetropolitan Washington State in the early 1990s, they do illustrate how the spatial organization of work is changing in urban and rural areas. See Priscilla Salant, Lisa Carley, and Don Dillman, "Lone Eagles among Washington's In-Migrants: Who Are They and Are They Moving to Rural Places?" *Northwest Journal of Business and Economics* (Bellingham, Wash.: Center for Economic and Business Research, Western Washington University, 1997), available in October 2001 at www.ac.wwu.edu/~cebr/loneagle.pdf.

The attractive power of natural amenities and its associated economic impact have been examined effectively by Ray Rasker. See, for example, Ray Rasker and Ben Alexander, *The New Challenge: People, Commerce, and the Environment in the Yellowstone to Yukon Region* (Washington, D.C.: Wilderness Society, October 1998), available at www.wilderness.org/ccc/northrockies/y2y_report.pdf. More recently, David McGranahan has shown a strong correlation between high levels of natural amenities and selective rural growth. See McGranahan, "Natural Amenities Drive Rural Population Change."

6. As of spring 2000, when the survey was taken, Bend and Deschutes County were officially listed as nonmetropolitan by the U.S. Census Bureau. The Census Bureau was expected to reclassify Bend as a metropolitan area based on the 2000 census.

7. Tom Kenworthy, "Peaks, Valleys Define Today's West," *USA Today*, May 18, 2001, 3A.

8. Jim Yuskavitch and Leslie D. Cole, *The Insider's Guide to Bend and Central Oregon* (Helena, Mont.: Falcon, 1999), v.

9. See data from the U.S. Census Bureau made available by the Texas A&M Real Estate Center, posted for Deschutes County at http://recenter.tamu.edu/

Data/popc/pc41017.htm and for Crook County at http://recenter.tamu.edu/
Data/popc/pc41013.htm.

10. For data on the Natural Amenities Scale, see www.ers.usda.gov/emphases/rural
data/amenities/, a Web site that complements the David McGranahan
report "Natural Amenities Drive Rural Population Change," available at
www.ers.usda.gov/Publications/AER781/. In ordinal ranking, Deschutes County
ranks 56th and Crook County ranks 116th among 3,111 U.S. counties (excluding
Alaska and Hawaii) surveyed for the study.

For an overview of data on recreational county rankings, see Kenneth M.
Johnson and Calvin L. Beale, "Recreational Counties in Nonmetropolitan
America" (revised May 14, 2002), available at www.luc.edu/depts/
sociology/johnson/p99webr.html. In Johnson and Beale's study, Deschutes
County (where most area ski and golf facilities and related revenues occur)
ranks "high" in recreational amenities. Crook County does not rank "high"
as a recreational county. Anecdotal reports of Crook County residents indicate
that they go to Deschutes County for various "paid" recreational outings, such
as skiing and golfing, but often stay in Crook County to fish, swim, or hike.

11. For climatic information on Bend and Eugene, including average rainfall and
temperature, see www.weather.com.

12. Yuskavitch and Cole, *The Insider's Guide to Bend and Central Oregon*, v. The
description of Bend in the book's preface goes on to emphasize a mix of local
amenities: "For others, [the physical attraction] isn't enough. They're in
search of a quality of life that they have not yet found. And when they discover Central Oregon, they discover the place they have been searching
for. . . . That quality is evident everywhere you look. It's in the small towns
where neighbors chat over the backyard fence or at church suppers. It's where
you can be on a wilderness trail or at a trout stream within a few minutes
drive of your house. It's where retirees find a quiet, peaceful life away from
crowded suburbs and families find a safe place to raise their children. It's
where entrepreneurs are creating new businesses and economic opportunities in science, technology, and information alongside the area's traditional
industries of ranching, farming, and logging. Most important of all, in the
grand tradition of the West, it's a place where people come to pursue their
dreams and make them real."

13. Yuskavitch and Cole, *The Insider's Guide to Bend and Central Oregon*, 6.

14. Gerald Williams and Stephen Mark, *Establishing and Defending the Cascade Range
Forest Reserve from 1885 to 1912: As Found in Letters, Newspapers, Magazines, and
Official Reports* (Portland, Ore.: U.S. Forest Service, Pacific Northwest Region,
and U.S. Park Service, Crater Lake National Park, September 1995), available
at http://fs.jorge.com/archives/Cascade_Range/Introduction.html.

15. For further information on the rivalry between Hill and Harriman, see historical information on the Crooked River Bridge at www.odot.state.or.us/region4/crooked/history.htm.

16. To search online for population information on the area, see the Web site of the Texas A&M University Real Estate Center, available at http://recenter.tamu.edu. Data are based on information provided by the U.S. Bureau of the Census.

17. U.S. Census Bureau, Texas A&M Real Estate Center, available at http://recenter.tamu.edu.

18. Several analysts, including Ray Rasker, of the Sonoran Institute, and Spencer Philips, of the Wilderness Society, have taken the lead in documenting this economic transition. For example, see Rasker and Alexander, "The New Challenge"; see also Spencer Phillips, *Windfalls for Wilderness: Land Protection and Land Value in the Green Mountains* (Washington, D.C.: Wilderness Society, 2000), available at www.wilderness.org/newsroom/pdf/windfallsforwilderness.pdf.

19. Dean Runyan Associates (DRA), *Oregon Travel Impacts, 1991–2001* (2002), available at www.dra-research.com/impactsORstate.html. In a report for the Oregon Tourism Commission, DRA reports that, including air travel, travel spending in Deschutes and Crook Counties grew from $219.6 million in 1991 to $334.4 million in 1999, an increase of more than 50 percent in eight years.

20. For further information on the relationship between travel and telecommunications, see John S. Niles, *Beyond Telecommuting: A New Paradigm for the Effect of Telecommunications on Travel* (Washington, D.C.: U.S. Department of Energy, Office of Energy Research, 1994), available at www.lbl.gov/ICSD/Niles.

21. Deschutes County, Oregon, "Draft Deschutes County Final RPS Report" (June 2, 1999).

22. Oregon Water Resources Department, "The Deschutes Ground Water Work Group" (December 13, 1999), 5, available at http://powder.wrd.state.or.us/programs/deschutes/overview.shtml.

23. A history and report on the current projects of the Deschutes Resources Conservancy are available at the organization's Web site, available at www.deschutesrc.org/drchist.htm. For additional information on the DRC and its role in the Camp Polk Meadow Project, see "Testimony in Support of H.R. 1787 by Ron Nelson, Chairman, Deschutes Basin Resources Conservancy, April 6, 2000," available at www.house.gov/resources/106cong/water/00apr06/nelson.htm.

24. "PGE Donation Protects Essential Steelhead Habitat in Deschutes Basin," available at www.portlandgeneral.com/about_pge/corporate_info/news/archives/pge_donation_protects_steelhead_habitat_deschutes_basin.asp.

25. We based our estimate that 50 percent of all residents had lived in the area for more than ten years on population growth estimates by the U.S. Bureau of the Census in 1999 as well as 1990 census data on mobility, or "churn."

26. The Hughes VSAT system installed for Les Schwab is described in a news story posted at www.lunacity.com/spacer/199905.html.

27. The telephone numbers were acquired from STS Inc. in random order and were prescreened for nonworking status. The sample frame consisted of all active 100-digit banks of telephone numbers in Deschutes and Crook Counties. The sample size was determined by four criteria: (1) there should be at least 150 complete interviews of in-migrant households; (2) there should be a total of at least 150 complete interviews with long-term households; (3) each number in the phone sample should have an equal probability of resulting in a complete interview; and (4) the sample size should be minimized subject to the foregoing three criteria. To satisfy the third criterion, we established a protocol for the number and timing of call attempts that would be made to each number. This resulted in a final sample size of 347, somewhat larger than the theoretical minimum.

A total of 2,942 dial attempts, not including busy signal calls, were made by Jennifer Reese, based in Cambridge, Massachusetts, to Deschutes and Crook Counties, Oregon, between March 7 and May 4, 2000. A total of 1,043 unique telephone numbers were called.

Of the numbers called, 322 numbers were ineligible (i.e., they were assigned to fax or data lines, business lines, second homes rather than primary residences, or lines that were not working). There were 214 calls to numbers that had an "undetermined status" (i.e., there was no response after six call attempts over a period of at least ten days).

There were 160 numbers that were "refusals" (this number includes one household that had already been interviewed through a call to a different phone line and ten households for which the interviewer judged the potential respondent to be incapable of completing the questionnaire—four potential respondents were hard of hearing, one potential respondent had a severe speech impediment, two potential respondents did not speak English, one potential respondent was a confused elderly person, and two potential respondents lived in nursing homes).

There were 347 numbers called that belonged to households that provided complete interviews—that is, 68 percent of eligible households gave completed interviews. At the beginning of each interview, the interviewer asked to "speak with a head of a household." If no "head of household or their spouse" was available, the interviewer asked to speak with another adult "capable of answering all questions." Among these 347 households interviewed, there were no incomplete interviews. All interviews that were started and eligible were completed. All but one of the respondents identified himself or herself as a householder (342 respondents) or a spouse of a householder (4 respondents). Having noted this exception, we describe the respondents to this survey as "householder-respondents."

28. The design of the questions on Internet use was informed by other surveys of Internet usage performed in the recent past, including the survey associated with U.S. Department of Commerce, National Telecommunications and Information Administration, *Falling through the Net: Defining the Digital Divide* (revised November 1999), available at www.ntia.doc.gov/ntiahome/digitaldivide/index.html, as well as the "1999 Pilot Survey of Oregon Telecommunications," described by Toshihiko Murata and Patricia Gwartney in *Oregon Economic and Community Development Department's Oregon Household Telecommunications Survey, Methodology and Results* (winter 2000), available at http://darkwing.uoregon.edu/~osrl/telecomoedd/frmtelecom.htm. The questionnaire covered the following topics:

1. Location and length of residence
2. Household computer presence and use
3. Household Internet presence and use (general)
4. Household composition (ages and relationships of household members)
5. Respondent-householder information

- Internet use, purpose, frequency, mode of access, and location
- Frequency of use and purpose of use (work or other) of other communications technologies (fax, cell phone, express delivery)
- Employment status and location
- Occupation
- Length and mode of commute to work
- Demographic information (age, gender, race, level of education)

6. Second householder-spouse

- Internet use, purpose, frequency, mode of access, and location
- Employment status and location
- Occupation
- Length and mode of commute to work
- Demographic information (age, gender, race, level of education)

7. Internet use by other adults in the household
8. Internet use by children in the household
9. Housing characteristics
10. Reasons for moving to the area and location of previous residence
11. Direct use of residential land, water and sewer connections
12. Affiliations with conservation organizations
13. Household income

29. The survey software used during the interview process was SPSS Data Entry, version V. The sample is valid subject to the unavoidable limitations of any sample survey. Sampling error is in the moderate range: +/–2.8 percent for a characteristic with 50 percent frequency in the full sample. The sample characteristics (e.g., age, household size, tenure, income, location within the region) are similar to comparable benchmark distributions from other sources, such as the 1990 Census and the Oregon Population Survey.

In the course of a small number of interviews, the respondent passed the phone to a second householder or adult who was more knowledgeable about the technology in question. These respondents are, of course, more likely to use the technology than the initial respondents. Any resulting response bias is at the "respondent" and "second adult" level only. Since the interviewer coded the alternate as the primary respondent, use of the Internet and perhaps other communications technologies by the respondent-householder is therefore likely biased upward, and conversely, their use by the "second adult" householder is biased correspondingly downward. However, at a household level, the only effect should be greater accuracy than if the initial, less knowledgeable respondents had been required to complete the interview. With the exception of this item, there are no apparent sampling biases.

One set of answers is a noteworthy indicator of the general increase in "connectedness." In response to a question about the number of phone lines in the household, forty-three (12 percent) of the household respondents reported more than one voice line. As noted above, one household was contacted twice on two different lines. To adjust for higher-than-average inclusion of these households in the survey, their responses were weighted down in inverse proportion to their number of voice lines.

30. Neilsen/Net Ratings and Media Matrix numbers reported in Ben Charny, "More U.S. Households Online Than Not," *ZDNet News*, August 16, 2000, available at www.zdnet.com/zdnn/stories/news/0,4586,2616761.html.

31. The category "to be near recreational activities/locations" includes such responses as "dry climate, ski and golf," "beautiful climate, great skiing," and

"to snowboard." The category "to be near the area's natural amenities or the environment" includes such responses as "like the outdoors," "the most beautiful place on earth," and "loved it, beautiful there."

32. To test the significance of other factors that might differentiate Internet households from non-Internet households, one of the authors, John Pitkin, built a model that takes household head age, income level, and household size into account. The model employs a binomial logistic regression methodology to test whether an independent variable has statistical significance after controlling for age, income, and household size.

33. Based on the estimated regression model described above.

34. Only a few respondents (i.e., six respondents) cited the Internet itself as a primary or secondary motivation for their recent in-migration to the Bend area. Regarding interstate highways in the 1960s and 1970s, it is more likely that middle-class urban residents chose to move to the suburbs for such reasons as better schools, less crime, and affordable homes with spacious yards. The interstate highways are more aptly characterized as "facilitating" suburbanization. See, for example, Judy Davis for Parsons Brinckerhoff Quade & Douglas, "Consequences of the Development of the Interstate Highway System for Transit," Transportation Research Board, National Research Council, Transit Cooperative Research Program, *Research Results Digest* 21 (August 1997): 3, available at http://nationalacademies.org/trb/publications/tcrp/tcrp_rrd_21.pdf. Davis reports that analysts take two main positions regarding the effects of the interstate highway system on transit. One is that "the interstate highway system facilitated the suburbanization of households and jobs, creating origins and destinations that were difficult for conventional transit to serve."

35. Information describing the Deschutes Basin Land Trust is posted at www.deschuteslandtrust.org. See the results of the Camp Polk Christmas Bird Count (CBC), part of the local CBC circle and the National Audubon CBC for North America, at the Web site posted by Paradise Birding available in August 2001 at http://paradisebirding.com/sys-tmpl/nss-folder/sisterscbc2000/SisCBC2000.html.

36. Information on Oregon Birders Online, see www.cyber-dyne.com/~lucyb/obol.html.

37. See the Web site for Wild Birds Unlimited of Bend at www.wbu.com/bend/.

38. Discussion by author James Levitt, with Paul Adamus, ornithology consultant, Corvallis, Ore., September 2001.

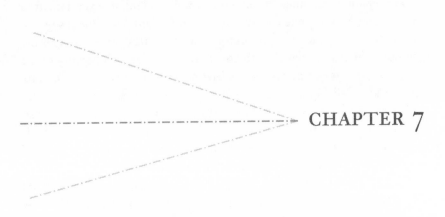

CHAPTER 7

Rural Development and Biodiversity

A CASE STUDY OF GREATER YELLOWSTONE

Andrew J. Hansen and Jay J. Rotella

Attracted by stunning scenery and abundant opportunities for outdoor recreation, many Americans are choosing to build primary or secondary homes in the forests and rangelands on the periphery of Grand Teton, Rocky Mountain, Glacier, and other national parks.[1] As in many areas of rural America, the human population living on the lands surrounding these and other nature reserves is growing dramatically. As explained in preceding chapters, it appears that myriad factors are influencing in-migration to such areas, including the availability of new information technologies and advanced transportation systems that allow knowledge workers and others to locate their homes and conduct their business in exurban locations.

Scientists have only recently begun to understand how rural residential development outside nature reserves may influence wildlife within the reserves.[2] However, early indications are that "nature seekers" moving to these areas may be degrading the very natural amenities and wildlife that attracted them in the first place—including biodiversity within protected areas.

This chapter examines the effects of rural residential development in the Greater Yellowstone ecosystem on native plant and animal species. Our central thesis is that although human densities are still low in the region, human activities tend to concentrate in the small portion of the landscape that is critical to many native species. This is because private lands under development tend to occupy fertile valley bottoms that are home to abundant wildlife. Yellowstone National Park, in contrast, encompasses mostly high plateaus and mountains with a harsher climate and poor soils.[3] Many plants and animals in the park may depend on foods and habitats in more favorable landscape settings outside the park. As rural residential development and other land uses intensify on these lands, the risk of extinction for wildlife populations within the park may grow.

Like Yellowstone National Park, many other nature reserves in the United States and around the world protect relatively harsh landscape settings. We speculate that intensifying land use in areas abutting these reserves is eroding biodiversity and ecological functioning in protected areas across the globe. If this thesis is correct, new conservation initiatives outside nature reserves will prove essential to maintaining native species within reserves such as Yellowstone National Park.

Rural Residential Development and Wildlife

Scientists have much to learn about how rural residential development affects biodiversity, including how these effects may vary with home density and spacing and how they compare with other land uses, such as agriculture, logging, and grazing. Nonetheless, a growing body of evidence indicates that rural residential development may substantially influence the populations of many native species. These effects can occur through several routes, including changes in habitats, ecological processes, and interactions among species of wildlife, as well as direct human disturbance of wildlife.[4]

Habitat Destruction

The creation of homes, lawns, and roads destroys the habitat of native species and may also alter the habitat's configuration, yielding smaller patches of native habitat, more abrupt edges, and greater distances between patches. Native species in such disrupted habitats may be more susceptible to pressure from human settlements and less likely to disperse among the habitat fragments that remain. Intense land use also creates new habitats that allow exotic species of weeds, invertebrates, predators, and diseases to invade more easily, crowding out native species. The combined effects of small populations of native species, fragmented habitats, more difficult movement, and invasion by exotic species can greatly increase the risk that native species will become extinct.

Alteration of Ecological Processes

Disturbing habitats modifies their energy flow and nutrient availability, pushing the types and rates of ecological processes outside the range required by native species. For example, new rural development often brings more emphasis on suppressing fires—which means that the fires that do occur are more severe. Both influences curtail the habitat of fire-dependent species. Efforts to control floods to protect homes similarly may mean that flooding is less frequent but more severe, and thus that native species cannot survive.

Sewage and landscaping-associated outflow from homes may alter aquatic communities and reduce water quality in streams and lakes by changing the flow of nutrients—for example, increased levels of nitrogen and phosphates associated with residential septic systems and lawn management practices may substantially change the chemical balance for relatively small (for example, algae) and large organisms (for example, trout) that live in these aquatic environments.

Alteration of Biotic Interactions

Growing numbers of pets, invasive plants, and other human-adapted species can significantly change the nature and intensity of interactions among species, boosting parasitism, predation, disease, and competition. For example, populations of predators such as coyotes and of species such as the brown-headed cowbird that prey on the broods of other species often rise with growing rural development because of the increased food supplies near rural homes. These predators dramatically reduce the number of small mammals and songbirds. Diseases may also spread more rapidly among wildlife that congregate at rural residences, including among

Figure 7.1: The Greater Yellowstone ecosystem includes Yellowstone and Grand Teton National Parks, several national forests, and privately owned property. (Map by David Deis, Dreamline Cartography.)

birds attracted to feeders. Contact with livestock and pets may also infect wildlife with disease.

Human Disturbance

People and pets can alter the population dynamics of native species by directly or indirectly displacing them from the habitats they require. Domestic dogs and cats may roam up to several miles from home and chase or prey upon a variety of native species. The more numerous roads and rising traffic associated with rural residential development can also kill significant numbers of wildlife. Also, home owners often shoot bears or cougars attracted to food sources linked to rural residences or contact wildlife control officials, who either remove or destroy the offending animals.

Rural Development and Biodiversity in the Greater Yellowstone Ecosystem

The Greater Yellowstone ecosystem encompasses Yellowstone National Park and the surrounding public and private lands that are ecologically and economically connected to the park (see Figure 7.1). This vast area embraces national forests, national wildlife refuges, and other public lands as well as private lands used for forestry, grazing, farming, or homes.

The ecosystem includes strong variations in topography, climate, and soils.[5] Yellowstone and Grand Teton National Parks occupy high elevations, including the Yellowstone Plateau and surrounding mountain ranges. Other public lands sit largely at mid-elevations, on the flanks of the plateau. In striking contrast, private lands primarily occupy valley bottoms and the plains surrounding public lands.

Because volcanic activity created the Yellowstone Plateau, soils at higher elevations consist largely of nutrient-poor rhyolites and andesites with low water-holding capacity. Many valley bottoms, in contrast, contain fertile alluvial soils. Because of the long, harsh winter at higher elevations, the summer growing season in the national parks lasts only about two months, while the growing season in valley bottoms extends for five or six months. Because of these variations in soil and climate, plant production is much higher in rich valley bottoms than in the national parks. More foods are thus available for herbivorous species and their predators at lower elevations.

The vast Greater Yellowstone ecosystem supports such wilderness species as wolves, grizzly bears, and cutthroat trout, and the area is also home to free-roaming herds of elk, bison, and pronghorn antelope. Many of these species respond to strong gradients in climate and foods by either specializing in low-elevation habitats or migrating seasonally between habitats at low and high elevations.

Although the human population of the Greater Yellowstone ecosystem is relatively small, it has grown by 55 percent since 1970, spawning large changes in land use.[6] People still concentrate in small cities, such as Bozeman, Montana, and Jackson, Wyoming, and in surrounding ranch

Figure 7.2: Many new residents in the Greater Yellowstone ecosystem have chosen to live out of town on "ranchettes" and in rural subdivisions, making sprawl a major issue. This new development is in Little Bear, Montana, just south of Bozeman, within the Greater Yellowstone ecosystem. (Courtesy of Andrew J. Hansen, Montana State University.)

lands, but the area these cities occupy has grown 348 percent over the past thirty years. Movie stars, corporate executives, and many other professionals are relocating here, as the local economy shifts from resource industries to a New West mix of traditional and new sectors such as high-tech firms, real estate, and recreational enterprises. Many new residents have chosen to live out of town on ranchettes and in rural subdivisions (see Figure 7.2); in fact, the number of residences outside city limits has grown nearly 400 percent since 1970. Rural homes are now located along nearly all the major rivers that flow from the Yellowstone Plateau, and sprawl has become a major issue.

Thus—although the Greater Yellowstone ecosystem remains a vast wilderness with relatively low human population density—private lands, human dwellings, and intense land use are concentrated in the low-elevation settings where climate and soils produce favorable foods and habitats for many native species. How might residential development of these privately owned lowlands influence wildlife populations in the region's nature reserves? Our studies of birds suggest that development on private lands may have a large effect on these populations. Other studies suggest that many mammals, fish, and plants also are at risk from growing rural development.

Biotic Interactions

Our detailed studies of birds in the Greater Yellowstone ecosystem provide an example of how changes in biotic interactions associated with rural residential development can strongly influence native wildlife.[7] We sampled the distributions of more than 100 bird species across the northwestern portion of the ecosystem (the "study site" delineated in Figure 7.1, also called here the "study area"). The goal was to determine how the abundance of individual species, as well as community diversity, varies across different elevations, soils, climates, and habitats. We also studied the reproduction of specific bird species to pinpoint areas where they

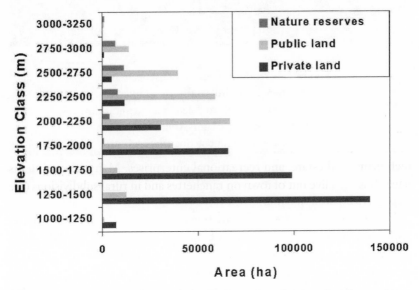

Figure 7.3: In the study area, private land is located predominantly at lower elevations. Like many species of breeding birds, private home owners who settle in the Greater Yellowstone ecosystem prefer to live in places at lower elevations with relatively mild climates, good soils, rivers, and deciduous forests. (Graph by Ute Langner, Montana State University.)

Figure 7.4: Yellow warblers in hot spots near homes suffered cowbird parasitism; American robins were better able to defend their nests against cowbirds and other predators. (Photo courtesy of the Massachusetts Audubon Society.)

were growing and declining.

We selected birds for study for several reasons. First, they provide important ecosystem services that benefit humans, such as controlling insect outbreaks. Second, the populations of many species of songbirds that migrate to and from the tropics have been declining in the United States, but the causes of these declines are poorly understood. Finally, birds rely on the types of foods and habitats that many other species also need and thus may be good indicators of the health of wildlife that are more difficult and expensive to study.

We found that bird species were not distributed randomly across different kinds of topography, climate, and other environmental gradients. Rather, they were associated with landscape settings that were warmer in June and lower in elevation and that contained rich soils and higher plant productivity. These sites were dominated by deciduous forest cover consisting of trees such as aspen, cottonwood, and willow.

We found that such landscape settings were relatively rare in the study area. "Hot spots"—where bird species were numerous and abundant—encompassed less than 3 percent of the study area and were mostly on or near private lands. Only about 7 percent of hot-spot areas fell within nature reserves. Hence, reserves like Yellowstone National Park—considered by many people primary habitat for native wildlife—contain relatively few hot-spot habitats for migratory birds, which are more abundant on private lands at lower elevations (see Figure 7.3).

How is land use distributed relative to these bird hot spots? Within private lands, we found that rural homes were placed disproportionately close to hot spots. Like many species of birds, home owners apparently prefer mild climates, good soils, rivers, and deciduous forests, typically found in the Greater Yellowstone ecosystem at relatively low elevations. This raises the question of how rural homes affect birds within these hot-spot habitats.

We found that human development influenced birds in hot spots primarily by favoring avian predators such as the black-billed magpie and the brown-headed cowbird, which feed on the broods of other species. These species were especially abundant near rural residences. Other bird species that are susceptible to nest parasitism and predation suffered very low rates of reproduction. For example, nearly half of all nests of yellow warblers in hot spots near homes suffered cowbird parasitism, and only about 20 percent of the nests successfully fledged young. The American robin, in contrast, was able to better defend its nests against cowbirds and predators; nest success in hot spots near homes for this species was nearly 50 percent higher than for the yellow warbler (see Figure 7.4).

Net population growth provides an important measure of these differences in reproductive success. Using computer simulation models extrapolated from field data, we estimated that the American robin population is growing substantially each year in these hot spots, while the yellow warbler population is declining. Apparently the population of yellow warblers is unable to sustain itself in the face of the higher nest parasitism and predation associated with rural residential development.

In addition, in projecting the population dynamics of yellow warblers across the study area, we found that changes in biotic interactions on small portions of private lands can raise the risk of extinction in nature reserves tens of kilometers away. This is because yellow warblers within the park have relatively low reproduction rates, owing to the harsh climate at higher elevations. These habitats are likely "population sinks," where mortality exceeds reproduction. When we modeled the viability of yellow warbler populations in the reserves, we found a high probability of extinction if we assumed that no immigrants arrived from outside the reserves. In fact, nearly half of the simulated populations within the reserves went extinct within fifty years.

These results led us to hypothesize that, in presettlement times, immigration from low-elevation hot spots—which acted as population sources—maintained populations in these high-elevation sinks. We further hypothesize that the expansion of rural residential development and other intense use of lowlands have likely converted biodiversity hot spots, which were source areas for species such as the yellow warbler, to population sinks. This has reduced the viability of subpopulations even in protected areas such as Yellowstone National Park.

It is important to point out that the yellow warbler is a widespread and abundant species in North America and is not threatened with extinction. The fact that the species is abundant allowed us to gather enough data to describe this source/sink mechanism of population extinction. The significance of our results lies in the implications of this mechanism for other species that may be more at risk nationally.

The pronghorn antelope provides a second example of how changes in biotic interactions outside Yellowstone National Park may raise the risk of extinction within it.[8] This grassland species historically migrated between highland summer habitats within the park and lowland winter habitats in Yellowstone Valley. Although the pronghorn antelope is protected in the park, the population has declined substantially in the past decade, to about 200 individuals. Scientists have not established the causes of this rapid decline. However, they speculate that farming and rural residential development in low-elevation habitats have favored coyotes and other predators, producing high mortality among antelope when the herd is in the winter range. The replacement of winter range with crops and periodic hunts to reduce crop damage are also possible contributing factors. This population is now considered highly vulnerable to extinction.[9]

Habitat Destruction

Human alteration of natural habitats may influence many other wildlife species in the Greater Yellowstone ecosystem. Rural residential development, farming, and livestock grazing have destroyed much of the habitat of grassland-dependent species. Mark R. Clawson has found that several grassland bird species that are endemic to Greater Yellowstone and declining regionally and nationally are not present in hayfields and croplands in the northwest portion of the ecosystem.[10]

Another lowland habitat of concern is the deciduous woodland, found largely in riparian areas. Like birds, several species of mammals, such as the water shrew; invertebrates, such as Gillette's checkerspot butterfly; and amphibians, such as the tiger salamander, live primarily in riparian woodlands.[11] Some 50 percent of the aspen, cottonwood, and willow forests on private lands within the Greater Yellowstone ecosystem are within two kilometers of a rural home. Homes within the woodlands have reduced this habitat, and homes adjacent to the woodlands may also jeopardize native species by changing biotic interactions.

Just as habitat destruction is undermining some native species, the creation of human habitats and other human activities are confronting Yellowstone and Grand Teton National Parks with serious problems from exotic species and diseases. Exotic plants and animals may establish along roads and in lawns and pastures near rural homes, serving as population source areas that allow non-native species to invade surrounding natural habitats. Weedy plants, such as knapweed and spurge, for example, are displacing native plants and reducing forage for wild grazing animals. The population of New Zealand mud snails is growing rapidly in some streams in Yellowstone National Park, displacing many native mollusks and possibly cutting off food for native trout. Further rural residential development in the Greater Yellowstone ecosystem likely will accelerate the invasion of exotics into the national parks.

Human Disturbance

Death and displacement of wildlife by humans and pets are especially likely among the many species of mammals in the ecosystem that move between higher and lower elevations seasonally or during their life cycles. Elk that summer in Yellowstone National Park, for example, migrate as many as 100 kilometers to winter in lowland valleys, often on private lands.[12] Bison,

Figure 7.5: Among the many species of mammals in the Greater Yellowstone ecosystem, grizzly bears may be the most at risk from human disturbance. (Photo courtesy of the National Park Service.)

moose, pronghorn antelope, and grizzly bear also make pronounced seasonal movements between high and lower elevations.

Among these species, grizzly bears may be most at risk from human disturbance (see Figure 7.5). As the population of the threatened grizzly bear recovers in Yellowstone National Park under protection from the Endangered Species Act, bears are increasingly observed in productive low-elevation habitats. When they are attracted to rural subdivisions, people often relocate or kill them to protect humans and pets. Grizzlies also die on roads and in encounters with hunters. Thus a disproportionate number of grizzly bear mortalities are occurring on private lands. In and around Glacier National Park, in Montana, for example, federal biologists report that more than 60 percent of conflicts between grizzlies and humans occur on private lands, even though such lands represent only 17 percent of the region.[13] Overall rates of human-induced mortality of grizzlies in the Northern Rockies have grown in each of the last three years to record high levels. Many conservationists argue that higher human densities on private lands will threaten the recovery of this species.

The migration of elk to lowlands provides an opportunity for game managers to reduce an overpopulation in Yellowstone National Park. Hunting seasons are designed to harvest animals after they move to unprotected lands at lower elevations in fall and winter.[14]

The migration of bison among elevations, in contrast, presents a substantial management challenge.[15] Some bison carry the cattle disease brucellosis. The policy of livestock agencies is to prevent bison from mixing with cattle and possibly transmitting the disease. But during severe winters, bison migrate out of Yellowstone National Park because it does not contain important low-elevation range. The primary response has been to kill bison that migrate out of the park, creating a substantial public controversy. Giving some of the productive low-elevation habitats protected status likely would minimize this problem.

Finally, two species of birds in the Greater Yellowstone ecosystem became endangered in the last century, probably because of human disturbance in lowland habitats. The trumpeter swan and bald eagle populations suffered high mortality owing to shooting and other human pressures. Since coming under the purview of the Endangered Species Act, these species have rebounded, and their populations are growing most rapidly in low-elevation riparian and other wetland habitats.[16] As with yellow warblers, climate limits the number of these species at higher elevations in the Greater Yellowstone ecosystem. Maintaining the quality of low-elevation riparian and other wetland habitats therefore may prove essential to the regional recovery of these two species. However, sustaining such quality will be increasingly difficult as rural residential development expands in these lowland areas.

Ecological Processes

Changes in water quality in streams and the introduction of exotic species have strongly affected native fish in the Greater Yellowstone ecosystem. In presettlement times, native west-slope cutthroat trout and Arctic grayling lived throughout the watershed, from headwaters down to major rivers.[17] Fish probably dispersed among headwaters through rivers lower in the watershed. The introduction of exotic trout into major rivers, and possibly habitat changes associated with irrigation and other intense land use, led to the extinction of cutthroat and grayling in those rivers. Grayling went extinct even in Yellowstone National Park, although they have since been reintroduced, while west-slope cutthroat were able to persist only in headwater streams. These headwater populations of cutthroat have a relatively high probability of extinction because they are no longer able to disperse across the major rivers.[18]

Protecting Biodiversity Hot Spots

Overall, these studies suggest that the wave of new residents in rural America may be strongly influencing biodiversity even in large wilderness areas such as the Greater Yellowstone ecosystem. Although human densities in such regions are still low, human activities concentrate in the small portion of the landscape that is critical to many native species. Because nature reserves tend to occupy higher elevations and have harsher climates and poorer soils than private lands, many species in these reserves may depend on more favorable landscapes outside them.[19] As rural residential development and other land uses intensify adjacent to protected areas, such as Yellowstone National Park, new conservation initiatives will prove essential to maintaining native species within those areas.

Strategies that could help achieve this sustainability include the following:[20]

- *Integrated assessment and management of public and private lands.* Administrative boundaries of the Greater Yellowstone ecosystem correspond poorly with ecological boundaries. This means that programs that aim to preserve biodiversity in the region are often costly and of limited effectiveness. To make them more effective, land managers need to assess land use and to link goals across jurisdictions. A successful example is the cooperative management of elk in the Montana portion of the Greater Yellowstone ecosystem. This management occurs across federal, state, and private lands that compose the high-elevation summer and low-elevation winter habitats of the elk.[21]
- *The development and use of decision-support tools for land management.* Field studies designed to assess biodiversity rapidly, remote sensing, geographic information systems, and computer simulation models are powerful tools for evaluating past ecological change and projecting future change under different land management scenarios. Unfortunately, many state agencies and local governments do not have access to such tools because they lack funds or technical expertise. Cooperative programs are needed to allow federal, state, and local agencies to share the development and use of such decision-support tools.
- *Prioritization of lands based on ecological and socioeconomic goals.* Sound human judgment along with decision-support tools must identify regions of the landscape best dedicated to ecological objectives ver-

sus socioeconomic development. Local governments can then use this information to decide where to encourage subdivisions and commercial activities and where to protect green space. Our analysis indicates that protecting only 2.7 percent of the land in the Greater Yellowstone ecosystem—the portion that represents biological hot spots—could substantially advance biodiversity goals. Land trusts and other organizations that protect fragile areas through purchase of conservation easements need to perform similar analyses to pinpoint hot spots. The federal government could also expand its Conservation Reserve Program, which pays farmers to avoid plowing marginal croplands, to protect agriculture lands that are important to biodiversity.

• *Public education.* Many home owners, real estate agents, and recreationists are poorly informed about the ecological consequences of their daily decisions. Rather than assuming that land managers and public lands can fully protect important ecosystems, residents and recreationists need to share in the pursuit of such community goals. To encourage this participation, educational programs are desperately needed to teach citizens how to live more lightly on the land. Elected officials would also benefit from training in managing growth, including analyzing fiscal impacts of different forms of development and land use, integrating scientific information into growth management plans, using conservation easements and other options for publicly financing protected areas, and conducting citizen-driven planning and zoning.

ACKNOWLEDGMENTS

This manuscript was improved through discussions with Robert Garrott, Bruce Maxwell, Cal Kaya, Tom McMahon, Charles Schwartz, and Brad Shepard. This research was supported by NASA's Land Cover/Land Use Change Program.

NOTES

1. Ray Rasker and Andrew J. Hansen, "Natural Amenities and Population Growth in the Greater Yellowstone Region," *Human Ecology Review* 7:2 (2000): 30–40.

2. See, for example, John M. Marzluff, F. R. Gehlbach, and David A. Manuwal, "Urban Environments: Influences on Avifauna and Challenges for the Avian Conservationist," in John M. Marzluff and Rex Sallabanks, *Avian Conservation: Research and Management* (Washington, D.C.: Island Press, 1998).

3. Andrew J. Hansen et al., "Spatial Patterns of Primary Productivity in the Greater Yellowstone Ecosystem," *Landscape Ecology* 15 (2000): 505–22.

4. Scott Powell et al., "Rural Residences and Biodiversity: How Where We Live Influences Native Species," *Conservation Biology* (in review).

5. Hansen et al., "Spatial patterns of primary productivity in the Greater Yellowstone Ecosystem."

6. Andrew J. Hansen, Jay J. Rotella, and Matthew L. Kraska. "Ecology and Socioeconomics in the New West: A Case Study from Greater Yellowstone," *BioScience* 52:2 (2002): 151–58.

7. Andrew J. Hansen et al., "Dynamic Habitat and Population Analysis: A Filtering Approach to Resolve the Biodiversity Manager's Dilemma," *Ecological Applications* 9:4 (1999): 1459–76. Hansen et al., "Ecology and Socioeconomics in the New West." *BioScience* 52:2 (2002): 151–168. Andrew J. Hansen and Jay J. Rotella, "Biophysical Factors, Land Use, and Species Viability in and around Nature Reserves," *Conservation Biology* (2002).

8. Yellowstone National Park, *Yellowstone's Northern Range: Complexity and Change in a Wildland Ecosystem* (Mammoth Hot Springs, Wyo.: National Park Service, 1997).

9. Daniel Goodman, "Viability Analysis of the Antelope Population Wintering near Gardiner," National Park Service report, Yellowstone National Park, Wyo., 1996.

10. Mark R. Clawson, "An Investigation of Factors That May Affect Nest Success in CRP Lands and Other Grassland Habitats in an Agricultural Landscape" (master's thesis, Montana State University, 1996).

11. Susan K. Skagen, E. Muths, and R. D. Amans, *Towards Assessing the Effects of Bank Stabilization Activities on Wildlife Communities on the Upper Yellowstone River, U.S.A.* (Fort Collins, Colo.: U.S. Geological Survey, Midcontinent Ecological Science Center, 2001).

12. Yellowstone National Park, *Yellowstone's Northern Range.*

13. Chris Serveen, personal communication with author Andrew J. Hansen, August 17, 2001.

14. Michael B. Coughenour and Francis J. Singer, "Elk Population Processes in the Yellowstone National Park under the Policy of Natural Regulation," *Ecological Applications* 6:2 (1996): 573–93.

15. Paul Schullery, *Searching for Yellowstone: Ecology and Wonder in the Last Wilderness* (Boston: Houghton Mifflin, 1997).

16. Terry McInenny, personal communication with author Andrew J. Hansen, July 16, 2000.

17. John D. Varley and Paul D. Schullery, *Yellowstone Fishes: Ecology, History, and Angling in the Park* (New York: Stackpole, 1998).

18. Brad Shepard et al., "Status and Risk of Extinction for Westslope Cutthroat Trout in the Upper Missouri River Basin, Montana," *North American Journal of Fisheries Management* 17 (1997): 1158–172.

19. J. Michael Scott et al., "Nature Reserves: Do They Capture the Full Range of America's Biological Diversity?" *Ecological Applications* 11:4 (2001): 999–1007.

20. Hansen et al., "Ecology and Socioeconomics in the New West."

21. Schullery, *Searching for Yellowstone.*

PART III

HARNESSING THE POWER OF
NEW NETWORKS TO ACHIEVE
CONSERVATION OBJECTIVES

Paradoxically, even as the Internet and its kin appear to be enabling demographic shifts, changes in land use patterns, and associated environmental disruptions, as discussed in previous chapters of this book, the new networks are also having profound constructive impacts across a wide array of conservation practice areas. Such practice areas include conservation science, conservation education, advocacy, resource management, and land use planning. The following chapters offer insight into the ways that leading conservation organizations are using new networked information technologies to make strategic, long-term differences in how they pursue their missions.

One effort, spearheaded by the University of Kansas Natural History Museum and Biodiversity Research Center, is described in Chapter 8. The initiative, which enables the integration of the records of natural history museums all over the world, allows conservation scientists to work with unprecedented speed and precision. In another case, through the BirdSource partnership between the Cornell Lab of Ornithology and the National Audubon Society described in Chapter 9, citizen science and conservation education are being taken to a new level. BirdSource engages tens of thousands of observers to gather data on bird population status and migrations at a continental scale. A case study of a Natural Resources Defense Council campaign presented in Chapter 10 details new tactics and strategies used by environmental advocates in achieving an important success in habitat protection. And in Chapter 11, the authors describe how the

Orton Family Foundation, recognizing the complex demographic and environmental pressures in states such as Vermont and Colorado, is developing new software applications to help rural communities plan for growth and change. These examples offer a window on a period of vital growth and change in the land and biodiversity conservation communities-of-practice.

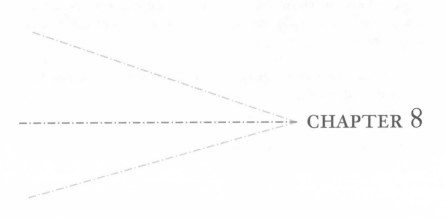

CHAPTER 8

The Green Internet

A TOOL FOR CONSERVATION SCIENCE

Leonard Krishtalka, A. Townsend Peterson,
David A. Vieglais, James H. Beach, and E. O. Wiley

The Grand Challenge

In the twenty-first century, the grand challenge for science is to understand global environmental systems in order to use them in a sustainable fashion.[1] This knowledge is critical to science and society—for managing natural resources, for improving human health, for ensuring economic stability, and for enriching the quality of human life. The urgent need for this knowledge increases daily as the ongoing disruption of natural systems accelerates the decline of the earth's natural environments and biological diversity.

At the current rate of species extinction, science has about fifty years to answer this challenge.[2] If medical science is concerned with the health of one species on the earth, biological science is concerned with the evolutionary and ecological pulse of the earth's other 10 to 50 million species and the ecosystems they compose. The twenty-first century, as E. O. Wilson predicts, will be the century of the environment.[3]

James Levitt, in this volume, has established three incisive points about collateral impacts of transportation and communications networks on the environment:

1. In the past 200 years, transportation and communications networks—from the railroad, telegraph, and telephone to electric power grids and the U.S. interstate highway system—have been associated with enormous shifts in demographic patterns and economic markets.
2. These shifts have had dramatic impacts on land use, associated with a relentless disruption of natural environments and their biological diversity.
3. The newest, fastest-growing, and most pervasive communications network is the Internet. By enabling people to live and work anywhere, it is associated with a new wave of changing land use patterns, especially the movement to and settling of outlying suburban, exurban, and rural locations. The proliferation of new networks, associated demographic changes, and land use shifts are helping to accelerate the human transformation of natural systems to artificial ones and the consequent shrinking of natural ecosystems and their biotas.

Ironically, the Internet may also be critical to saving and conserving the earth's remaining ecosystems and biological diversity. The Internet can mediate knowledge discovery of complex environmental phenomena that were hitherto intractable—knowledge that is vital to informing the stewardship and conservation of natural environments. Specifically, the Internet and its sibling information technologies can enable natural history institutions to tackle one essential component of this century's grand challenge: harnessing knowledge of the earth's biological diversity and how it shapes the global environmental systems on which all of life depends.

How do we deploy the Internet to help meet this goal?

A Key Element of the Solution

Biological diversity, or simply *biodiversity*, is a critical component of the earth's natural capital. It sustains our economy and way of life; cleans our air, water, and soil; and provides us with food, fiber, fuel, pharmaceuticals, and aesthetic enjoyment. Biodiversity comprises the life of the planet—all plants, animals, and microbes and their levels of biological organization, from genomes to species to ecosystems. Approximately 1.8 million species are known; 10 to 50 million more await discovery, especially among biotas in tropical rain forests, soils, and oceans. Almost all known species are variable across individuals and populations, and each has a complex biology and chemistry, the result of almost 4 billion years of evolution.

We Have the Biodiversity Information

For more than 300 years, humans have conducted a systematic biological exploration of the earth. The resulting knowledge of biodiversity is documented by vast collections of animals and plants and their associated biotic data archived in natural history institutions—particularly museums—worldwide. From the 7 million specimens at the University of Kansas Natural History Museum and Biodiversity Research Center to the 500 million held in all U.S. museums to the 3 billion estimated globally, these collections document the identity, global composition, spatial distribution, habitats, ecology, systematics, and history of known life on Earth.

This documentation is twofold: first, natural history institutions and others ensure that the physical specimen of an animal or a plant collected at a locality is cataloged, curated, and conserved for research and education for current and future generations. Second, these organizations typically also hold some biotic and environmental data recorded for each specimen of animal or plant when it is collected—for example, its taxonomic identity (that is, its species name), its geographic provenance, the date it was collected and by whom, its morphology, and ambient climatic and ecological conditions. These data are recorded in field notes, ledgers, specimen labels, and, ultimately (in institutions with adequate expertise and budgets), electronic databases augmented by research on the physical specimens.

The data associated with each of these 3 billion specimens derive from the biotic, geologic, oceanic, atmospheric, and geospatial sciences. The data are multimodal, ranging from numbers and text to images, sound, and video (for example, recordings of behavior and vocalizations). Cumulatively, the collections and associated data provide the raw research material for revealing the patterns, processes, history, and causes of evolutionary and ecological phenomena.

Natural history institutions and their biocollections are our libraries of life. They harbor an invaluable information commodity—authoritative data on Earth's biotas, past and present. Use of the economic term *information commodity* is deliberate here because, in the end, stewardship of global biodiversity and ecosystems is an economic necessity. Yet, at the close of the twentieth century, after 300 years of systematic collection, this biotic information commodity remained largely untapped and inaccessible for research, education, or public policy. Now, with the Internet and information technologies, natural history institutions can finally deploy their libraries of life for the benefit of science and society.

The hurdle, however, was overcoming history. After 300 years of species inventory, the biodiversity science community lacked the means—an information architecture and set of common practices—for the discovery, retrieval, and integration of data. From one collection to the next—often within the same institution—underlying specimen data are heterogeneous and incompatible. The data are recorded and stored in thousands of idiosyncratic, independently developed information systems and are dispersed worldwide across academia, government agencies, conservation organizations, research institutes, and private museums.

Therefore, our ability to engage biocollections information in automated systems was severely limited. Lacking interoperability—the ability to integrate data across species of animals and plants and across research domains—museum biocollections lacked synthetic power. Without a community information infrastructure, the vast libraries of life in our museums remained largely unseen and unread. To paraphrase Umberto Eco, the good of biocollections and their data lies in their being read; without an eye to read them, they contain signs that produce no concepts—therefore they are dumb.

It is not surprising then to have study after study recognize that the digital acquisition, integration, and application of biocollections data are fundamental to the solution of complex biodiversity challenges. Recent examples of such studies include the Australian government's *Darwin Declaration*, the U.S. President's Council of Advisors on Science and Technology's *Teaming with Life*, the National Biodiversity Information Center's *Consensus Document*, the systematics and biocollections communities' *Systematics Agenda 2000*, the National Science Foundation's *Loss of Biological Diversity*, and the President's Office of Science and Technology Policy's *Strategic Planning Document—Environment and Natural Resources*.[4] At the international level, the OECD Megascience Forum recommended establishment and support of "a distributed system of interlinked and interoperable modules (databases, software and networking tools, search engines, analytical algorithms, etc.), that together will form a Global Biodiversity Information Facility."[5]

How can natural history institutions achieve the goal of harnessing their vast, authoritative, collection-based information of the planet's biodiversity to inform the environmental management of the planet? The enabling medium is the Internet; the enabling engine is biodiversity informatics.

Biodiversity Informatics

Biodiversity informatics is an interdisciplinary field uniting earth systems sciences, computational science, and software engineering. Informatics deals with biotic data; their storage, integration, and retrieval; and their use in analysis, prediction, and decision making. The National Science Foundation (NSF) identifies bioinformatics as having the highest priority for knowledge creation in the biological sciences, whether it is mining neuroscience data, genomic data, or biodiversity data.[6]

By employing current biodiversity informatics technology, natural history institutions can furnish instant, powerful, and shared electronic access to their specimen-based archives of biodiversity data. Single collections of birds from Mexico or fish from the Caribbean rarely contain enough information (that is, specimen data points) for comprehensive biodiversity analyses of an evolutionary lineage or the biota of a geographic region or geologic period. It is generally true, when dealing with census records, that pooling data from multiple sources vastly increases the reach and usefulness of the data. In the field of biodiversity informatics, pooling all of the records of plants and animals from multiple collections across institutions yields a much denser record across space and through time, which researchers can use to address complex biotic questions.

Further, these data can be integrated across biotic, geospatial, atmospheric, and genomic domains. For example, they can be combined with terrain, land cover, climate, and gene sequence data to create new classes of biotic information for computational analysis and modeling. Such efforts have the potential to yield impressive results: imagine a computational model that imports genetic, geospatial, geographic, and ecological information to provide a predictive simulation of the potential spread of a genetically engineered organism across a complex ecological landscape.

The Species Analyst

In an initiative begun at the University of Kansas in 1997, a consortium of biodiversity researchers and computer scientists at the Natural History Museum and Biodiversity Research Center developed an Internet-based informatics architecture that helps transform a museum's isolated biodiversity data archives into networked knowledge. The organizing principle for this architecture is the quantification, analysis, and prediction of biodiversity, which is achieved by integrating biocollections data and biological modeling with earth systems science research.

This system, dubbed the Species Analyst, is a software architecture that enables any user to query multiple collection databases simultaneously.[7] Once retrieved, the information is assembled and integrated in seconds with geospatial, atmospheric, and computational tools and data, permitting analysis, modeling, and prediction of species occurrences based on locality data and underlying environmental and climatic variables.

The Species Analyst uses the ANSI/NISO Z39.50 standard for information retrieval from dispersed and idiosyncratic information systems. This approach to achieve interoperability among disparate databases has proven highly successful in enabling data sharing and knowledge networking in the bibliographic and geospatial domains.[8] Applications and computational algorithms for predicting species distributions, called GARP (the Genetic Algorithm for Rule-set Production), developed at the San Diego Supercomputer Center (SDSC), provide an Internet-mediated facility for creating distribution maps from biocollections data on species occurrences and from electronic maps of climate, land cover, and soils. With the collaboration of other institutions, the Species Analyst has become the infrastructure for NABIN (the North American Biodiversity Information Network) and provides a direct connection with GARP's predictive modeling capability at SDSC.[9]

The Species Analyst Technology

The Species Analyst consists of two key components: (1) a common interface for each of the data providers (that is, biodiversity databases at natural history museums and other institutions), and (2) analytical tools that enable the use of such data in research and education. The common database interface ensures retrieval of data records from the data providers in a common format without requiring translation by the user. A data profile, known as the Darwin Core, describes the data fields and record structures produced by each database and is intended for use with the Z39.50 standard protocol for information retrieval, which is employed by the Species Analyst for machine-to-machine communication. Because all data providers support a common interface, an end user employing a client application can broadcast a query to any number of data providers and combine the results into a single set of

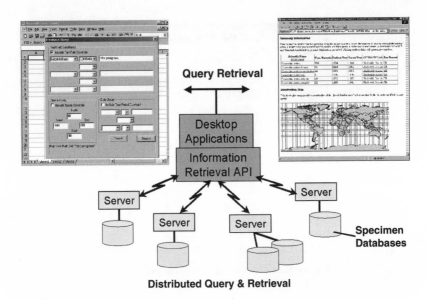

Query Retrieval

Desktop Applications

Information Retrieval API

Server

Server

Server

Server

Specimen Databases

Distributed Query & Retrieval

Figure 8.1: The Species Analyst allows users to query databases of biodiversity information that are supported by servers all over the world, though the data is presented to the end user as an integrated record.

records. To the end user (or client), the data providers to the Species Analyst appear to be providing a very large, single database (see Figure 8.1).

The Species Analyst World Wide Web interface allows users to (1) perform broadcast queries of any number of selected data providers (see Figure 8.2); (2) view the results in a number of text and map formats (see Figure 8.3); (3) use the results to model and predict the species' distribution via a link to the SDSC; and (4) find associated taxonomic and genomic information on the particular species via links to the federal government's Integrated Taxonomic Information System and Genbank, respectively (see Figure 8.4).[10] For instance, in the latter example, if a user knows only the common name of a species of plant or animal the user wishes to query, such a link automatically provides the scientific name or names for that species.

Current Applications of the Species Analyst

As an Internet-based application for public use, the Species Analyst has been used by researchers to address a spectrum of research challenges in

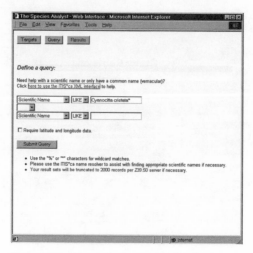

Figure 8.2: Through the Species Analyst's World Wide Web interface, users can search for records of a species by entering either its scientific or its common name.

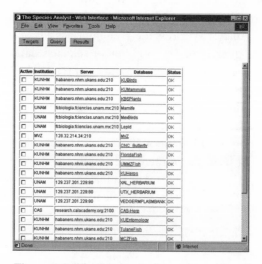

Figure 8.3: The Species Analyst returns a list of specimens from biocollections around the world.

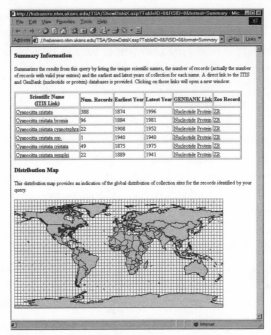

Figure 8.4: The Species Analyst's World Wide Web interface provides users with a summary of the results of a query, including a distribution map, which indicates the locations of collection sites for the records identified in the query.

biodiversity patterns, processes, and conservation. Described below are applications of the Species Analyst available in early 2001 that employ interfaces generally suitable for the scientific research community; in future applications, such as Backyard Biodiversity and Lifemapper, scheduled for release in 2002, a more public, user-friendly Web interface will be introduced.

REPATRIATION OF BIODIVERSITY INFORMATION

Collection-based biodiversity information is highly unevenly distributed around the world, a legacy of the history of the biological exploration of the earth during the past 300 years. Although the tropics are the most biodiverse part of the world, information about tropical diversity is concentrated in North American and European institutions.[11] With the ability to access, integrate, and freely share biodiversity information via the Internet, at least the information component of a given nation's biodiversity patrimony is being made available for use by science and society in that nation and throughout the world.

CONSERVING BIODIVERSITY

The ability to integrate authoritative biodiversity data with new predictive analytical tools enables researchers to develop quantitative, optimal schemes for conserving biodiversity on regional scales.[12] This approach is a dramatic improvement over previous approaches to biodiversity conservation because it focuses broadly across species and can be used to identify optimal additions of geographic areas to existing reserve systems.[13] For example, researchers used the Species Analyst to predict the geographic ranges of native bird species in the southwestern dry forest of Mexico. Having this predictive capability helped determine how a reserve system might have been expanded to include critical concentrations of certain bird populations.

REDESIGNING AGRICULTURAL LANDSCAPES

Agroecosystems are integral components of the biological world. Farmers and researchers have a particular interest in the myriad pests that can dramatically reduce agricultural productivity. Recent research with a prototype application of the Species Analyst to this problem has demonstrated that biocollections data can be used to predict the distributions of pest rodents in Mexico.[14] In turn, these rodent pest distributions correlate well and may account for 5 to 7 percent of the variation in yield for seven major crops— that is, the rodent pest species may be responsible for this much of a drop in crop yield. The ability to maximize crop productivity by predicting (and then controlling) the geographic extent of potential pest species can have important economic implications—especially if these results are tested and replicated in other regions and for other pest species and crops.

PREDICTING SPECIES INVASIONS

Invasive species are recognized as damaging threats to natural and human-managed ecosystems, an example being the Asian long-horned beetle, which recently invaded the United States from China in wood-based pack-

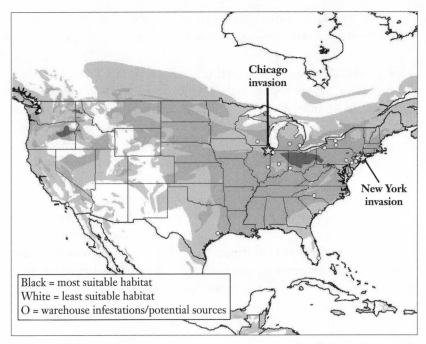

Figure 8.5: Using information on the ecological niche and geographic distribution of the Asian long-horned beetle in China, scientists simulated the possible spread of the beetle across the United States.

ing material.[15] The Asian long-horned beetle is destroying trees in urban areas, such as New York City and Chicago, as well as in other areas across the Northeast. At the University of Kansas, we used biocollections records of specimens of the Asian long-horned beetle from China to model its ecological niche and geographic distribution in that nation. We compared that model to a similar model of the United States and simulated the beetle's potential invasive spread across this country, with considerably greater predictive success than traditional methods used by U.S. government agencies (see Figure 8.5).[16]

The current approach to dealing with invasive species is reactive: intervene only after the invasion has occurred and determined to be threatening. Given sufficient computer time, this predictive approach instead enables science to simulate the invasion of any (and all known) species from one country or region to another *before* it occurs, archive the results, and inform policy planning and strategic, proactive solutions.

PREDICTING BIOLOGICAL CONSEQUENCES
OF GLOBAL CLIMATE CHANGE

That rapid climate change is occurring is now rarely disputed.[17] However, the effects of climate change on elements of the biotic world are only beginning to be understood.[18] For example, scientists are using biocollections data and the current climate regime to model the current geographic distribution of specific birds in Mexico. If the climate parameters are changed—for example, if temperature is increased several degrees centigrade (as global climate change models predict for this region in the next 50 to 100 years)—the predicted geographic distribution of these birds is dramatically fragmented and reduced, a harbinger of eventual extinction. As with invasive species, massive use of biocollections data to model the current and possible future distributions of all species of animals and plants under different scenarios of climate change can build an archive for planning policy regarding biodiversity and natural resources.[19]

CENSUS OF MARINE LIFE

A census of marine life is an initiative started by the Sloan Foundation and funded through the NSF and a variety of other agencies represented by the National Oceanographic Partnership Program. The Fishnet consortium— currently consisting of twenty-three institutions: in the United States (seventeen), Canada (two), Europe (two), Africa (one), and Australia (one)— is a response to this initiative.[20] Fishnet is one example of the collaboration of a biodiversity research community formed around the power of Internet-based distributed information systems such as the Species Analyst.

Initially, Fishnet will serve biotic data associated with more than 40 million specimens of marine and freshwater fish worldwide to provide ichthyologists, fisheries biologists, and other parties interested in biodiversity with an electronic atlas of fish distributional data that can be viewed on the Web or downloaded into a variety of formats for analysis and predictive modeling. Eventually, Fishnet aims to become a virtual, Internet-based, worldwide museum of fish biodiversity that may also include fish databases from noncollection institutions, such as government agencies.

Fishnet enables us to (1) map where species of fish occur based on the authoritative geographic locality data associated with the fish collections of participating partners, and (2) identify critical geographic gaps—important marine or freshwater areas with few or no fish records—in order to plan strategic biotic surveys. Preliminary research using the Species Analyst toolbox has integrated fish biodiversity data with environmental information and used GARP to predict the distributions and ecological niches of different fish species. We are now testing the accuracy of the predictions. One near-term goal is to add a third dimension, oceanic depth, to GARP's predictive modeling of geographic distribution, which is critical to analyzing and predicting marine biodiversity.

Remaining Barriers

Although the Species Analyst has achieved enormous success, biodiversity informatics still faces formidable sociological and technological barriers.[21] First, too many museums have not yet grasped the first principle of the information age—namely, that access to their authoritative biotic information for knowledge creation and decision making is as valuable as the information itself.

Second, even if some of their collection data have been captured in electronic databases, many museums still have policies that discourage or do not permit data sharing. Ironically, such policies deliberately quarantine museum collections and their essential specimen information from research *on the very biodiversity phenomena* that those collections were intended to help elucidate. These institutions need to redefine their missions to incorporate and invest in biodiversity informatics if the millions of specimens they house are to be relevant to understanding biological diversity and sustaining the earth's plants, animals, microbes, and natural environments.

Third, many biocollections personnel worry that successful use of electronic biocollections data for biodiversity conservation will convince administrators and government authorities that new biotic surveys and collections are no longer needed. On the contrary, integration and visualization of collection data can both demonstrate the power of existing collections for research, education, and informing policy and reveal the geographic, ecological, and taxonomic gaps in our knowledge of worldwide biodiversity that need to be filled by additional biotic surveys and collections. Integration of existing museum biocollections data may be the best tool for biodiversity scientists and funding agencies to plan, prioritize, and support global, strategic, cost-effective biotic surveys.

Fourth, most collection-based biotic data, especially those associated with millions of insect and plant specimens, are still not captured in electronic databases. Many specimen records need to be checked for accuracy in species identification and geographic location, and a large fraction of many collections needs to be geocoded (that is, text-based localities must be replaced with latitude and longitude coordinates) for data retrieval and use with geospatial tools. In fact, geospatial mapping and modeling of collections data can expose outliers in databases for correction (if the records are erroneous) or exciting research (if they are real). Clearly, this will require a concerted investment of resources by museums and funding agencies and a collaborative approach to efficient, cost-effective technological solutions by collection-based systematic communities. Models here are Fishnet for the ichthyology community, MaNIS for the mammal community, and HerpNET for the collections of amphibians and reptiles.

Future Technology

Evolving standards and ongoing innovations in information technology are integral to the future of any distributed information system. They will enable continued development of the Species Analyst and other tools to best utilize the shared resources of the Internet. Adoption of these standards will drastically reduce current bottlenecks in the rapid deployment of databases worldwide for the Species Analyst network.

One Internet innovation is XML (eXtensible Markup Language), which advances the encoding of information for Internet transfer and will be used to re-implement the Species Analyst's current Z39.50 protocol for data transport. Another significant Internet innovation for the future of the Species Analyst is UDDI (Universal Description, Discovery and Integration). It provides a single, global, distributed directory for automatically searching and retrieving business services. Everything from a florist's phone number to a museum database of beetles or frogs can be registered in the UDDI directory. Finally, a new Internet initiative called IGIR (Interface for Generic Information Retrieval) will enhance access to and retrieval of information from museum collections databases. In combination, these technological implementations will permit any user of the Species Analyst to discover relevant biological data sources and analytical toolkits through the global UDDI directory and query any number of them to retrieve and integrate a set of records for further analysis.

On a larger scale, these innovations will foster rapid coordination of the Species Analyst with similar biodiversity informatics initiatives in Europe (ENHSIN: the European Natural History Specimen Information Network), Mexico (REMIB: the World Information Network on Biodiversity), and other regions, ultimately forming a distributed global biodiversity network.

Future Applications

The current implementation of the Species Analyst provides a glimpse of the future capabilities of a global system for accessing biodiversity information. Although originally intended as a prototype for combining these data and analyses into a single package, the Species Analyst is evolving into a model for a global facility for integrating and serving biodiversity information (GBIF: Global Biodiversity Information Facility).[22]

The greatest potential of the Internet for understanding complex environmental phenomena and informing conservation is the integration of information across disciplinary domains to create new classes of knowledge. Especially important will be applications that integrate three systems: natural systems (for example, genomics, species, communities, and ecosystems), physical systems (for example, climate, geology, topography, and hydrodynamics), and human systems (for example, agroeconomics, socioeconomics, demographics, land use, and human health). Examples of such future applications include the following areas.

TRACKING HUMAN DISEASE

Diseases generally are systems of interacting species among which one becomes a threat to human health. One example is the recent invasion of North America by the West Nile Virus. First, the virus was transported to the continent; then, species interactions propagated the virus in avian populations, from which it infected humans; and finally, the virus spread rapidly across the continent. Other diseases can invade a new area and spread as a result of climate change (for example, malaria and Chagas' disease reaching the southern United States) or evolutionary change (for example, HIV). These systems are ripe for assessment, modeling, simulation, and proactive intervention using new, Internet-mediated, integrative biodiversity informatics tools and data.

FOSTERING NEW INDUSTRIES AND NEW PRODUCTS

Many commodities of economic importance, such as penicillin and bio-engineered crop strains, begin as biological products. Currently, however, the search for such products takes little advantage of our authoritative knowledge of biodiversity. New informatics tools can be used to combine phylogenetic (tree of life) information on evolutionary kin groups of animals or plants that might, for example, share pharmaceutical properties, with geospatial predictive modeling based on biocollections data to identify the most promising areas to prospect for such plants or animals.

UNDERSTANDING MIGRATION SYSTEMS

Animal migration, and in particular the factors governing species' movements, is poorly understood. Recent groundbreaking work has shown a tight correspondence between species' movements and their precise climatic context.[23] It appears that some species follow particular climatic regimes through the course of a year. Internet-mediated modeling of species migrations using biocollections and other biodiversity and environmental data is a promising approach to simulate the migration routes of species for planning conservation areas and policies.

MODELING 1 MILLION SPECIES: LIFEMAPPER

As discussed above, using the Species Analyst, GARP, and other algorithms, we have successfully created credible, verifiable models of species distributions on a continental and global scale from as few as twenty specimen records from natural history collections. But no species on the earth occurs in geographical isolation from all other species. Perhaps the greatest potential for biodiversity informatics is to compare known and predicted distributions of assemblages of species, rather than single taxa, which would enable us to address large-scale research questions in biodiversity and ecosystem science, conservation management, and informal science education for the public.

To accomplish this, scientists at the University of Kansas Biodiversity Research Center are planning to complement our Internet-distributed query architecture with two new products: a new computational framework—a distributed screen saver—for continuous calculation of geographic distribution models of *all* of the earth's species, and an enormous, Web-accessible archive of those models. The archive, development of which is being funded by the NSF, will allow us to combine the spatial distribution models of animal and plant species into communities for analysis of biodiversity–ecosystem relationships.

The project has two major hurdles. First, we estimate that only 1 million of the earth's known 1.8 million species are represented by specimens archived in natural history institutions. Worse, the majority of these specimens, especially the arthropods and plants, have yet to be cataloged and geocoded in an Internet-accessible database.

The second hurdle is the computational power required to compute 1 million species models derived from querying distributed museum databases via the Species Analyst, particularly when these databases are being continually updated with new georeferenced records of specimens and observations. Because species distribution models require six hours of GARP processing on Intel Pentium III processors, serial processing of one species at a time, or even fifty concurrent analyses, is insufficient to tackle the predictions for a million species. For example, to compute a complete set of distributional models for 100,000 species would require 500 days of Intel servers running fifty GARP simulations simultaneously. Models for 1 million species would require roughly 13.7 years.

The solution lies in parallel computing, with each species model being calculated independently and asynchronously by tapping the tremendous unused resources of desktop computers worldwide. For example, the Search for Extra Terrestrial Intelligence @ Home Project scans for signals of life beyond the earth by packaging aliquots of radio astronomy data and distributing them for analysis to desktop screen savers around the world. We have developed Lifemapper to document life on the earth using the same screen saver-parallel computing architecture—and we will be launching this program on the Web soon.

Technically, our Lifemapper users interact with an Internet data server, which gathers all of the available distribution data for a particular species within the Species Analyst network, temporarily caches those data, and then scatters them to PC clients (screen savers) for computing when those PCs are ready for a new job. The PC screen saver then runs GARP in the background or as a viewable screen saver and returns the species distribution model to the server when the computation is complete.

Parallelizing GARP analysis over the Internet can increase the computational throughput by three orders of magnitude. The Lifemapper project sustains about 500,000 active PC users at any one time. With that many clients, Lifemapper is expected to generate 1 million species distribution models per week, each of which can be continuously updated and served through a geographic information system Web server for the public and via XML data streams for integration with other kinds of data and tools. Screen saver options will allow users to choose their favorite plant or animal group for modeling, and each user will be credited with a "life-list" of species models computed on his or her PC.

Just as important as the computing architecture is the extent to which Lifemapper will enable scientists, educators, students, and citizens worldwide to participate in the predictive modeling of the earth's species of animals and plants for conservation and environmental planning. Ultimately, it is the connectivity and bandwidth of the Internet that will make possible the collaboration of hundreds of thousands of citizens around the world in this project.

Conclusion

The Species Analyst uses the Internet to enable knowledge networking of biodiversity information. Via the Internet, natural history institutions are now bringing the intellectual content of the world's biocollections into currency for science and society and are fostering the use of that information across the domains of research, education, commerce, and government.

The Internet allows biodiversity science to break new intellectual ground technically and conceptually. Technically, the Internet enables the retrieval and integration of the information resources of the broader biocollections, ecological, library, genomic, and geospatial data communities. Conceptually, it enables the mining, visualizing, and analysis of a 300-year legacy of biological inventory data for research challenges across earth systems disciplines and computational biology. Only through such interdisciplinary integration will science make tractable the modeling and simulation of complex biodiversity phenomena that were hitherto impossible. The ability to do so on the fly through Internet II backbones using local and distributed computational resources has revolutionized the speed, timeliness, and accuracy of biodiversity modeling and its potential to inform policy. Essentially, the Internet is the chassis on which biodiversity science is evolving from a traditionally descriptive enterprise to a predictive, prescriptive one.[24]

This use of the Internet to advance knowledge discovery in what Rita Colwell, director of the NSF, calls "biocomplexity" is, to paraphrase Jacques Monod, "the actual." More daring and perhaps utopian is the "possible"—the equivalent of a Weather Channel or Napster for all knowledge. The Species Analyst and its Internet kin beg the possible: the seamless, rapid integration and analysis of data for simulation of large-scale phenomena across earth and human systems by the next generation of supercomputers. In 2001, according to officials at the Los Alamos National Laboratory, the best supercomputers were capable of 3 trillion operations per second. In 2010, they will be capable of 1,000 trillion operations per second. Not only will these supercomputers be better calculators, as has been their evolution until now, but they will be new kinds of thinking machines. They will provide, for example, the capability of large-scale simulation: a living object put together atom by atom; an organism's brain assembled neuron by synapse; or the biosphere and its composition, structure, and processes simulated across space and through time, beginning 4.5 billion years ago.

Knowledge networking across earth and human systems is needed to inform society, from prudent resource management to desktop science homework, from the scientific issues governing the life of the planet to the implications for the quality of human life. It is also the ultimate democratization of knowledge. Ever since the species of *Homo* evolved on Earth, its core, singular adaptation has been information management, from rock carvings to language to the printing press to the Web. The Internet—and its evolutionary descendants—should become humanity's knowledge server for guiding planetary management.

NOTES

1. Erich Bloch et al., *Impact of Emerging Technologies on the Biological Sciences* (Arlington, Va.: National Science Foundation, 1995).
2. *Systematics Agenda 2000* (New York: Charting the Biosphere, 1994); Edward O. Wilson, "Integrated Science and the Coming Century of the Environment," *Science* 279 (1998): 2047–48; Peter H. Raven and Edward O. Wilson, "A Fifty-Year Plan for Biodiversity Surveys," *Science* 258 (1992): 1099–1100.

3. Wilson, "Integrated Science and the Coming Century of the Environment."
4. Environment Australia, *The Darwin Declaration* (Canberra: Australian Biological Resources Study, Department of the Environment, Environment Australia, 1998); President's Council of Advisors on Science and Technology (PCAST), *Teaming with Life: Investing in Science to Understand and Use America's Living Capital* (Washington, D.C.: PCAST Panel on Biodiversity and Ecosystems, 1998); Advisory Planning Board of the National Biodiversity Information Center, *Strategic Planning Document—Environment and Natural Resources* (Washington, D.C.: National Science and Technology Council's Committee on Environment and Natural Resources, 1994), available in July 2000 at www.whitehouse.gov/White_House/EOP/OSTP/ NSTC/html/enr/enr-3a.html; *Systematics Agenda 2000*; Craig C. Black et al., *Loss of Biological Diversity: A Global Crisis Requiring International Solutions* (Washington, D.C.: National Science Foundation, National Science Board, 1989), 89–171; Advisory Planning Board of the National Biodiversity Information Center, *Strategic Planning Document.*
5. Organisation for Economic Co-Operation and Development (OECD), *Final Report of the OECD Megascience Forum Working Group on Biological Informatics* (Paris: OECD Publications, 1999), available at www.oecd.org/ehs/icgb/biodiv8.htm.
6. Bloch et al., *Impact of Emerging Technologies on the Biological Sciences.*
7. See http://speciesanalyst.net.
8. Jocelyn Kaiser, "Searching Museums from Your Desktop," *Science* 284 (May 7, 1999): 888.
9. Collaborating institutions include NAFTA's Commission on Environmental Cooperation; Mexico's CONABIO (Consejo Nacional para el Uso y Conocimiento de la Biodiversidad) and UNAM (Universidad Nacional Autónoma de México); Canada's CNC (Canadian National Collection in Agriculture Canada); and the U.S. Department of the Interior.

   For information on NABIN, see A. Townsend Peterson, Adolfo G. Navarro-Siguenza, and Hesiquio Benitez-Diaz, "The Need for Continued Scientific Collecting: A Geographic Analysis of Mexican Bird Specimens," *Ibis* 140 (1998): 288–94.
10. Viewing formats provided include XML (eXtensible Markup Language), raw text, ESRI Arcview Geographic Information Systems shape files, and Excel spreadsheets.
11. Peterson, Navarro-Siguenza, and Benitez-Diaz, "The Need for Continued Scientific Collecting."
12. Mandaline E. Godown and A. Townsend Peterson, "Preliminary Distributional Analysis of U.S. Endangered Bird Species," *Biodiversity and Conservation* 9 (2000): 1313–22.

13. A. Townsend Peterson, L. G. Ball, and K. M. Brady, "Distribution of Birds of the Philippines: Biogeography and Conservation Priorities," *Bird Conservation International* 10 (2000): 149–67; A. Townsend Peterson, "Predicting Species' Geographic Distributions Based on Ecological Niche Modeling," *Condor* 103:3 (2001): 599–605.

14. Victor Sanchez-Cordero and Enrique Martinez-Meyer, "Museum Specimen Data Predict Crop Damage by Tropical Rodents," *Proceedings of the National Academy of Sciences USA* 97 (2000): 7074–77.

15. James T. Carlton, "Pattern, Process, and Prediction in Marine Invasion Ecology," *Biological Conservation* 78 (1996): 97–106.

16. A. Townsend Peterson and David A. Vieglais, "Predicting Species Invasions Using Ecological Niche Modeling," *BioScience* 51 (2001): 363–71.

17. Thomas R. Karl et al., "Indices of Climate Change for the United States," *Bulletin of the American Meteorological Society* 77 (1996): 279–92; Karen O'Brien and Diana Liverman, "Climate Change and Variability in Mexico," in Jesse C. Ribot, Antonio Rocha Magalhaes, and Stahis S. Panagides, eds., *Climate Variability, Climate Change and Social Vulnerability in the Semi-Arid Tropics* (Cambridge: Cambridge University Press, 1996), 55–70; Camille Parmesan, "Climate and Species' Range," *Nature* 382 (1996): 765–66; Craig D. Allen and David D. Breshears, "Drought-Induced Shift of a Forest-Woodland Ecotone: Rapid Landscape Response to Climate Variation," *Proceedings of the National Academy of Sciences USA* 95 (1998): 14839–42; David W. Inouye et al., "Climate Change Is Affecting Altitudinal Migrants and Hibernating Species," *Proceedings of the National Academy of Sciences USA* 97 (2000): 1630–33.

18. Robert L. Peters and Joan D. S. Darling, "The Greenhouse Effect and Nature Reserves," *BioScience* 35 (1985): 707–17; Andrew Dobson, Alison Jolly, and Daniel I. Rubenstein, "The Greenhouse Effect and Biological Diversity," *Trends in Ecology and Evolution* 4 (1989): 64–68; Robert D. Holt, "The Microevolutionary Consequences of Climate Change," *Trends in Ecology and Evolution* 5 (1990): 311–15; Robert L. Peters and John Peterson Myers, "Preserving Biodiversity in a Changing Climate," *Issues in Science and Technology 1991–1992*, 8(2): 67–72; Warren P. Porter, Srinivas Budaraju, Warren Stewart, and Navin Ramankutty, "Calculating Climate Effects on Birds and Mammals: Impacts on Biodiversity, Conservation, Population Parameters, and Global Community Structure," *American Zoologist* 40 (2000): 597–630.

19. A. Townsend Peterson, L. G. Ball, and K. C. Cohoon, "Predicting Distributions of Mexican Birds Using Ecological Niche Modelling Methods," *Ibis* 144:1 (2002): E27–E32.

20. See http://habanero.nhm.ukans.edu/fishnet/.

21. On the success of the Species Analyst, see Elizabeth Pennisi, "Taxonomic Revival," *Science* 289 (2000): 2306–8; and James L. Edwards et al., "Interoperability of Biodiversity Databases: Biodiversity Information on Every Desktop," *Science* 289 (2000): 2312–14.

22. See www.gbif.org/relafram.htm.

23. Leo Joseph, "Preliminary Climatic Overview of Migration Patterns in South American Austral Migrant Passerines," *Ecotropica* 2 (1996): 185–93; Leo Joseph, "Towards a Broader View of Neotropical Migrants: Consequences of a Re-examination of Austral Migration," *Ornitologia Neotropical* 8 (1997): 31–37.

24. Leonard Krishtalka and Philip H. Humphrey, "Fiddling While the Planet Burns," *Museum News* 77:2 (2000): 29–35; Pennisi, "Taxonomic Revival"; Edwards et al., "Interoperability of Biodiversity Databases."

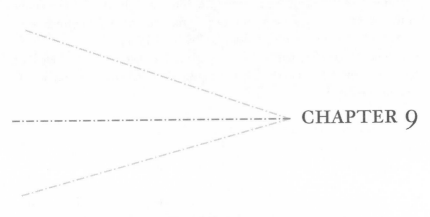

CHAPTER 9

BirdSource

USING BIRDS, CITIZEN SCIENCE, AND THE INTERNET

AS TOOLS FOR GLOBAL MONITORING

John W. Fitzpatrick and Frank B. Gill

Many chapters in this book emphasize the liberating impact of the Internet on how humans disperse themselves across the landscape. The new freedom to live and work anywhere is increasing the rate at which humans, accompanied by their associated ills, can colonize natural areas that once were too remote to be ruined. In this context, the Internet is predicted to be a force that may exacerbate threats to natural areas and counteract efforts to conserve biodiversity.

Here we propose that the Internet also presents a powerful and constructive opportunity to unite a dispersed human populace with its natural landscape. The power of the Internet—like the telephone before it— is its capacity for fast, meaningful, *two-way* transfer of information between dispersed consumers and a centralized server. We contend that the noncommercial aspects of this two-way capacity have until now remained largely unexplored but that they are profoundly important. For the first time in human history, individual diary keepers living at all corners of the globe have the means to contribute to a collective and intricately detailed "memory" of the earth. Such use of the Internet will reveal much about both the day-to-day functioning of natural systems and the long-term impact of humans on these systems. By contributing to our collective knowledge, diary keepers will be empowered to take action on behalf of their place on the planet. Soon, we suggest, this could catalyze a new culture of conservation.

The Internet as Two-Way Communication

During its first decade in the public marketplace, the World Wide Web, apart from highly popular e-mail applications, was largely used as a device to distribute relatively static information, or "brochureware." With the presence of sophisticated search engines aiding increasing numbers of curious end users, Web sites sponsored by nonprofit and public mission-driven organizations flourished alongside Web sites hosted by for-profit enterprises. Most mission-driven and profit-driven Web sites during the 1990s worked largely to transport information from a central source (for example, servers clustered in a large-scale commercial computing facility) to dispersed points (for example, client personal computers in residential living rooms). Even as e-commerce pioneered genuine client-to-server information flow (for example, point-and-click shopping carts and credit card purchasing), most people continued to use the Web principally as recipients of a one-way flow of information. Mission-driven applications spread knowledge about their sponsor organizations and the world to their Web-browsing clientele, and profit-driven applications distributed knowledge, largely in the form of advertising regarding commercial goods and services, to the online public (see Figure 9.1).

By the year 2000, numerous applications had emerged that use the World Wide Web to collect information from the dispersed public. Not surprisingly, however, most of these applications were commercial, collecting information about the consumer primarily to aid in marketing decisions. To the chagrin of privacy advocates, much of this information flow was involuntary.

In 1997, the Cornell Laboratory of Ornithology and the National Audubon Society launched BirdSource, available at www.birdsource.org, an innovative application of Internet-based information exchange designed to further the missions of our nonprofit and educational organizations. Our goal was to offer individual citizens an opportunity to play an active role in

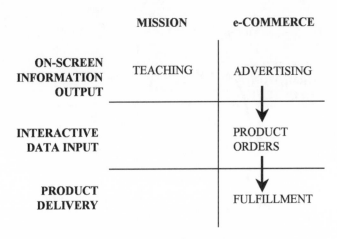

Figure 9.1: In the 1990s, most Web sites facilitated only a one-way flow of information, from suppliers to consumers.

a massive information feedback system focused on how the natural world works and how it is changing (see Figure 9.2). In this model, dispersed individuals feed the system by supplying data that are used as the basis of new layers of knowledge about the world. The scale at which this interactive monitoring system could be implemented was wholly new in human society, and we believe that the model has revolutionary implications for humans' ability to conserve and manage what remains of the earth's natural systems.

Birds as Nature's "Hook"

It is commonly observed that humans generally invest in landscape protection only after achieving some degree of understanding about that landscape. Whatever it means to "understand" a landscape, certainly the initial steps include asking questions about the landscape and seeking answers to those questions. In short, understanding begins with "wondering." If this is so, then one of the crucial social challenges of biological conservation is to motivate individual citizens in large numbers to wonder about their own local landscapes rather than to treat these landscapes simply as backdrops for their day-to-day activities. Fostering natural curiosity about what surrounds us lies at the heart of any good environmental education program.

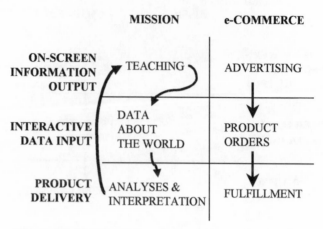

Figure 9.2: BirdSource is a mission-driven application that uses the Internet to facilitate a two-way flow of information about the natural world.

As many advertisers and journalists can explain, attracting the attention of a disinterested public requires a "hook." Talking lizards can attract us to a brand of beer, beautiful models can focus us on almost anything, and an intriguing opening vignette can make us want to keep reading a story. Like the profit-motivated marketers of commercial goods and services, conservationists are also in the advertising business. Our product is nature, and the currency we seek is investment by humans in nature's protection. Our advertising challenge is to induce motivation to invest among the masses. For this we, too, need a "hook."

Of all the hooks provided by nature, our experience suggests that birds are the most enjoyable, convenient, and informative—indeed, this fact explains the very existence of our two organizations. Birds are plentiful and colorful, they come in myriad styles and sizes, their songs are wonderfully musical to our ears, they can fly (who has not dreamed of having this skill?), and we see them outside our windows at home as well as anywhere we travel. Spectacular, semiannual global migrations of birds remind us that we live on a small planet, on which all of our natural systems are closely intertwined. In short, birds are enormously powerful communicators to humans.

Birds are also sensitive indicators of a fragile world. In the 1960s, the mere thought of a "Silent Spring," a scenario brought to the public's attention by Rachel Carson's masterful book on that idea, induced even the most cynical among us to consider the need to change our behavior.[1] In the United States, peregrine falcons and bald eagles declined toward extinction throughout the 1950s and 1960s, poisoned by our overuse of organochloride pesticides. Understanding these declines led directly to bans on the worst pesticides, including DDT. Ultimately, the dialogue regarding the widespread use of toxic chemicals helped catalyze the creation of the U.S. Environmental Protection Agency (EPA) in 1972. By 1999, with the help of national regulations regarding pesticide use and ongoing EPA enforcement actions, the peregrine falcon was eligible to be removed from the Endangered Species List, and the bald eagle followed suit shortly thereafter.

Birds can produce environmental success stories even in the bleakest circumstances, such as on the windward slopes of East Maui. No place on earth better reveals the environmental havoc wrecked by humans than Hawaii, a tropical archipelago undergoing ecological collapse after 1,000 years of perturbation by humans. In the degraded cloud forests of East Maui, several species of lobelia, among the rarest plants in the world, are eaten by feral pigs—an invasive species introduced by human settlers. As a consequence, some of the rarest birds in the world—five species of Hawaiian honeycreeper—die from introduced malaria borne by introduced mosquitoes breeding in pig-felled tree fern stems. By drawing attention to the honeycreepers' plight as an endangered species indigenous to East Maui, as well as to the area's critical importance as a water supply for the Hawaiian Islands, conservationists led by the Nature Conservancy were able to galvanize financial and political support for comprehensive local conservation management efforts. As a result, the East Maui Watershed Partnership—a remarkable consortium of federal, state, and local government agencies, landowners, private irrigation companies, municipal watersheds, and nonprofit organizations—is at present restoring to Maui one of the world's most spectacular forested slopes.[2]

Keeping Common Birds Common

For decades now, endangered species such as peregrine falcons, bald eagles, and Hawaiian honeycreepers have been recognized widely for their flagship roles in conservation. Literally, endangered birds "sing" to us about the plights of the endangered ecosystems amid which they perch. Humans do not respond easily to the plants, frogs, or beetles of the great old-growth forests in the Pacific Northwest, but we do respond to the deep and sorrowful brown eyes of the spotted owl. It bears emphasis, however, that the Endangered Species Act (the ESA) passed into law by the U.S. Congress in 1973 was not written primarily to save the owls, as most believe, but *to save the ecosystems beneath the owls*. Stated clearly in Section 2B of the ESA is the law's primary biological rationale:

> To provide a means whereby the ecosystems upon which endangered species and threatened species depend may be preserved.

It was not lost on the authors of the ESA that saving ecosystems required paying attention to a few of its key species. Indeed, the biological validity of this rationale has been repeatedly upheld.[3]

If birds are such good indicators of ecosystem health, then why should we limit our attention to those few that are endangered? They are, after all, just a tiny fraction of the flora and fauna of the world, and the ecosystems they represent are by no means the only systems being degraded or destroyed as a consequence of human activity. What if we had a way to measure in detail how *all* bird populations were faring through time throughout their ranges? Could such a monitoring tool supply a "finger on the pulse" for local ecosystems, presenting early warning signals as ecological changes take place?

In fact, monitoring bird populations has become a common enterprise, and several continental-scaled databases now exist from which to draw conclusions about ecosystem trends. For example, the North American Breeding Bird Survey (BBS) reveals that twenty-two of twenty-eight bird species in our native grasslands have declined since the mid-1960s, many of them precipitously.[4] Yet not one of these species appears on the U.S. Endangered Species List. In the woodlands of eastern North America, the BBS reveals that once-common species such as the cerulean warbler, the northern (yellow-shafted) flicker, and the wood thrush have declined steadily over the past thirty years at annual rates of 3 to 7 percent.

In these and most other such cases, the ecological causes of population declines are poorly understood. Without doubt, they involve mixes of conflicting and compounding influences that differ from species to species—and it should be pointed out that more than half of the woodland bird species in North America have actually increased during the same period. Where we can identify significant factors causing population declines, however, it is clear that human land use plays an overwhelmingly important role.[5]

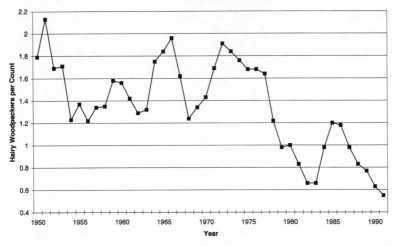

Figure 9.3: Although the hairy woodpecker is still common in other parts of the country, in peninsular Florida the species is becoming quite rare, indicating an ecological imbalance in America's southeastern forests. (Data from the Audubon Christmas Bird Count.)

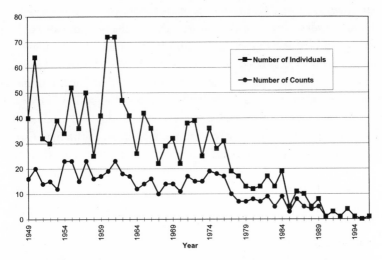

Figure 9.4: Bewick's wren is nearing extinction in the eastern United States, another signal of a regional ecological imbalance. (Data from the Audubon Christmas Bird Count.)

Population trends for wide-ranging bird species are seldom uniform across their range. Many species show stable or increasing numbers in some areas even as they decline precipitously in others.[6] Patterns of increase or decrease among wide-ranging species can signal ecological problems at local levels, despite the species as a whole being in no danger of extinction. For example, a century of data from the Audubon Christmas Bird Count (CBC) reveal that the hairy woodpecker has become extremely rare in peninsular Florida (see Figure 9.3) and that the Bewick's wren is all but extinct east of the Ozark Mountains (see Figure 9.4). Both species remain common elsewhere across North America, but they signal significant ecological imbalances in the forests of the southeastern United States. In Florida, hairy woodpeckers depend on pine forests subject to periodic wildfire, a natural disturbance that was prevalent prior to human colonization. A century of fire suppression and a generation of human population growth are now causing fire-dependent species such as the hairy woodpecker to decline throughout the Florida peninsula. Bewick's wrens are abundant in scrublands of western North America, but the eastern populations depended on natural and man-made forest clearings that were free of the competitively dominant house wren.[7] Because of the spread of humans (and their associated wren houses) along with altered forest management practices that eliminated large, early successional clearings in the Appalachians, the eastern Bewick's wren is functionally extinct today.

Because birds respond so closely to landscape changes, the idea of keeping common birds common throughout their natural ranges could be a useful proxy for successful landscape conservation. Yet, historically, we have had no way to measure locally relevant indicators in detail and no standards against which to compare today's numbers. How many other hairy woodpecker or Bewick's wren stories exist amid the wild populations of birds around the world? How sensitive might we become to the management needs of local habitats if we could equip ourselves with (1) detailed population data on birds, (2) easy access to these data, (3) an engaged public wondering what these data mean, and (4) a conservation community empowered by data that track the landscape changes in real time?

The New Army

All data discussed in this chapter—indeed, virtually all the graphs, maps, tables, and trend lines available for analysis of bird population trends the

Table 9.1: Trends in Outdoor Recreational Habits of Americans, 1982–95 (in millions)

	1982–83	1994–95	% change
Fishing	60.1	58.3	–3%
Hunting	21.2	18.8	–11.4%
Birdwatching	21.2	54.1	+155%

Source: Survey data from H. Ken Cordell, ed., *National Survey on Recreation and the Environment* (Washington, D.C.: U.S. Forest Service, 1997).

world over—draw on information gathered by a unique group of research assistants. Because people enjoy looking at birds—and at least a few record what they see in organized ways—conservation biologists now have a few quantifiable, if rudimentary, measures of long-term changes in the biological landscape.

During the late 1800s, demand skyrocketed for exotic feathers to adorn high-fashion ladies' hats. During this period of unprecedented exploitation of wild bird populations, a hardy band of bird enthusiasts emerged.[8] The birdwatchers' enthusiasm was fed through the 1900s by brilliantly illustrated field guides, increasingly good and lightweight binoculars, and a steadily growing drumbeat of environmental consciousness. By the year 2000, this legion of birders had grown into a full-fledged "citizen army."

The number of birdwatchers appears to be rapidly expanding. Throughout North America, citizens today watch birds in numbers that rival professional sports as a leisure pastime. Survey after survey reveals that while certain outdoor hobbies such as hunting and fishing actually have declined somewhat over the past twenty years, birdwatching has exploded (see Table 9.1).[9] Birding is the fastest-growing outdoor recreation in North America, estimated now to be second only to walking as a way in which people enjoy the outdoors. As of 1995, an estimated 50 to 60 million people in the United States described themselves as at least casual birdwatchers.

Where does the information go as all these people watch birds across the landscape? Historically, a few dedicated scientists and hobbyists kept careful logs, often filling shoeboxes with three-by-five cards recording their lifetime sightings and occasionally pooling their observations to obtain broad-based snapshots of bird life across the continent. For instance, in 1900, a small band of twenty-seven observers, led by Frank Chapman, an early leader of the Audubon movement, launched the Christmas Bird Count tradition, keeping detailed year-end, place-based observational records at sites stretching from New Jersey to California and Ontario, Canada.

For most of the twentieth century, however, the larger part of the great birdwatching army let all its information vanish with the wind, because it had little access to a collective memory. Beginning in the 1980s, however, an increasing number of largely citizen-based projects began to demonstrate that amateur naturalists, linked together with increasingly powerful and affordable communications networks, could provide enormous quantities of scientifically valuable information about birds.[10] Acting as "citizen scientists," such individuals, most without formal scientific training, continue to this day to dramatically enlarge the scope and scale of ornithology efforts around the world. As only one indication of a widespread movement, by the year 2000 there were more than 51,000 individuals—the vast majority active as volunteer naturalists—involved in the 101st CBC, at sites throughout the Western Hemisphere stretching from the Arctic Circle to southern South America.[11]

The Internet Age and BirdSource

As we progress into the twenty-first century, we believe that the Internet will rival field guides and binoculars in its revolutionary impact on our understanding of nature. Properly configured with modern data management software and mass digital storage, the Internet makes it feasible to quickly collect, organize, archive, and display vast amounts of information, collected by professional ornithologists and citizen scientists alike. The potential applications of such a vast common resource for the study and teaching of nature are legion.

Combining Resources

One of the keys to the Internet's power is its ability to combine the resources of widely distributed organizations and individuals. Beginning in 1995, three trusted colleagues engaged in an extended brainstorming effort to figure out how to leverage that power for conservation purposes. At that time, John Fitzpatrick, a coauthor of this essay, had just been appointed to the director's position at the Cornell Laboratory of

Figure 9.5: The BirdSource Web site facilitates a
two-way exchange of information on bird populations.

Ornithology. His friend, John Flicker, a twenty-year veteran of the
Nature Conservancy, had just accepted an offer to become the new president of the National Audubon Society. Soon afterward, Audubon hired
Frank Gill (the second coauthor of this essay), a veteran ornithologist,
as senior vice president for science. Flicker, Fitzpatrick, and Gill conceived of an Internet-based tool for citizen science that would combine
the impressive scientific resources at Cornell, which included access to
the university's highly sophisticated supercomputer center, with the
broadly distributed organizational resources linked to Audubon, itself a
national nonprofit with some 550,000 members, spread out across the
United States and abroad, with a clearly defined interest in birds and their
habitat.

The tool we invented, as noted in the introduction to this chapter, is
a Web site and database structure called BirdSource, first launched online
in 1997 (see Figure 9.5). We believe that BirdSource is helping to change
fundamentally the way birds are observed and understood across the
globe.[12]

Making Historic Records Widely Available for the First Time

The first application of BirdSource was to capture retroactively a pair of
existing citizen-gathered databases, Audubon's CBC (initiated by Frank

Chapman in 1900 as an alternative to the annual gentlemen's contest determining which team could kill the most birds in a given time period, known as the "Christmas hunt") and the Cornell Lab's Project FeederWatch (begun in 1988 to track populations at bird feeders throughout each winter).

Both data sets were digitized and converted into relational databases and have subsequently been updated on a regular basis to reflect the results of ongoing CBC and FeederWatch activities. When initially offered online, the BirdSource Web site provided public access to these data sets for the first time in history. A century of previously inscrutable citizen observations about birds was suddenly available for inspection, providing interested researchers and curious visitors unlimited opportunity to make rapid search-and-sort queries. Published analyses of changes in bird populations based on BirdSource databases began appearing almost immediately.[13] A number of additional projects, including a search for early signals of global warming, are currently under way.

The CBC and FeederWatch portions of the BirdSource Web site undergo regular upgrades to improve their usefulness and level of interactivity. Indeed, the version of the BirdSource Web site now available on the Web, several years after the site's launch, features maps, charts, and customizable queries as well as artfully written stories that illustrate some dramatic changes in bird populations through the twentieth century. Web visitors appear to be pleased; they continue to visit the site in increasing numbers.[14]

Backyard Birding as Citizen Science

Another application of BirdSource, known as the Great Backyard Bird Count (the GBBC), was launched in February 1998 to invite the general public to make something quite useful out of what had been an unrecorded pastime. By engaging birders to systematically record their observations, effectively censusing birds via the Internet, the GBBC on BirdSource has lived up to its ambition of fostering "Birding with a Purpose."

The structure of the GBBC is straightforward. During a three-day period in mid-February, participants are encouraged to spend any amount of time they choose observing birds in their backyard, schoolyard, municipal park, or favorite natural area. They then report, using the BirdSource Web site (or on paper, if they choose), the following information: the

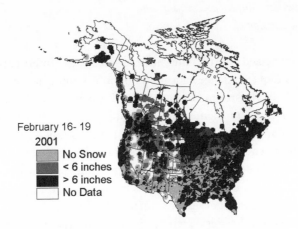

February 16- 19
2001
No Snow
< 6 inches
> 6 inches
No Data

Figure 9.6: The Great Backyard Bird Count resulted in the collection of significant data on the relative abundance of hundreds of bird species across North America. The population distribution could then be correlated with other data, such as the amount of snow cover, as shown on this map.

observation site location, identified by zip code; the length of time spent counting birds; their experience level in identifying birds, characterized as "beginner, intermediate, or expert;" and the highest number of individual birds of a given species observed over a specified period.

The GBBC has quickly gained popularity. It received 14,300 submissions in 1998. That number grew to 42,000 in 1999 and to some 67,000 in 2000. Data gathered in the course of annual GBBCs now provides the basis for a remarkable set of maps depicting relative abundances of 420 species of birds across North America (see Figure 9.6). Based on comparisons with data gathered by other means, these maps are remarkably accurate and detailed, supplying a vital new source of information about the distribution of bird populations in late winter, a perilous period for survival, just prior to the first stages of spring migration.

Besides its scientific value, the GBBC has struck a chord among the general public. Feedback from participants has been extremely positive. Many adult participants have taken the GBBC as an opportunity to engage computer-savvy children and grandchildren in watching birds for the first time. Others have reported it as their first experience using the Internet for truly interactive purposes.

Conservation Education: Training the
Next Generation of Citizen Scientists

A widely publicized but still underutilized value of the Internet is its potential role as a tool for science education not only in informal, family settings but also in the classroom.[15] Once again, birds prove to be a wonderful "hook," this time for elementary and middle school teachers, many of whom today introduce their students to nature study by mounting and watching birdfeeders outside the classroom window. Simple, inquiry-based science curricula using birds can encourage students to observe, identify, record, and—most important—ask their own questions, pose their own hypotheses, and construct tests for their ideas.[16]

The Classroom FeederWatch program, sponsored by the Cornell Lab of Ornithology with funding from the National Science Foundation, provides teachers with a curriculum on ornithology that turns students into citizen scientists.[17] Classroom FeederWatch helps teachers organize data collection, observational aids, data analysis tools, and ancillary information about the subjects outside the window. The Internet supplies a ready opportunity for students to learn that carefully collected data can become part of a much larger community of information, as local classroom results can be compared immediately with those coming in from around the country that are posted on BirdSource.

By providing access to large, communal databases, the Internet becomes a vital tool that science teachers can use to encourage their students to ask questions about their own data. Through the Classroom FeederWatch curriculum and BirdSource, students have the opportunity to investigate their own research questions, and they can submit their findings for publication in *Classroom Birdscope*, which is distributed to all classes participating in the program. The Internet increases access in the classroom to the sense of wonder and thrill of discovery that are at the foundation of scientific enterprise.

Resource Management: Helping to
Change the Way We Use the Earth

A principal goal of BirdSource is to make available information about avian distribution, abundance, and population trends to help identify landscape problems and establish conservation priorities. Making population data

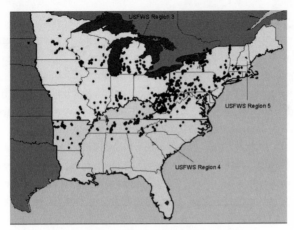

Figure 9.7: The Internet facilitated the coordinated mapping of breeding localities of cerulean warblers. (Map courtesy of Cerulean Warbler Atlas Project [CEWAP].)

available over the Internet allows the world at large, including independent investigators at widely dispersed locations, to dream up a wealth of conservation-related questions not conceivable by the original data providers alone. Many such questions, of course, relate to long-term population trends—for example, consider the stories regarding Bewick's wren and the hairy woodpecker discussed earlier in this chapter.

Other questions simply focus attention on where a particular species is most abundant. For example, soon after the CBC database was posted on BirdSource, we received a query from a biologist interested in red-throated loons. The researcher was particularly concerned about the problem of unintended by-catch of loons by fishing boats off the U.S. Atlantic coast. A quick query of BirdSource revealed a previously unnoticed concentration of red-throated loons off Cape Hatteras, North Carolina—the wintertime "mother lode" for this pelagic species (that is, a species living for at least part of the year in the open ocean). As a consequence of this discovery, conservation steps are now being taken in North Carolina to reduce loon mortality by changing offshore fishing practices.[18]

With proper guidance, citizens can be engaged to spend considerable time focusing on individual species in order to answer questions relevant to the conservation of such species. This fact is well demonstrated by the results of Project Tanager, a pre–Internet era project of the Cornell Lab of

Ornithology that used a simple field protocol to ask sophisticated questions about the response of certain bird species to forest fragmentation. Volunteers were recruited to count tanagers, search for nest sites, and document the presence of nest predators in a variety of forest patch sizes. The volunteers also recorded habitat characteristics, such as the dominant tree species and canopy height. Statistical analysis of the results allowed researchers to demonstrate that scarlet tanagers respond to landscape fragmentation in different ways across their range, an important finding because it suggests that best practices for land management may vary from region to region.[19] The project demonstrated that data gathered by citizens from their backyards and local nature preserves can prove essential in making sound land use decisions.[20]

The Internet facilitates even more detailed analysis of focal species of conservation concern. Scientists at the Cornell Lab, for example, are using the Internet to coordinate atlas projects to map the breeding localities of certain bird species that are undergoing sharp population declines, such as the cerulean warbler and the golden-winged warbler (see Figure 9.7). The Golden-winged Warbler Atlas Project (the GOWAP) will use BirdSource as a tool to help gather data on the species' habitat requirements; identify sites for acquisition, protection, or management; and determine potential general and site-specific threats to the species.

Resource Protection: Identifying and
Protecting Important Bird Areas

This social transition—getting new people to watch and wonder about the landscape—is fundamental not only to the advancement of conservation science, education, and resource management efforts but also to the actual protection of habitat critical to birds and other species. For example, by focusing attention on key localities where conservation action might help protect golden-winged warblers, the rate at which such places are being protected is already beginning to increase. On an international scale, such attention could make a major impact on worldwide efforts to protect habitats important to birds. This idea is the basis of the Important Bird Area (IBA) program, a wide-ranging effort of BirdLife International that includes National Audubon as the U.S. Partner Designate.

The IBA program represents the efforts of an international coalition of more than 100 partner organizations that collaborate under the aegis of

BirdLife International to identify, monitor, and protect sites important to birds in 156 countries around the world. For example, as a result of IBA-associated efforts in Europe that began in the 1980s, some 3,600 sites on or near that continent have been identified for continued attention. In part because of IBA activity, hundreds of such sites and millions of acres have received better protection.

In the United States, the National Audubon Society launched its IBA initiative in 1995. Through the National Audubon system of state offices, as well as in collaboration with independent conservation organizations, more than 1,200 IBAs in forty U.S. states had been identified as of November 2001. A key tool in establishing these sites has been data provided through BirdSource on WatchList birds (avian species that are rapidly declining, threatened, or endangered across their historic ranges) as well as on species that are currently more common. With data provided via BirdSource, local bird enthusiasts, in association with a National IBA Technical Committee, continue to identify habitats, at state, national, global, and continental levels, that are critical to birds' survival.[21] The current plan is to have IBA programs up and running in all fifty states by the end of 2002.

Through the IBA program, local birdwatchers are given specific suggestions for how they can help to protect birds and their habitats; they become participants in the conservation of species that they see on a regular basis. In some cases, sites designated as IBAs are able to obtain significant resources for long-term habitat protection—for example, conservationists working to protect the Montezuma Wetlands Complex in New York state, in cooperation with the IBA program, were able to obtain $2.5 million from the federal Land and Water Conservation Fund in 2001 for habitat acquisition and restoration.

In brief, the IBA program encourages people to invest in critically important bird monitoring projects throughout the year, around the nation and the world. In doing so, volunteer participants acting as citizen scientists significantly increase the probability that important habitat will be protected and properly managed over the long term.

Conclusion: The Opportunity to Involve People in Local Places

Myriad examples through the ages—including a few mentioned here that are already emerging from Internet-mediated exercises—demonstrate that when people are asked for information about their place, when they are encouraged to get out and study what is out there in their own backyard, schoolyard, workplace, or favorite recreational landscape, then their thoughts and behaviors toward those places begin to change. The act of watching, observing, counting, and recording things about our own place simply makes us more interested in our place; it makes us want to ask questions about our place. We propose that this shift also makes us want to save our place.

Today we collect information from more than 50,000 birdwatchers in North America with the help of BirdSource. Imagine how much information and insight we could gain if each of the 50 million people that are reported to be watching birds in North America spent just a few minutes each month logging in to provide information about the birds that they see (or do not see) at their favorite spot. Conducted at massive scales, focal area monitoring provides an unprecedented opportunity for reading the pulse of the landscape and gaining detailed scientific knowledge about geographically specific, long-term biological changes.[22]

Such a collective enterprise was unimaginable just a few years ago. As we enter the twenty-first century, enabled with networked technologies such as the Internet, such an enterprise—on a growing but still relatively modest scale—has become an everyday reality. This two-way interaction between individuals at their place, on the one hand, and a collective memory about the landscape, on the other, has already produced singular advances for conservation. The opportunities for constructive applications of this capability appear limitless. We have only just begun.

As the history of science is written in decades and centuries to come, we believe that the twentieth century will be remembered as the age of physics. In contrast, the twenty-first century appears destined to be known as the age of biology. Through biotechnology and genomics, we can now engineer biological systems that affect our daily lives, with a broad spectrum of applications in fields ranging from disease prevention to medicine, agriculture, and pollution control. Equally important, humans are coming to grips with the fact that we are the ecological managers of Planet Earth. This is the most complex system in our known universe, and we humans are "stationed at the bridge." For centuries, we have been managing the earth without really knowing how. Most important, perhaps, we have been proceeding entirely without an organized system of biological measures, monitors, gauges, barometers, or warning signals. How effective can we be as "systems managers" without such system gauges?

We believe that the Internet and related technologies will augment our ability to manage the earth in three fundamental ways. First, by allowing us to draw on the observational power and abilities of millions of people and sensors dispersed across the landscape, the new networks will allow us to capture real-time biological monitoring information at scales varying from single organisms to woodlots and entire continents. We will have the capacity to receive enormous amounts of information about natural systems, from a variety of sources and geographic locations, simultaneously.

Second, new communications and computing technologies will allow us to integrate and display environmental data instantly, permitting real-time comparisons against long-term patterns to detect subtle changes and, as necessary, to sound warning signals. Third, by drawing individuals into the process, not only will we get the closest possible glimpse of the "heartbeats" of natural systems, but we will also engage a great number and variety of individual human spirits in the process. It is with those spirits that we have our best opportunity to pass on our irreplaceable natural heritage to countless generations to come.

NOTES

1. Rachel Carson, *Silent Spring* (Boston: Houghton Mifflin, 1962).
2. For further information on the East Maui Watershed Partnership, see the Nature Conservancy, "East Maui Watershed Partnership Wins Achievement Award," press release (n.d.), available at http://nature.org/wherewework/northamerica/states/hawaii/preserves/art2363.html.
3. National Research Council, *Biological Basis for the Endangered Species Act* (Washington, D.C.: National Academy Press, 1995).
4. Bruce G. Peterjohn and John R. Sauer, "Population Status of North American Grassland Birds from the North American Breeding Bird Survey, 1966–1996," *Studies in Avian Biology* 19 (1999): 27–44.
5. For example, see Peter D. Vickery and James R. Herkert, eds., "Ecology and Conservation of Grassland Birds of the Western Hemisphere," *Studies in Avian Biology* 19 (1999).
6. For example, see Francis C. James, C. E. McCulloch, and David A. Wiedenfeld, "New Approaches to the Analysis of Population Trends in Land Birds," *Ecology* 77 (1996): 13–27.
7. E. Dale Kennedy and Douglas W. White, "Bewick's Wren," in Alan Poole

and Frank Gill, eds., *The Birds of North America*, no. 315 (Washington, D.C.: American Ornithologists' Union, and Philadelphia: Academy of Natural Science, 1997).

8. Mark V. Barrow Jr., *A Passion for Birds: American Ornithology after Audubon* (Princeton, N.J.: Princeton University Press, 1998).

9. H. Ken Cordell, ed., *National Survey on Recreation and the Environment* (Washington, D.C.: U.S. Forest Service, 1997).

10. See, for example, Stanley A. Temple, John R. Cary, and Robert E. Rolley, *Wisconsin Birds: A Seasonal and Geographical Guide* (Madison: University of Wisconsin Press, 1997); Keith L. Bildstein, "Long-Term Counts of Migrating Raptors: A Role for Volunteers in Wildlife Research," *Journal of Wildlife Management* 62 (1998): 435–45; Bruce V. Lewenstein, Rick Bonney, and Dominique Brossard, "Citizen Science: Measuring Scientific Knowledge and Attitudes in a New Type of Public Outreach Project," Texas A&M University, Center for Science and Technology Policy and Ethics, Discussion Paper Series #98-1, College Station, Tex., 1998; and Bruce V. Lewenstein, Rick Bonney, and Dominique Brossard, "Measuring Scientific Knowledge and Attitudes in Citizen Science," in Jon Miller, ed., *Perceptions of Biotechnology: Public Understanding and Attitudes* (Cresskill, N.J.: Hampton, 2002, in press).

11. For more information on the Christmas Bird Count, see "CBC History," available on the National Audubon Society Web site at www.birdsource.org /cbc/hist.htm.

12. For more information on BirdSource, see www.birdsource.org.

13. Wesley M. Hochachka and Andre A. Dhondt, "Density-Dependent Decline of Host Abundance Resulting from a New Infectious Disease," *Proceedings of the National Academy of Sciences USA* 97 (2000): 5303–6; Wesley M. Hochachka et al., "Irruptive Migration of Common Redpolls," *Condor* 101 (1999): 195–204.

14. For example, Steve Kelling, information technology manager at the Cornell Lab of Ornithology, reports an increase in Web traffic on the Lab's collection of citizen science–related sites of 15 to 20 percent between 2000 and 2001.

15. Rick Bonney and André A. Dhondt, "Project FeederWatch," in Karen C. Cohen, ed., *Internet Links to Science Education: Student Scientist Partnerships* (New York: Plenum, 1997).

16. Deborah J. Trumbull, *The New Science Teacher: Cultivation of Good Practice* (New York: Teachers College Press, 1999); Deborah J. Trumbull et al., "Thinking Scientifically during Participation in a Citizen-Science Project," *Science Education* 84 (2000): 265–75.

17. In addition to funding Classroom FeederWatch, the National Science Foundation supports a number of other Cornell Lab of Ornithology citizen science activities. See, for example, Allison C. Wells, "Lab Awarded $2.25 Million for Citizen Science Online," *Birdscope: News and Views from Sapsucker Woods*, newsletter of the Cornell Lab of Ornithology (Autumn 2001): 4.

18. Dominic F. Sherony, Brett M. Ewald, and Steve Kelling, "Inland Fall Migration of Red-Throated Loons," *Journal of Field Ornithology* 71 (2000): 310–20.

19. Kenneth V. Rosenberg, James D. Lowe, and André A. Dhondt, "Effects of Forest Fragmentation on Breeding Tanagers: A Continental Perspective," *Conservation Biology* 13 (1999): 568–83.

20. Kenneth V. Rosenberg et al., *A Land Manager's Guide to Improving Habitat for Scarlet Tanagers and Other Forest-Interior Birds* (Ithaca, N.Y.: Cornell Lab of Ornithology, 1999).

21. For further information on the National Audubon Society's participation in the Important Bird Area program and on its WatchList activities, see "What Is the Status of the IBA Program," available at www.audubon.org/bird/iba/prog_status.html. See also various linked Web pages available at www.audubon.org/bird/iba/index.html.

22. For example, Jeffrey V. Wells et al., "Feeder Counts as Indicators of Spatial and Temporal Variation in Winter Abundance of Resident Birds," *Journal of Field Ornithology* 69 (1998): 577–86.

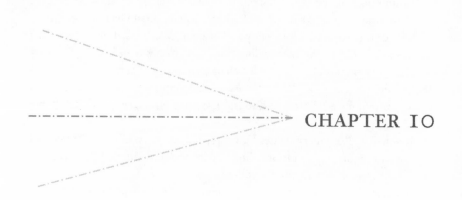

CHAPTER 10

Conservation Advocacy and the Internet

THE CAMPAIGN TO SAVE LAGUNA SAN IGNACIO

S. Jacob Scherr

Run your finger down a map of the West Coast of North America. Passing San Diego, you will touch the long narrow peninsula of Baja California, Mexico. About halfway down, you will see a spur of land reaching out into the Pacific, and just beyond that a small finger of ocean labeled Laguna San Ignacio pushing into the desert. The nearest paved road—the transpeninsular highway—lies some fifty miles to the east. Even on the most detailed maps, you will find the names of just a handful of small fishing camps on the laguna's shores. This tiny lagoon—just seventeen miles long and on average three miles wide—appears insignificant and remote; indeed, it has no telephone or electrical lines.

Yet Laguna San Ignacio is very much connected to the rest of the world—environmentally, economically, and most recently, electronically. Every winter, the lagoon is a critical end point for the longest annual mammal migration in the world. For three months, it is home to hundreds of gray whales, which seek out its calm waters to mate, give birth, and nurture their young until they are strong enough to begin the 4,000-

mile journey back to the food-rich seas off Alaska. During the whale season, many of the lagoon's fishers set aside their nets and lobster traps to cater to the ecotourists who arrive from around the world to watch—and often touch—the whales. The tourists come to savor the beauty and riches of one of our planet's true biological gems.

The largest community in the area, Punta Abreojos, sits just six miles to the northwest of the lagoon's mouth. Although too distant from the temporary whale-watching camps set up on the east side of the lagoon to benefit from tourism, the people of Punta Abreojos are very much tied to the global market. This village of just 1,000 people has prospered from its rich harvests of abalone and lobsters, which are shipped worldwide.

The laguna also caught the eye of one of the world's largest companies, Mitsubishi Corporation of Japan. In 1994, Mitsubishi—in a joint venture with the government of Mexico—announced its intentions to build the world's largest salt factory on the laguna's shores. This announcement triggered immediate expressions of deep concern from local communities and scientists working in the area. Over the next five years, the battle to save the laguna developed into a major international environmental campaign. This campaign was one of the first to make full use of the emerging power of the Internet to enable concerned citizens around the world to become intimately involved in this struggle. The Web, in turn, helped empower the people of Punta Abreojos and other communities around the laguna to stand up against this project.

A columnist for the *Japan Times* opined that this campaign, as it played out on the World Wide Web, provided a "glimpse" of how environmental battles would be fought in the twenty-first century.[1] In the end, as one Mexican newspaper rejoiced, "The Whales Won."[2] On March 2, 2000, President Ernesto Zedillo announced the permanent cancellation of the plans for the saltworks. This victory was important not only because it achieved its immediate goal of saving a precious natural area, but also because it provides significant lessons and precedents for using the Internet as a new tool for advocacy to conserve special places around the world.

Laguna San Ignacio and the Proposed Saltworks

The laguna's best-known visitors are the gray whales. The area is one of three such shallow saltwater lagoons on the Pacific coast of Baja

California sought by the whales each winter as a breeding and nursery ground. Laguna San Ignacio and the surrounding area are also home to many other species, including endangered pronghorn antelope and sea turtles. The laguna is the only one of the three that has not been subjected to extensive development.

Over the past twenty years, the Mexican government has taken steps to protect and preserve the region. In 1979, Mexican president José López Portillo declared the lagoon a "whale sanctuary." It later became a part of El Vizcaino Biosphere Reserve—the largest protected natural area in Latin America—established by President Miguel de la Madrid's government in 1988.[3]

Five years later, the United Nations recognized the Mexican government's efforts to protect the lagoon by designating it part of a World Heritage site, one of less than 140 natural areas on the World Heritage list. Under the 1972 World Heritage Convention, nations can request that their most important natural and cultural resources be recognized and listed as part of the heritage of all humankind. The national governments pledge that they will protect and preserve these treasures, which include the great pyramids of Egypt and Yellowstone National Park.[4]

Mitsubishi's proposal in 1994 to construct an industrial facility twice the size of Washington, D.C., on the shores of the lagoon thus came as a jarring surprise. The company's joint venture with the Mexican government—Exportadora de Sal, S.A. (ESSA)—already operated a massive saltworks facility approximately eighty miles north of San Ignacio. This facility, built in the 1950s, sits on the shores of another whale lagoon, Ojo de Liebre, or "Scammon's Lagoon."[5] Since the facility's construction, the adjoining town of Guerrero Negro has grown from a tiny settlement of 50 people to a sprawling town of more than 12,000 residents.

Employing some 800 people, the saltworks produces more than 7 million tons of salt annually. After being collected, the salt is loaded into massive trucks, then washed and loaded onto barges. It is first shipped to Isla de Cedros, an island forty miles off the coast, which is the nearest deepwater port. It is then loaded onto large tankers destined primarily for Japan. In Japan, the salt is broken down to produce chlorine and sodium hydroxide, which are further processed to make plastics and other industrial chemicals.

In the early 1990s, ESSA began developing a plan to build a second salt-processing facility at San Ignacio Lagoon. This facility would have a slightly larger capacity (8 million tons of salt per year), use modern equipment, and be almost fully mechanized. The new facility would employ only 200—far fewer than the existing plant. Plans for the new saltworks also included the construction of a new 1.2-mile-long pier at which the large salt tankers could dock. This would avoid the costly barging required at Guerrero Negro. Juan Bremer, CEO of ESSA, stated in an interview in 1994 that the main objective for the new facility was to enhance the "productivity and competitiveness of [ESSA] by reducing the high costs associated with double handling of salt loading and shipping operations at the Guerrero Negro facility."[6]

To describe the new facility as massive is an understatement. Its footprint would extend 116 square miles. Water would be pumped out of the lagoon at some 6,600 gallons per second—approximately an Olympic-sized swimming pool every two seconds—to fill open-air artificial evaporation ponds formed with earthen dikes. The expanse of evaporation ponds would be larger than Laguna San Ignacio itself. Over a two-year period, water would be pumped from pond to pond, as the salt concentrated, until it reached crystallization ponds. The remaining liquid, called bittern brine waste, would be drained before graters scraped the salt off the bottom. The bittern brine—which is toxic to marine life—was to be dumped back into the mouth of Whale Bay, at the entrance of the lagoon, at a rate of 22,000 tons per day. The 1.2-mile-long shipping pier would be built in the pathway of migrating whales and right in the middle of the region's most productive abalone and lobster fishery.

Beyond its immediate impact on tidal flats, wetlands, and estuaries, the saltworks would have transformed the sensitive region with an influx of thousands of people. Residents of the communities around the laguna foresaw little or no economic benefit from the saltworks and feared its impact on their livelihoods from fishing and tourism.

The Campaign

The dispute over the proposed saltworks evolved from a matter of local and limited concern in Mexico into a pitched battle in fora around the world. The five-year fight proceeded through seven rounds which are described here.

Round One: Proposal and Rejection

Along with its proposal, ESSA submitted the required environmental impact statement (EIS) to the Mexican National Institute of Ecology (INE), which is part of SEMARNAP (Secretaria de Medio Ambiente Recursos Naturales y Pesca), the Mexican environmental ministry.[7] Environmentalists countered that it would be both inappropriate and illegal to allow such a massive industrial project in the buffer zone of a federally protected biosphere reserve, in which economic activities are strictly limited and new human settlements prohibited.[8] Critics also pointed out that the EIS failed to address some of the most basic questions regarding the potential impact of the salt plant. The 465-page document devoted only 21 pages to an analysis of such impacts, just 5 pages to possible mitigation measures, and just 24 lines to the gray whale.[9]

In early 1995, well-known author and poet Homero Aridjis first called public attention to the project. His organization, Grupo de los Cien (Group of One Hundred), which represents Mexico's most prominent artists and writers concerned about the environment, pressured the government to open up the environmental review process. On February 21, 1995, Aridjis published the first major piece on the project in his column in *La Reforma*, one of Mexico's largest newspapers.[10]

Six days later, INE rejected the saltworks plan, calling it "incompatible with the area's conservation objectives and . . . also incompatible with land-use, zoning, and other legal provisions." The decision determined that "[in] the area of the project, one finds plant and animal species under various categories of protection . . . 14 plant species (4 rare, 2 threatened, 2 under special protection and 6 endemic) and 72 animal species (15 rare, 39 threatened, 6 in danger of extinction, 7 under special protection and 5 endemic). These species could be directly or indirectly harmed by habitat alteration and construction and operation of the project in question." Therefore, INE concluded, "there are no valid reasons which justify the loss of the natural environment in such as extensive area and within a biosphere reserve."[11]

Round Two: The Project Is Revived

ESSA was determined to see that the saltworks would be approved and, in March 1995, appealed the decision and asked INE to reconsider its rejection of the project.[12] There was also strong political pressure to proceed with the project from Herminio Blanco, the Mexican secretary of commerce. Serving also as chair of ESSA's board of directors, Blanco argued that Mexico could not afford to pass up the $120 million foreign investment to construct the facility and the some $80 million a year in expected hard currency revenues when the plant became operational.

Fearing that Mexican environmental officials would be overwhelmed, the Grupo de los Cien reached out for support from environmentalists in the United States and worldwide. Aridjis, the group's president, contacted Jacob Scherr at the Natural Resources Defense Council (NRDC). He asked Scherr if NRDC would be willing to sign onto an advertisement in the *New York Times* highlighting the threat Mitsubishi posed to San Ignacio Lagoon. The May 1995 advertisement put the San Ignacio controversy on the world map. The debate over the fate of the lagoon began to take shape in the *Times* as ESSA/Mitsubishi responded with its own advertisements claiming that its plans would not harm Laguna San Ignacio.

In June 1995, SEMARNAP succumbed to ESSA's pressure and agreed to allow the company to submit a new proposal and a new environmental impact assessment (EIA) for the project. To her credit, Julia Carabias, secretary of SEMARNAP, decided on an unprecedented process for ensuring the quality and credibility of the new EIA by appointing an independent committee of Mexican and international scientists. This committee was to set out the terms of reference (TOR) for the EIA and have a role in its review. The chair of the committee was Dr. Gonzalo Halfter Salas, from the Institute of Ecology, A.C., in Mexico. Its members included three well-known whale biologists—Drs. Bruce Mate, Steven Swartz, and Stephen Reilly, all of the United States—as well as Dr. Exequiel Ezcurra Real de Azua, a researcher at the National University of Mexico, and Dr. Victor H. Marin Briano, a researcher in the Department of Ecological Sciences at the University of Chile.[13]

In late February 1996, the committee held hastily arranged public hearings in La Paz, Mexico, in which a broad range of agencies, institutions, organizations, and experts participated. On July 12, 1996, the final detailed ecological and biological TORs were published. At the same time, INE issued a TOR on the socioeconomic impacts of the project.[14] These guidelines called for extensive information gathering, research, and analysis on some seventy different issues concerning ecological and social conditions around the laguna and on the proposed project, its impacts, and potential mitigation measures. As just two examples, the new EIA was asked to provide an "explanation and development of a plan for the effective handling of the brine resulting from production processes, considered toxic effluent," as well as inventories of land and marine species in the area and analyses of how the project might affect them.[15] INE indicated that the scientific committee would be asked to review the completed EIA and to present its own evaluation, which would be of great importance to INE's final decision.[16]

Round Three: An International Campaign Emerges

For many versed in Mexican politics, any struggle against the saltworks seemed fruitless. The Mexican government, which had been controlled by a single party for decades, was viewed as corrupt and willing to use pressure tactics to get its way. Many residents of San Ignacio Lagoon appeared to be afraid of the repercussions—such as the denial or delay of fishing or ecotourism (whale watching) permits—that might ensue if they spoke out against ESSA, the largest company on the lower Baja peninsula. Although having achieved a number of admirable successes, the environmental community in Mexico still was not seen as a powerful force. Unlike in the United States, environmentalists could not count on the courts to order government ministries to halt illegal projects. Allied with Mitsubishi, one of the world's largest corporations, ESSA would eventually and inevitably prevail. While publicly announcing that they were awaiting the results of the environmental studies, ESSA officials began quietly purchasing choice land along San Ignacio Lagoon in anticipation of the approval of the saltworks plan.

Over the next eighteen months, opposition to the project began to grow and become better organized. NRDC began to mobilize its own supporters. In November 1995, we mailed out an Earth Advocate Update, a periodic mailing to our members, on the saltworks controversy. We also began to encourage other international environmental groups to become involved in the debate. We soon developed a strong partnership with the International Fund for Animal Welfare (IFAW), an organization with a long history of work on whales. NRDC and IFAW retained Mark Spalding, one of the leading U.S. experts on Mexican environmental law, to begin work with his extensive contacts in Baja. Pro

Esteros, a group that had already been active in protecting sensitive coastal areas on the peninsula, then decided to take the lead in organizing a coalition of regional groups to oppose the saltworks.

Even at this early stage, NRDC began to use its new Web site to enhance communication and advocacy on this campaign.[17] Visitors to the site could send e-mails directly to Mitsubishi and Minister Carabias. NRDC also began to post more detailed information about the proposed saltworks and its impacts. Meanwhile, SEMARNAP posted its terms of reference for the EIA on its own Web site, while Mitsubishi laid out its response to the growing number of letters and e-mails on ESSA's Web site.

Electronic mail also proved critical among the dozen individuals scattered around Mexico and the United States who were working actively on the campaign. E-mail enabled us to share information and coordinate activities with an effectiveness and efficiency that would not have been possible just a few years earlier.

In February 1997, NRDC and IFAW joined with the Grupo de los Cien and Pro Esteros in a "mission" to the laguna that took the campaign to a higher level. This was the first chance for all the opposition organizations to meet together and with representatives of the communities around the lagoon. We also invited scientists, celebrities, and media. The journalists were particularly interested in interviewing actors Pierce Brosnan and Glenn Close and NRDC senior attorney Robert F. Kennedy Jr. This visit signaled that the battle to save Laguna San Ignacio had truly become international and involved a number of powerful individuals and organizations.[18] As media interest in the campaign began to rise, the Web became a valuable way to make news and information about the saltworks easily available to journalists.

The Mexican government and Mitsubishi began to take real notice and respond. Mitsubishi ran an ad in U.S. newspapers lauding its project and claiming that its existing saltworks had operated in harmony with nature for more than forty years. SEMARNAP prepared and distributed its own bilingual (Spanish-English) report on the government's consideration of the project "to provide the general public with the basic information on the steps being taken."[19] On October 3, 1997, Mitsubishi and ESSA announced that they had entered into a contract for between $1 and $2 million with the Universidad Autonima de Baja California Sur to undertake the new environmental assessment. The Universidad, in turn, hired scientists from the well-respected Scripps Institute of Oceanography.

The proponents of the saltworks wanted to limit consideration of the project to the formal environmental review—the process over which they had the most control. In Mexico, the United States, and many other countries, the proponents of a project have initial responsibility for preparing the EIA. In the case of the saltworks, ESSA contractors would manage various scientific and technical studies, which they would then synthesize and edit for presentation in the EIA. Not surprisingly, such EIAs invariably conclude that environmental impacts are manageable and that the projects should proceed.

Saltworks proponents asserted that the decision on the saltworks would be based on science. The detailed research called for by the TORs would indeed improve scientific understanding of the laguna's ecology and inform the discussion of various potential environmental impacts, safeguards, and mitigation measures. Yet even scientists who intensely study an aspect of a proposed project may at best have only an opinion as to its wisdom, given the inherent uncertainties, immense complexity, and inevitable trade-offs. The decision on whether to proceed with the saltworks would inevitably depend on judgments and values.

While critics accepted the environmental review process, we wanted to make sure that people living around the lagoon, as well as concerned scientists and citizens around the world, had a voice in the decision making. Our goal was to open up the debate on the future of Laguna San Ignacio as much as possible. We also made it clear that we would continue to oppose the project even with modifications or mitigation, because we agreed with the initial rationale for its rejection: a major industrial facility is irreconcilable with the preservation and protection of this special natural area.

As a result of NRDC's outreach to the public through both direct mail and our Web site, Mitsubishi began to receive thousands of petitions and e-mails opposing their saltworks plans. Many people took the time to set out their views in their own words and in detail. Even schoolchildren across the United States sent postcards and drawings. Familiar refrains among these children's cards were "How would you feel if someone built a huge saltworks in your home?" and "Let the whales be."

NRDC was not alone in encouraging the public to communicate with Mitsubishi. Other groups began their own electronic petitions using our Web-based materials. One group provided the e-mail address for James Brumm, vice president of Mitsubishi's U.S. subsidiary. Brumm later said that he received dozens of e-mails every day and that he had taken these messages seriously and replied to each one. He even engaged in extended exchanges with some of the e-mail protesters.[20] The Internet gave individuals a chance to participate in this campaign with an immediacy that did not exist just a decade ago.

Perhaps the weakest element of our argument against the new saltworks was the lack of available evidence to counter Mitsubishi's claims that its existing saltworks in Guerrero Negro, next to Laguna Ojo de Liebre, another gray whale lagoon, had operated in "harmony with nature" for more than forty years. Owing to its remote location, the Guerrero Negro saltworks had received little scrutiny until the campaign began. But Mitsubishi's defense began to unravel in December 1997, when ninety-four endangered sea turtles were found dead on the shores of the laguna. SEMARNAP undertook a technical investigation of the turtles' deaths. In its report issued six months later, it concluded that the deaths had been caused by the release of toxic bittern brine by the saltworks.

Investigators also witnessed a spill of brine in the spring that killed thousands of fish, and they found some 294 batteries discarded at the bottom of the lagoon. Under pressure from attorneys working with the campaign, Mexico's environmental prosecutor eventually fined ESSA about $20,000 for these violations.[21]

Round Four: Taking the Campaign to the United Nations, Japan, and Europe

On June 23, 1998, our growing coalition of organizations opposed to the saltworks filed an unprecedented petition asking the U.N. World Heritage Committee to declare the whale sanctuary of El Vizcaino—which includes Laguna San Ignacio—a World Heritage site "in danger."[22] The World Heritage Convention includes provisions for responding to "serious and specific threats to these sites," including "large scale public or private projects or rapid urban or tourist development projects."[23] Although the World Heritage Committee, administered by UNESCO, can place such sites on an "in danger" list, it does not have the authority to punish nations that fail to protect their own sites and has limited resources to assist nations. Still, such a listing carries the weight of international concern.

Although no formal process for citizens to urge the committee to declare sites "in danger" exists, we nonetheless filed a detailed document calling upon the committee to do so in regard to Laguna San Ignacio. NRDC also used the mails and our Web site to generate e-mails to the committee, and the flow of electronic messages throughout the fall attracted the secretariat's attention. This effort put this issue on the agenda for the December 1998 meeting of the full World Heritage Committee, being held that year in Kyoto, Japan.

Representatives from NRDC, IFAW, and Pro Esteros traveled to Japan to meet with the committee and present its president with some 30,000 petition forms submitted to NRDC in response to a direct mail letter. The Mexican government's representative to the committee argued that the site was in no danger. However, under pressure from other members, Mexico did agree to accept a technical mission from the World Heritage Committee to review the status of the site.

While in Japan, we also presented the controversy to the Japanese media, which had not covered the story until that point. We held press conferences and published advertisements in major Japanese newspapers. (Opinion advertising is rare in Japan, and one newspaper would not accept our ads unless we took out "Mitsubishi" and instead mentioned only "a major Japanese corporation.") We were heartened by the response. The Japanese public is very aware of the U.N. World Heritage system and proud of their country's own heritage sites, including the historic monuments of Kyoto and Yakushima Island, home to ancient cedars. The media was intrigued that a Japanese company showed so little concern about degrading a World Heritage site in Mexico when it would never contemplate doing so at home.

In Tokyo, we held a remarkable face-to-face meeting with the member of the Mitsubishi corporate board responsible for its chemicals division, of which its salt-making operation in Mexico is a very small part. He confessed that he had never before met with environmentalists during his forty-year career. It was clear to us that our message was at least being heard at the highest levels of the company.

By now, tens of thousands of people had become deeply interested in our campaign, and wanting to give them a firsthand sense of our successful mission to Japan, we produced a daily diary posted on NRDC's Web site.[24] This diary gave many people a sense of direct involvement in our campaign. Recognizing the new interest in Japan, we later added Japanese-language pages to our Web site.

Meanwhile, IFAW—through its offices in Belgium—lobbied the

European Parliament to oppose the saltworks. A number of parliamentarians raised the issue during discussions surrounding the proposed European Union–Mexico trade agreement. Additional support came from international parliamentarians representing more than sixty countries at the fourteenth-anniversary meeting of the Global Legislators Organization for a Balanced Environment. These legislators passed a resolution requesting that "the Government of Mexico withdraw its support for the proposed saltworks without delay."[25]

Round Five: Taking the Scientific High Ground

By late 1998, ESSA's new EIA was well under way and Mitsubishi was eager to constrain the debate until its own studies were complete and could be presented to the Mexican government, the media, and the public as the voice of science. In a full-page ad in the *New York Times* on December 14, 1998, Mitsubishi took issue with our petition to the World Heritage Committee and charged that we were undermining the formal Mexican EIA process "designed to protect the environment."

The Mexican environmental groups and their international partners were prepared to participate fully in a detailed review of the EIA. However, we felt that the question of whether the saltworks was acceptable could and should be addressed by scientists in forums other than those favored by the proponents.

In an effort that would have been difficult if not impossible without electronic communications, NRDC attorney Joel Reynolds began to work with noted whale expert Dr. Roger Payne of the Whale Conservation Institute/Ocean Alliance on a scientists' statement on the laguna. Payne has argued strongly that it is not sufficient to study nature: scientists have a moral obligation to stand up for its protection. Payne contacted colleagues in the scientific community to explain the threats posed by the saltworks to San Ignacio Lagoon and ask for their support for the statement. Payne sent the statement by e-mail to these scientists, asking for their endorsement. NRDC and IFAW also circulated the statement widely by e-mail and fax to scientists with whom they had worked, including NRDC board members Dr. Sylvia Earle, a preeminent marine biologist, and Dr. George Woodwell, a noted ecologist. Mexican environmental groups contacted atmospheric scientist, Nobel Prize winner, and Mexican native Dr. Mario Molina, Mexican biologist

Dr. Arturo Gomez-Pomba, and others, who agreed to endorse it.

In the end, the statement was signed by thirty-four of the world's top scientists from a broad range of fields, including nine Nobel Prize winners. The statement noted that the saltworks would introduce three important threats to the gray whales: loss of habitat, ship collisions, and bioaccumulation of contaminants. The statement went on to say that "the industrialization of this undisturbed breeding habitat is contrary to the principles and values that sanctuaries, biosphere reserves, and World Heritage Sites were created to uphold." The scientists concluded: "For these and other reasons, we believe that ESSA's proposed saltworks would pose an unacceptable risk to significant biological resources in and around Laguna San Ignacio" (for the full text of the statement, see the appendix to this chapter).[26]

The statement ran as an advertisement in newspapers in Mexico and the United States in July 1999, setting off a flurry of news coverage. Faced by a strong statement from leading scientists opposed to the project, Mitsubishi was now clearly on the defensive. The company had lost the advantage of having its scientists speak first and set the terms of the debate. Mitsubishi challenged the credibility of the scientists' statement by pointing out that the signatories had not themselves studied the lagoon—a fact that the statement's sponsors readily acknowledged. However, Mitsubishi never took issue with the content of the statement itself, which had been very carefully drafted.

Round Six: Stepping Up the Campaign in Mexico

During 1999, many of the Mexican groups that had opposed Mitsubishi's original saltworks proposal in 1994 joined with NRDC and IFAW to form the international Coalition to Save Laguna San Ignacio (Coalición Para la Defensa de Laguna San Ignacio). The coalition involved more than fifty Mexican organizations, including Grupo de los Cien, UGAM (the Union of Environmental Groups of Mexico), CEMDA (the Mexican Environmental Law Center), and Pro Esteros. Andrés Rozental, a former top-ranking official in the Mexican Ministry of Foreign Affairs who had served as Mexican ambassador to Great Britain and Mexico's representative to the International Whaling Commission, took responsibility for coordinating the activities of the coalition. In addition, Alberto Székeley, one of Mexico's leading environmental lawyers, began working for the coalition on a series of lawsuits designed to compel disclosure of the environmental problems at the existing saltworks. The coalition also encouraged and participated in a set of hearings on ESSA's activities conducted by a special commission of the Mexican Chamber of Deputies.

The coalition soon gained a high profile in Mexico City, organizing educational events for children and a well-attended music festival. The coalition's hard-hitting billboard ads ("Laguna San Ignacio: Heritage of the World, or of a private company?") had a strong impact in establishing the campaign as a defense of Mexico's patrimony. Beginning in summer 1999, the coalition began to implement a well-coordinated media strategy, which included daily summaries of campaign news coverage sent by e-mail to all members. With more than fifty Mexican organizations involved, it was possible to draw in a large variety of spokespeople for the campaign. Radio programs from children's shows to political news carried three or four interviews a week, with many different coalition members.

Starting in March 1999, the coalition's attorney, Alberto Székeley, began to file a series of legal actions regarding the sea turtle deaths and other environmental problems at the existing Guerrero Negro saltworks facilities. One important result was the disclosure that during summer 1995 a government audit showed 298 violations of Mexican environmental law, many of which still had not been resolved.

The Web was also an important element of the coalition's effort to stimulate public concern in Mexico about the saltworks. In 1999, the coalition launched a sophisticated Spanish-language site (www.coalicionsanignacio.org), which included background information on the saltworks, the legal challenges in Mexico, and an extensive section for children. In response to hundreds of letters from children in Mexico, Juan Bremer, director of ESSA, wrote an open letter to the "children of Mexico," explaining ESSA's position on the saltworks.[27]

The campaign's strength was built, in part, on its broad geographic reach. The groups that composed the Coalition to Save Laguna San Ignacio were located throughout the United States and Mexico, while European—and to a lesser degree, Japanese—activists also became engaged in the campaign. Pro Esteros, a Baja California–based wetlands conservation group, was able to both represent the concerns of local environmentalists on a world stage and provide information to the coalition about local reaction to the growing international controversy. NRDC, with then some 400,000 U.S. members, and IFAW, with its offices worldwide, were able to mobilize tremendous international support for Mexican groups and also to participate directly in decision making in Mexico. For example, Mark Spalding, a San Diego–based environmental lawyer and lecturer at the University of California at San Diego, represented NRDC and IFAW in Mexican congressional hearings on the environmental consequences of the proposed saltworks.

The geographic distribution of coalition members created a need from the outset for strong communications between Baja California, the many Mexico City–based groups, Mark Spalding in San Diego, NRDC's offices in Los Angeles and Washington, D.C., and IFAW headquarters on Cape Cod, Massachusetts. This coordination became ever more important as the coalition developed public education and media strategies for both U.S. and Mexican audiences. Although e-mail, fax, telephone, and, to some degree, Web sites were important for daily communications, face-to-face meetings—including an annual mission to San Ignacio Lagoon during the winter whale-watching season—were also critical to coordinating the ever-expanding approaches to the campaign.

In one task that could not be conducted over the Internet, NRDC dispatched Ari Hershowitz, a young, Spanish-speaking scientist, to the communities around the lagoon early in 1999 to understand the residents' views of the saltworks. Local opinions regarding the saltworks had been ignored or entirely misrepresented by proponents of the project. ESSA officials had made numerous visits to Punta Abreojos, the site of the proposed pier and much of the saltworks' infrastructure. ESSA would fly in reporters and government officials by company plane, land in Abreojos, point out the site of the proposed saltworks, and fly out quickly, sometimes as local community leaders ran after them to explain their views.

Over the course of more than two months, Hershowitz spoke with hundreds of people around the lagoon and was able to pull together the first comprehensive picture of the concerns of local residents regarding the saltworks and their notions of how to achieve a sustainable economic future. Residents were interested in exploring ways to expand their incomes from fishing, ecotourism, and oyster aquaculture, and they were eager for help in improving educational, health care, and energy services. Most of the residents saw the saltworks as providing little or nothing in the way of jobs or benefits. At the same time, the project would bring many new residents into the area and threaten the natural resources on which the communities now depended. Hershowitz's efforts provided the basis for a strong working relationship between the coalition and the community of Punta Abreojos.

While local fishers were gaining a growing role in the debate, pressure from the international community was coming to a head. The coalition had anticipated that the World Heritage Committee mission would visit Laguna San Ignacio and Laguna Ojo de Liebre during the whale season (January to March). However, the government of Mexico began to balk at the proposed membership and terms of reference for the mission. The government insisted that the team include not only international experts on landscape ecology, marine mammals, and salt production but also three of its own scientists. Extensive negotiations on the TORs for the mission were held, which in the end amazingly did not include provisions for considering potential impacts of the proposed saltworks on the site. The Mexican government had prevailed with its rather surreal argument that the mission could not determine the potential for danger from the proposed saltworks, because "the project proponent had not yet made its submission to the proper authorities."[28] The Mexican officials were obviously worried that the mission would find the site to be "in danger" from the proposed saltworks. The officials were taking steps to make certain that this did not happen.

The World Heritage mission finally arrived in Mexico at the end of August 1999, visiting with government officials in Mexico City and then touring both whale lagoons. When the mission arrived in Guerrero Negro on Laguna Ojo de Liebre, hundreds of ESSA workers and their families anxious about any criticism of the existing saltworks greeted them with banners and signs supporting the company.[29] In contrast, fishers and tour operators at Laguna San Ignacio were almost unanimous in their opposition to the proposed new facility. The mission visit was widely covered in the Mexican media and was front-page news in a number of newspapers.[30]

Round Seven: "Mitsubishi: Don't Buy It!"

At a September 1999 press conference in Los Angeles, the international coalition, joined by actor Pierce Brosnan, announced an all-out effort in California to force Mitsubishi to drop the project. We chose to focus our limited resources on California for a number of reasons, including the state's high level of public awareness and concern about the gray whales, which migrate just off the California coast. We asked consumers not to purchase Mitsubishi products, and we asked mutual fund managers not to hold the company's stock. We encouraged cities and counties throughout California to pass resolutions opposing the saltworks and business with the company.

An important element of the "Mitsubishi: Don't Buy It!" phase of the campaign was a dedicated Web site (www.savebajawhales.com). Far more sophisticated than the Web sites we had used in the early days of our campaign and the Web, this site enabled people to target their advocacy. By entering their zip code, people in California could send an e-mail directly to their local Mitsubishi car dealer. By late in the year, Mitsubishi Motors was feeling the pressure, and top officials at the company were worried about the impact of the campaign on their reputation. A few Mitsubishi dealers called to indicate their willingness to break ranks and go on record opposing the saltworks.

Over the course of the campaign, many individuals came forward to volunteer their assistance. Often, their first contact with us was through the Web or e-mail. In Los Angeles, NRDC worked closely with Imaginary Forces, a motion graphics studio, to produce a compelling public service announcement against the saltworks. When TV stations were unwilling to carry the provocative announcement, we made it available on our Web site.

NRDC assisted IFAW in a major effort to encourage more than forty municipalities, counties, and major pension funds throughout California to pass resolutions against the saltworks.[31] We also asked a number of labor unions to take a stand on the issue, and the Service Employees International Union, California Building and Construction Trades, and California AFL-CIO all passed resolutions opposing the saltworks. Many of these resolutions called for a review of all existing contracts with companies within the Mitsubishi keiretsu and mandated no new contracts until the proposed saltworks was abandoned.

In October 1999, fifteen mutual funds handling more than $14 billion announced that they refused to buy Mitsubishi stock until the company abandoned its plans for the new saltworks.[32] A full-page advertisement in German newspapers highlighted this move, asking "Mitsubishi, Eine Gute Anlage?" ("Mitsubishi, a wise investment?").

In late 1999, the saltworks project suffered another blow when the World Heritage Committee, preparing to meet in Marrakesh, Morocco, received a detailed report on the August mission to Mexico.[33] SEMARNAP tried to "spin" the report by putting out a press release before the report was available to the media or the public. The release highlighted the mission's recognition of "the outstanding efforts of the Mexican government in protecting the El Vizcaino World Heritage Site."[34]

Once we were able to obtain the full report, we found that the mission had in fact signaled deep concern about the proposed saltworks. Despite the Mexican government's effort to force the team to turn a blind eye to the proposed plant at Laguna San Ignacio, the mission felt compelled to examine the site because everyone they met, including governmental officials, raised the issue. The mission found that Laguna San Ignacio "is in relatively pristine condition" and warned that the proposed saltworks "could threaten" the site's integrity, since evaporation and crystallization ponds would transform a large area within World Heritage boundaries.[35] The mission also pointed to the secondary impacts of the project, which would involve human encroachment, pollution, and waste disposal.

The coalition used the Internet, including e-mail and Web sites, to disseminate the report's real message to journalists worldwide. The media agreed with our assessment of the mission's findings. *Newsweek* went so far as to lead off its article as follows: "It's looking more like Mitsubishi against the world."[36]

In January 2000, we continued to pressure Mitsubishi in California, bringing a resolution before the state's Coastal Commission condemning the saltworks project owing to its effects on shared marine species. After presentations by NRDC, Mexican members of the coalition, and Mitsubishi representatives, the commission, in a nearly unanimous decision, adopted the resolution. A press release by Mexican members of the coalition regarding this decision resonated throughout Mexico and was reported in most major national newspapers (*El Dia, Excélsior, La Prensa, El Economista, La Jornada*, and others). Speculation that the U.S. federal government might get involved stirred considerable concern in the Mexican press.

Knockout

In February 2000, we were anxiously awaiting the release of ESSA's now-completed EIA on the project. We had started to identify scientists in Mexico and the United States who could assist us in a detailed review, because we worried about the boost to the project from the report's inevitable conclusion that the saltworks would have little or no impact on the whales or the lagoon.

As we were in the final stages of preparing for another mission to the laguna, we heard that President Zedillo was visiting the laguna, accompanied by his family and some close friends.[37] On March 2, as we began traveling to the site, we received word from Mexico City. President Zedillo had just announced that the project had been canceled.

In his statement, President Zedillo pointed out that the EIA had shown that the project would not harm the whales and was environmentally sustainable. Nonetheless, he noted that responsible environmental organizations, UNESCO, and the public had raised concerns about the alteration of the landscape, and he agreed that it was more important to preserve this area than to proceed with the project. According to President Zedillo, "It's a unique site on a world wide scale, for species habitat and for its natural beauty, which is also a value to be preserved. . . . We Mexicans are generating a new culture of appreciation, respect, and care for the natural resources of our nation."[38]

Mikio Sasaki, president of Mitsubishi, expressed his agreement with the decision.[39] ESSA never released a copy of its EIA but did make a thirty-four-page summary available for a while on its Web site. As expected, the EIA asserted that the saltworks could proceed with limited and acceptable impacts to the environment.[40]

The announcement of the saltworks' cancellation reverberated around the world. All major U.S. newspapers carried the story of the whales' victory. Victory messages were quickly distributed worldwide over the Internet, posted on NRDC's Web site as well as the "Mitsubishi: Don't Buy It!" and coalition sites. NRDC and IFAW placed advertisements in major U.S. newspapers, and the coalition published ads in Mexican newspapers congratulating President Zedillo on his decision. The director of the World Heritage Committee sent a letter of congratulations as well and posted this on the World Heritage Committee Web site.[41]

The *Boston Globe* reported that the campaign had "changed the shape of environmental politics in Mexico."[42] Andrés Rozental summarized it this way: "Its most important contribution is that it strengthens the [Mexican] environmental movement enormously by giving it a victory. . . . In these battles, whether it's the monarch butterflies or the Lacandon forest, a victory gives a renewed impetus to continue and go ahead into other campaigns."[43]

While the threat of the saltworks has subsided, NRDC and IFAW are continuing to work to protect and preserve Laguna San Ignacio. We have established a $100,000 fund to further sustainable economic alternatives, and we have already initiated a number of economic and social projects with the communities. For example, we are assisting Punta Abreojos in expanding their oyster aquaculture operations and in improving their ability to protect and manage their fisheries.

Lessons from the Laguna and the Future of Internet Advocacy

The campaign to save Laguna San Ignacio would not have succeeded, nor would effort of this magnitude and depth have been possible, without the Internet. Instant electronic communication was fundamental to coordinating the activities of scores of widely dispersed individuals and organizations. The Web also enabled tens of thousands of people around the world to monitor the progress of the controversy over the saltworks on their computer screens and to participate in the campaign. We received numerous e-mails from people with suggestions and offers of help.

The Web clearly made it easier to send formal expressions of concern about the saltworks directly to Mitsubishi, the Mexican government, and the World Heritage Committee. The company took its e-mail traffic seriously and engaged in extended electronic dialogues with some individuals. We also knew that if we posted material about the campaign on our Web site, Mitsubishi would see it. The Web also greatly facilitated media coverage, affording journalists ready access to detailed information on all sides of the controversy. We readily gave reporters links to other sites.

Of course, while the Internet was important, it was one of many different communications tools used in the campaign. Direct mail, earned and paid media, and even telephone and fax were essential in stimulating public awareness and concern about the project in Mexico, the United States, and worldwide, as well as in securing the significant resources needed to conduct this major campaign.[44] Face-to-face meetings were also critical to developing strong working relationships within the coalition and with the communities around the lagoon. Direct debates and discussions with Mexican officials and Mitsubishi were similarly important.

Finally, there was no virtual substitute for the actual experience of visiting the laguna and watching its whales. Over the course of five years, we organized visits involving more than 100 scientists, officials, journalists, celebrities, and activists. The area is extraordinarily beautiful, and visitors came away resolved that this special place should not be sacrificed for common salt.

The campaign to save Laguna San Ignacio provided a model for a new form of international environmental campaigning. Throughout the world, biological gems are threatened by ill-conceived development schemes, and local communities and environmentalists struggle against great odds to protect them. In January 2001, NRDC launched a Web-based BioGems initiative to save such threatened natural areas, with an initial focus on sixteen sites in the Americas (see www.savebiogems.org). These sites include the Arctic National Wildlife Refuge and the Everglades in the United States, the Macal River in Belize, and the Olivillo forest in southern Chile. The Web site provides brief overviews, including videos and slideshows, on the importance of each of these areas and the threats to them. Visitors can use e-mail to take action, send e-mail to their friends suggesting that they visit the site, and make donations. They can also sign up to be BioGem defenders and accumulate points for their activism on personalized Action Visas.

NRDC made a major effort to drive traffic to the BioGems site using "viral marketing" tools. "Viral marketing" employs word of mouth and a variety of electronic media, including person-to-person e-mail, to quickly and efficiently spread the word, "like a virus," about a new cause, idea, product, or service among a group of people. In early May 2001, we sent an e-mail from Robert Redford about the Arctic National Wildlife Refuge to some 75,000 people on NRDC's electronic activist lists and to some 250 environmental online mailing lists, newsgroups, and bulletin boards. Redford asked recipients to take action at the BioGems Web site and to forward the email to their friends. As a result, the number of visitors to the site soared from some 60,000 in April to 272,000 in May, while the number of BioGems defenders grew from some 18,000 to almost 145,000.

NRDC can target e-mail alerts and updates to defenders who have taken action on a particular BioGem. For example, on a Friday evening in early June, we sent an e-mail to 21,000 defenders suggesting that they visit the site to send a message urging Chilean officials to cancel plans for a road through the Olivillo coastal forest. This appeal generated some 8,000 messages over a single weekend. One of the targeted officials was quoted the next week complaining about the deluge, while our Chilean partner groups were pleased with the rapid and strong show of international support.[45]

Today, most environmental debate still occurs through conventional forums, such as administrative proceedings, diplomatic negotiations, and newspaper op-ed columns. But the Web may become more central to environmental decision making through "cyber-summits" and "cyber-tribunals."[46] Governments, corporations, and citizen organizations would present their views, debate issues, make commitments, and monitor their implementation—all through the Web.

Noted author and *New York Times* columnist Thomas Friedman recently observed in a speech that we are moving from a "world of walls" to a "world of webs."[47] As the campaign to save Laguna San Ignacio shows, the Internet is indeed helping break down barriers to the free flow of information and ideas worldwide. Governments and corporations are finding it more difficult to control and constrain decision making. The Internet is giving voice to communities whose natural resources are being threatened, while enabling people to express their concern about environmental problems all over the world and hold governments and corporations accountable for their promises. This globalization of citizen action parallels today's economic globalization. In time, the development of the Internet may be regarded as a turning point in the struggle to conserve and protect the earth.

ACKNOWLEDGMENTS

The author acknowledges with thanks the assistance of Ari Hershowitz in researching and drafting this chapter.

Appendix: Text of Scientists' Statement

The following letter was published in July 1999, as an advertisement, in the *New York Times*, the *Los Angeles Times*, and other major newspapers in the United States and Mexico.

"An unacceptable risk . . ."
Leading International Scientists Urge Mitsubishi to
Abandon Proposed Baja Saltworks at Laguna San Ignacio

We, the undersigned, are scientists united in our concern over a proposal
to build the world's largest saltworks on the shores of Laguna San
Ignacio, in Baja California, Mexico, a lagoon designated as a whale sanctuary by Mexico in 1976. In 1988 it was also included within the largest
international biosphere reserve in Latin America (the Vizcaino Biosphere
Reserve) and in 1993 was listed by UNESCO as a World Heritage Site.
It is the last undisturbed gray whale breeding and calving area on earth,
and for that reason may be of unique importance for the survival of that
species.

As proposed by Exportadora de Sal, S.A. (ESSA), a joint venture of
Mitsubishi Corporation (49%) and the Mexican government (51%), the
saltworks would create a massive 116 square mile industrial landscape of
evaporation ponds—larger than Laguna San Ignacio itself—a million-ton
salt stockpile, fuel and water tanks, a 1.25-mile long pier with a shipping
dock and conveyor belts running from crystallization ponds to the pier's
end, workshops, headquarters buildings and the facilities necessary to
support 200 employees while onsite. The upper end of Laguna San
Ignacio would be invaded by 17 pumps operating 24 hours a day to draw
6,600 gallons of saltwater per second from the lagoon into the evaporation ponds.

We believe that the industrialization of this undisturbed breeding
habitat is contrary to the principles and values that sanctuaries, biosphere
reserves, and World Heritage Sites were created to uphold. To build
major industry here, especially when it is constructed on, extracts its
water from and pumps its wastes into a UNESCO World Heritage Site,
will create a dangerous precedent—a precedent at odds with the broad
scientific consensus that life in the world's estuaries and coastal waters is
increasingly threatened by loss and degradation of habitat through physical alteration of ecosystems and by pollution.

More specifically, building the saltworks at Laguna San Ignacio will
risk introducing to the area the top three present threats to whales
besides whaling: loss of habitat, accidents involving collision with ships,
and the slow but inexorable bio-accumulation of contaminants in the
whales' bodies. A larger human population will be attracted, crowding
the whales in the lagoon with more boats, noise and waste. Large ocean-
going vessels will be introduced to the region. Large quantities of toxic
contaminants, such as oil, diesel fuel and concentrated brine wastes will

be present. The brine wastes contain toxic concentration of magnesium sulfate, potassium chloride, bromine, iodine, and boron, which, as proposed by ESSA, will be dumped into the adjacent Bahia de las Ballenas— Bay of the Whales.

The Vizcaino Reserve contains critical habitat for both terrestrial and marine species, including a highly endangered pronghorn antelope. These species would also be impacted by the saltworks project. According to Mexico's Federal Attorney General for Environmental Protection, a study ordered by his agency showed that a die-off in December 1997 of 94 endangered black sea turtles (Chelonia aggasiz) was possibly caused by brine waste contamination from an existing ESSA saltworks at nearby Laguna Ojo de Liebre. In the course of its scientific investigation, the agency also observed a fish kill caused by a spill of over four million gallons of brine waste from this same ESSA plant in May 1998.

There may be important regional impacts as well. The flooding of 116 square miles of coastal tidal flats and mangroves would certainly disturb the existing habitat of terrestrial species, altering tidal and runoff patterns and potentially affecting migratory birds. Because there is no way to extract the quantities of sea water needed without extracting fish fingerlings along with it, pumping of 6,600 gallons per second from the lagoon could adversely affect fisheries in the region. Finally, increased human population is inevitably accompanied by increased poaching of wildlife.

For these and other reasons, we believe that ESSA's proposed saltworks would pose an unacceptable risk to significant biological resources in and around Laguna San Ignacio. We respectfully urge Mitsubishi to abandon the project and trust that the Mexican government will stand by its original decision denying ESSA permission to construct a saltworks at Laguna San Ignacio.

"We, the undersigned, are scientists united in our concern . . ."

Roger Payne, (Drafter), Founder/President, Ocean Alliance; MacArthur Fellow; Lyndhurst Prize Fellow; United Nations Environment Programme Global 500 Laureate; Knighted by Netherlands: Order of the Golden Ark.

Philip Anderson, Professor, Princeton University; Nobel Prize (Physics); National Medal of Science; Dannie Heineman Prize; Bardeen Prize; Foreign Member, Japan Academy; Foreign Member, Royal Society of London.

George Archibald, Founder/Director, International Crane Foundation; MacArthur Fellow; United Nations Environment Programme Global 500 Laureate; World Wildlife Fund Gold Medal; Knighted by Netherlands: Order of the Golden Ark.

David Baltimore, President, California Institute of Technology; Nobel Prize (Medicine); Member, National Academy of Sciences; Former President, Rockefeller University; Foreign Member, Royal Society of London; Honorary Member, Japanese Biochemical Society.

Lester Brown, Founder, President, and Senior Researcher, Worldwatch Institute; MacArthur Fellow; United Nations Environment Prize; United Nations Environment Programme Environmental Leadership Medal; Gold Medal—Pro Habitability Award, King of Sweden; Blue Planet Prize, Asahi Glass Foundation.

Richard Dawkins, Professor, New College, Oxford University; Nakayama Prize for Human Science; Royal Society of London Michael Faraday Award; Medal of the Zoological Society of London.

Irven De Vore, Professor, Harvard University; Lifetime Achievement Award, Institute of Human Origins; Former President, Anthropology Section, American Association for the Advancement of Science.

Jared Diamond, Professor, University of California, Los Angeles School of Medicine; MacArthur Fellow; Pulitzer Prize; Cosmos Prize (Japan); Coues Award, American Ornithologists' Union; Burr Award, National Geographic Society; Member, National Academy of Sciences.

Rene Drucker-Colin, Chair, Department of Physiology, School of Medicine, Autonomous National Univ. of Mexico (UNAM); Vice-President and President-Elect, Mexican Academy of Sciences; National Sciences Prize; Guggenheim Fellow; Mexican Foundation for Health Prize; National Prize for Sciences and Arts (Mexico).

Sylvia Earle, Former Chief Scientist, U.S. NOAA; National Geographic Society Explorer-in-Residence; President, National Marine Sanctuaries Fndtn.; Knighted by Netherlands: Order of the Golden Ark; Olguin Marine Environment Award; Stratton Leadership Award.

Paul Ehrlich, Professor, Stanford University; Craoord Prize (Royal Swedish Academy of Sciences); MacArthur Fellow; Tyler Prize for Environmental Achievement; United Nations Environment Programme Sasakawa Environment Prize; Volvo Environment Prize; Heinz Award for the Environment; Member, National Academy of Sciences.

Thomas Eisner, Professor, Cornell University; National Medal of Science;

Tyler Prize for Environmental Achievement; Guggenheim Fellow; Centennial Medal, Harvard University; Member, National Academy of Sciences; Foreign Member, Royal Society of London.

Murray Gell-Mann, Professor and Co-Chair, Santa Fe Institute; Nobel Prize (Physics); Member, President's Committee of Advisors on Science and Technology; Member and John J. Carty Medallist, National Academy of Sciences; United Nations Environment Programme Global 500 Laureate; Foreign Member, Royal Society of London.

Arturo Gomez-Pompa, Distinguished Professor of Botany, University of California, Riverside; Tyler Prize for Environmental Achievement; Guggenheim Fellow; Knighted by Netherlands: Order of the Golden Ark; Herrera Medal (Mexico); Advisor to President of Mexico on tropical ecology issues.

Stephen Jay Gould, Professor and Museum Curator, Harvard University; MacArthur Fellow; President, Society for Study of Evolution; Former President, Paleontological Society; Member, National Academy of Sciences; Medal of the Zoological Society of London; Gold Medal for Service to Zoology, Linnaean Society of London.

Donald Griffin, Professor, Harvard University; Professor, Rockefeller University; Former President, Frank Guggenheim Foundation; Eliot Medal, National Academy of Sciences; Member, National Academy of Sciences.

Roger Guillemin, Distinguished Professor, the Salk Institute; Nobel Prize (Medicine and Physiology); National Medal of Science (USA); Lasker Foundation Award; Member, National Academy of Sciences (USA); Honorary Member, Japan Biochemical Society.

Sidney Holt, Founder/CEO, International League for Protection of Cetaceans; Marine biologist and environmental consultant; United Nations Environment Programme Global 500 Laureate; Knighted by Netherlands: Order of the Golden Ark; World Wildlife Fund Gold Medal.

Sir Andrew Huxley, Professor, Trinity College, Cambridge; Nobel Prize (Medicine); Member, Royal Society of London; Copley Medal; Order of Merit (United Kingdom); Grand Cordon of the Sacred Treasure (Japan).

Brian Josephson, Professor, Cavendish Lab, Cambridge University; Nobel Prize (Physics); Hughes Medal; Faraday Medal; Holweck Medal.

Donald Kennedy, Former President and currently Bing Professor of Environmental Science, Stanford University; Former Commissioner, U.S. Food and Drug Administration; Member, National Academy of Sciences.

Sir Aaron Klug, President, Royal Society; Nobel Prize (Chemistry); Honorary Fellow of Peterhouse and of Trinity College; Former Director, Medical

Research Council Laboratory of Molecular Biology, Cambridge; Former Director of Studies in Natural Sciences, Peterhouse, Cambridge.

Masakazu Konishi, Professor, California Institute of Technology; International Prize for Biology, Japan Society for the Promotion of Science; Dana Award for Achievement in Health; Member, National Academy of Sciences.

Donella Meadows, Former Director, Sustainability Institute and Professor of Environmental Studies, Dartmouth College; MacArthur Fellow; Pew Scholar; Co-Author, Club of Rome Report, "Limits to Growth" and "Beyond the Limits."

Mario Molina, Institute Professor, Massachusetts Institute of Technology; Nobel Prize (Chemistry); Member, President's Committee of Advisors in Science and Technology; Member, Secretary of Energy Advisory Board; Member, National Academy of Sciences; Board Member, US-Mexico Foundation of Science.

Giuseppe Notarbartolo di Sciara, President, Istituto per la Ricerca Scientifica e Tecnologica Applicata al Mare (ICRAM); Former Director, European Cetacean Society; Tridente d'Oro Prize.

Fernando Nottebohm, Distinguished Professor, Rockefeller University; Director, Rockefeller Field Research Center for Ethology and Ecology; Dana Award for Achievement in Health; Pattison Award for Distinguished Research in Neurosciences; Coue's Award, American Ornithologists' Union.

Peter Raven, Director, Missouri Botanical Gardens; Engelmann Professor, Washington University; MacArthur Fellow; Tyler Prize for Environmental Achievement; Volvo Environment Prize; United Nations Environment Programme Sasakawa Environment Prize.

Jorge Reynolds Pombo, Electrical and Bio-Engineer (including pioneer of pacemaker and whale electro cardiograms); Director, Grupo Whales Heart Satellite Tracking; Member, 34 Colombian and international scientific societies; Miembro Asociado por Invitacion, Sociedad Mexicano de Cardiologia; La Orden de Boyaca, Grado Gran Offical; Silver Medal of Scientific Merit, United Kingdom.

John Terborgh, Director, Duke University Center for Tropical Conservation; Founder, Manu Tropical Research Station, Manu, Peru; MacArthur Fellow; Guggenheim Fellow; Pew Fellow; Member, National Academy of Sciences; Daniel Giraud Elliot Medal, National Academy of Sciences.

James Watson, Director, Cold Spring Harbor Lab; Nobel Prize (Medicine); Presidential Medal of Freedom; Member, National Academy of Sciences;

Carty Medal, National Academy of Sciences; Member, Royal Society of London; Copley Medal, Royal Society.

Edward O. Wilson, University Research Professor, Harvard University; Crafoord Prize (Royal Swedish Academy of Sciences); National Medal of Science; International Prize for Biology (Government of Japan); Tyler Prize for Environmental Achievement; Pulitzer Prize (twice); Benjamin Franklin Medal of the American Philosophical Society.

George Woodwell, Director, Woods Hole Research Center; Founder/ Director, Ecosystems Center of the Marine Biological Laboratory; Heinz Award for the Environment; Member, National Academy of Sciences; Fellow, American Academy of Arts and Sciences; Former Chairman, World Wildlife Fund-U.S.

Richard Wrangham, Professor, Harvard University; MacArthur Fellow; Rivers Medal, Royal Anthropological Institute; Board Member and Former President, Dolphins of Shark Bay Research Foundation; Fellow, American Academy of Arts and Sciences.

Organizations listed for identification purposes only.

NOTES

1. Stephen Hesse, "Can Environmental Dialogue Prevent a Whale of a Disaster?" *Japan Times*, December 13, 1998.
2. Jorge González Torres, "Ganaron las Ballenas" (The Whales Won), *El Universal* (Mexico), March 4, 2000.
3. Serge Dedina, *Saving the Gray Whale: People, Politics, and Conservation in Baja California* (Tucson: University of Arizona Press, 2000), 60–61. Dedina provides a detailed overview of the history of the Mexican government's exploitation and more recent protection of gray whales.
4. United Nations Educational, Scientific and Cultural Organization (UNESCO), *Convention Concerning the Protection of the World Cultural and Natural Heritage* (Paris: UNESCO, 1972). For information regarding the convention and its implementation by the UNESCO World Heritage Committee, see http://whc.unesco.org. The convention language is available at http://whc.unesco.org/nwhc/pages/doc/main.htm.
5. Scammon's Lagoon was named after a nineteenth-century whaler whose rediscovery of the gray whale breeding areas nearly led to the species' extinction.
6. Serge Dedina and Emily Young, *Conservation and Development in the Gray Whale Lagoons of Baja California Sur, Mexico,* final report to the U.S. Marine

Mammal Commission, October 1995.

7. Secretaria de Medio Ambiente Recursos Naturales y Pesca (SEMARNAP), *San Ignacio Saltworks: Salt and Whales in Baja California*, SEMARNAP Working Papers (1997), 16.

8. Mark J. Spalding, *Laguna San Ignacio: A Briefing Book*, prepared for the Natural Resources Defense Council and the International Fund for Animal Welfare (January 1999), 1–2.

9. Dedina, *Saving the Gray Whale*, 88. Dedina writes that this document is "a model of how not to write an environmental impact statement."

10. Homero Aridjis, "El Silencio de las Ballenas" (The Silence of the Whales), *La Reforma* (Mexico), February 21, 1995.

11. Letter from Gabriel Quadri de la Torre, INE, to Juan Bremer Gonzalez, February 27, 1995. SEMARNAP would later ignore this letter and claim that the reason the saltworks project was rejected was that the environmental assessment "was deficient as regards identification, assessment, and description of the environmental impacts." SEMARNAP, *San Ignacio Saltworks*, 16.

12. SEMARNAP, *San Ignacio Saltworks*, 16.

13. SEMARNAP, *San Ignacio Saltworks*, 16–17.

14. See www.semarnap.gob.mx/noticias/reportajes/si_proyecto.htm.

15. SEMARNAP, *San Ignacio Saltworks*, 21–22.

16. SEMARNAP, *San Ignacio Saltworks*, 24.

17. NRDC launched its Web site on June 16, 1995.

18. For a report on that mission, see www.nrdc.org/wildlife/marine/hbaja.asp. On March 12, 1997, the Mexico City newspaper *La Reforma* carried a prominent story on the mission with a focus on Pierce Brosnan entitled "El 007 Viene a Mexico." The Associated Press and Knight Ridder News Service widely distributed stories on the controversy, and *Newsweek* carried a brief story in its March 24, 1997, issue.

19. SEMARNAP, *San Ignacio Saltworks*, 11.

20. Conversation with James Brumm, vice president, Mitsubishi International Corp., April 5, 2000.

21. "Sancionan Empresa $206 Mil 700" (Company Is Fined $20,700 [Pesos]), *La Reforma* (Mexico), December 8, 1999.

22. Letter from Pro Esteros, Cooperativa Ecoturismo Kuyima, Grupo de los Cien, NRDC, and IFAW to World Heritage Committee, UNESCO World Heritage Centre, June 23, 1998. The World Heritage site also includes a second whale lagoon, Ojo de Liebre, or Scammon's Lagoon, adjacent to ESSA's existing saltworks at Guerrero Negro.

23. UNESCO, *Convention Concerning the Protection of the World Cultural and Natural Heritage*, Article 11.4.

24. See www.nrdc.org/wildlife/marine/hkyoto2.asp.

25. PRNewswire release, August 27, 1999.

26. See the scientists' statement in the appendix to this chapter.

27. Juan Ignacio Bremer, "To All the Children of Mexico," open letter published in *La Reforma* (Mexico), January 8, 2000, and in *El Universal* (Mexico), January 9, 2000.

28. United Nations Educational, Scientific and Cultural Organization (UNESCO), *Report of the Mission to the Whale Sanctuary of El Vizcaino, Mexico, 23–28 August 1999*, report presented to the Bureau of the World Heritage Committee, 23rd Extraordinary Session, Marrakesh, Morocco, November 26–27, 1999, available at www.unesco.org/whc/archive/99-209-inf20.pdf.

29. The mandate of the World Heritage Committee's mission included examining the impacts of the existing saltworks on Laguna Ojo de Liebre. Rumors that the United Nations might shut down the facility floated around Guerrero Negro.

30. "La UNESCO Analizará los Daños que Causan las Salineras a las Ballenas de México" (UNESCO Will Analyze the Damage Caused by Saltworks to the Whales of Mexico), *La Crónica*, August 23, 1999. Reports on the World Heritage mission also appeared in *El Universal, Excélsior,* and other national Mexican newspapers.

31. By December 1999, the following local governments had passed resolutions opposing the saltworks: Alameda County Board of Supervisors, Humboldt County Board of Supervisors, Marin County Board of Supervisors, San Francisco County and City Board of Supervisors, Sonoma County Board of Supervisors, Berkeley City Council, Cotati City Council, Culver City Council, Davis City Council, Eureka City Council, Folsom City Council, Imperial Beach City Council, La Cañada–Flintridge City Council, Los Angeles City Council, Malibu City Council, Pismo Beach City Council, Poway City Council, Richmond City Council, Sacramento City Council, Sand City Council, Saratoga City Council, Sebastopol City Council, Santa Cruz City Council, Trinidad City Council, Vista City Council, Watsonville City Council, and West Hollywood City Council. The final total of local resolutions in California exceeded forty.

32. Michael Shea, "Beating Mitsubishi," *Campaigns and Elections* (July 2000): 44. The mutual funds that participated in this statement included Calvert Group, Citizens Funds, Crown Futures, Domini Social Investments, Everest Asset Management, First Affirmative Financial Network, Green Century Fund, Miller Howard Investments, MMA Praxis, Parnassus, Pax World Fund, Prentiss Smith & Co., Trillium Asset Management, Walden Asset Management, and Winslow Management.

33. UNESCO, *Report of the Mission to the Whale Sanctuary of El Vizcaino.*

34. For information on SEMARNAP's efforts to protect the gray whale, see "Bellena Gris, Accoón Institucional" at www.semarnap.gob.mx/naturaleza/especies/ballengris/accion.htm.

35. UNESCO, *Report of the Mission to the Whale Sanctuary of El Vizcaino.*

36. Paul O'Donnell, Seth Stevenson, and Victoria S. Stefanakos, "A Whale of a Whale Problem," *Newsweek*, November 29, 1999.

37. James Smith, "Activists Break New Ground to Help Shake Off Saltworks Project," *Los Angeles Times*, April 23, 2000, sec. A1, p. 10.

38. Zedillo's speech is summarized and referenced at www.semarnap.gob.mx/quincenal/qui-57/destacado.htm.

39. See www.mitsubishi.co.jp/En/press/release40.html.

40. Some of our colleagues were disappointed that the formal environmental review process was not completed and a formal decision rendered by INE on the project. They saw this as a lost opportunity to create an important precedent in Mexico for a fully transparent and participatory EIA process. However, the proponent of a project also has the option of withdrawing its proposal, which is what the Mexican government and Mitsubishi did in this case.

41. "Letter from the Chair of the World Heritage Committee to the President of Mexico concerning El Vizcaino," ref: WHC/74/Mexico.nat/MR, March 7, 2000. Available at www.unesco.org/whc/nwhc/pages/news/main2.htm.

42. Richard Chacon, "Whales Sink Plans for Mexican Salt Plant," *Boston Globe*, April 2, 2000.

43. Quoted in Chris Hardman, "Saving the Song of Sea Giants," *Américas* (April 2001): 38.

44. Only a tiny fraction of donations for the campaign came directly from the Web. Most were secured through traditional solicitations, although the Web and e-mail provided a new mechanism for keeping donors and supporters up to date.

45. The impact of e-mail petition campaigns may decline as they become more common. However, our experience suggests they will continue to be an important mechanism for putting corporate and government officials on notice regarding public concerns.

46. For a potential model of such a cyber-proceeding, see www.earthsummit-watch.org/shrimp, the site of the now-inactive "Shrimp Sentinel." The Sentinel addressed the sustainability of shrimp trawling and farming worldwide using information from governments, industry, and citizen groups.

47. Thomas W. Friedman, "Responsive Governance in the 21st Century," *The Mansfield American-Pacific Lectures* at International House of Japan, Tokyo, September 21, 2000 (Washington, D.C., Missoula, Mont. and Tokyo: Mansfield Center for Pacific Affairs, 2000), 3. Available at www.mcpa.org/programs/friedmancomplete.pdf. Friedman discussed his concept at several speeches delivered in 2000. In Tokyo, he is quoted as saying, "so basically, in the space of ten years, we've gone from a world of division and walls to a world of integration and webs."

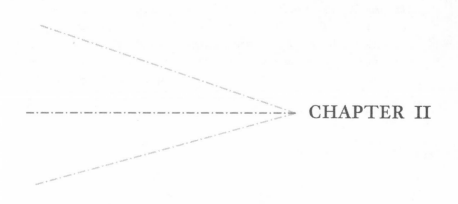

CHAPTER II

Envisioning Rural Futures

USING INNOVATIVE SOFTWARE FOR
COMMUNITY PLANNING

William Roper and Brian H. F. Muller

The family of Lyman Orton, proprietor of the Vermont Country Store in Weston, Vermont, has a long history as Vermont merchants. Orton's grandfather had a country store in North Calais, Vermont, where Lyman's father, Vrest, grew up in the storekeeping atmosphere. After a journey that included a stint in New York working for H. L. Mencken's *American Mercury*, and a position at the Pentagon during World War II, Vrest returned to Vermont. He opened the Vermont Country Store in Weston in 1946—a village that he described as having been "a ghost town all right"—helping to spark the town's renewal.[1] In 1952, the *Saturday Evening Post* ran a story about Vrest called "The Happy Storekeeper of the Green Mountains," which earned the Vermont Country Store national visibility. Skilled as a publicist, Vrest became chair of the Vermont Historic Sites Commission, was a frequent contributor to *Vermont Life* magazine, and became a prominent advocate for protecting the special qualities of small-town life.

Lyman Orton took over the store in the late 1970s. During the Vermont boom of the 1980s, he became increasingly concerned at the inability of communities to effectively define and shape their future. Growth occurred in vastly different ways than the citizens envisioned. Following through on such concerns, Lyman and his sons, Cabot, Gardiner, and Eliot, have significantly advanced Vrest's interest in preserving the quality of the Vermont landscape. The family, along with their friend Noel Fritzinger, founded the Orton Family Foundation in 1995, with the mission of helping rural communities cope with rapid economic, social, and environmental change—change that appears to be intensifying as "today's technologies permit many Americans to earn their living farther away from urban centers, and more people are drawn to the quality of life that smaller communities offer." The Ortons have expressed a particular concern for citizens "struggling to manage these pressures in ways that promote and enhance the beauty, community and quality of rural life."

Accordingly, the foundation's first efforts "were focused on creating awareness of the issues rural communities face, defining the challenges and exploring potential solutions." As a result of such efforts, the foundation's leadership decided to help develop a "toolbox" that would allow people in rural places to "participate in fully informed decision-making processes about their common future."[2]

Lyman Orton is quite clear in his conviction that if ordinary citizens are to participate in the town planning process, the process itself must become more tangible and transparent. As he explained in a recent editorial contribution to the *Vermont Country Store Catalog*, the community planning process is "so dense and brain-damaging that it's no wonder we leave it up to the 'experts.' The result? Others are defining how our very own towns will grow, and we're not even particularly aware of it."[3]

In the interest of making the community planning process more open and understandable, the Orton Family Foundation committed early in its existence to help develop, as part of its toolbox for rural communities, a new generation of planning software. The software was intended to produce graphically vivid output that could make clear, to anyone interested, what the result of a given set of community land use and planning policy choices might be.

Launching the software development effort in 1997, the foundation assembled a development team composed of leaders in land use planning, information technology, simulation modeling, and 3-D visualization. The resulting software suite, called CommunityViz™, is an innovative set of integrated tools that allow users to interactively sketch land use scenarios;

evaluate them against community objectives and constraints; view com-prehensive information on the past, present, and future impacts of their choices; and walk through realistic three-dimensional simulations of those scenarios.[4]

CommunityViz™ is based on ArcView, a market-leading geographic information system (GIS) software used by public, private, and nonprofit organizations around the world. GIS expands the notion of a traditional database to include layers of information developed for different purposes, including satellite-generated images of roads and other land uses and maps depicting property boundaries; zoning rules; water, sewer, and utilities; police, fire, and emergency response data; and historic and natural resources. GIS provides a powerful mechanism for layering these types of information on top of one another, facilitating sophisticated analysis and decision making. CommunityViz™ applies GIS capabilities in a relatively novel fashion by converting two-dimensional geographic information into three-dimensional visualizations that can be interactively manipulated.

The software suite developed by the Orton Family Foundation has, since 1997, been through initial development, early prerelease "beta" test-ing, and a "limited release" testing through 2001. The software suite's ini-tial general release commenced in the fourth quarter of 2001. This chap-ter briefly reviews how the software works as well as some of the lessons learned by the Orton Family Foundation during software development, prerelease beta testing, and limited release testing. The CommunityViz™ software has significantly benefited from rigorous field testing, and it should continue to gain strength as a planning tool as it is used in a vari-ety of settings in the United States.

How It Works

The CommunityViz™ decision-making software consists of three com-ponents. The core component, Scenario Constructor, is designed to enable users to evaluate the immediate impact of different development proposals. It was developed for the CommunityViz™ application by Fore Site Consulting, an advanced developer of GIS-based spatial spreadsheet and impact analysis. The Policy Simulator component, created by the technology consulting group at PricewaterhouseCoopers, uses new developments in simulation theory to facilitate long-term evaluation of the cumulative impact of alternate land use policy decisions. The final component, Sitebuilder3D, developed by MultiGen Paradigm—a firm specializing in three-dimensional visualization techniques—is designed to allow communities to evaluate the visual impact of changes in zoning rules and various development proposals. The three components are integrated into the full CommunityViz™ suite of software, which allows user-inputs in any one of the three components to be reflected in the outputs of all three components.

Planning with the Scenario Constructor Component

The Scenario Constructor component of CommunityViz™ was designed by Fore Site to allow users to customize their framework for making land use decisions by defining and specifying several basic planning elements:

Indicators—the objectives or desired outcomes
Alternatives—the options available
Attributes—the measurable characteristics that are the basis for comparing
  alternatives
Variables—the assumptions used to assess potential solutions
Constraints—the policies and requirements that all viable alternatives must
  satisfy
Analysis—how well an alternative satisfies the constraints and fulfills the
  objectives

As users experiment with different assumptions and land use layouts, Scenario Constructor updates attributes, monitors constraints, and tracks indicators.

For example, a community may wish to evaluate the impact of different land development scenarios on public services such as water and wastewater services, open space and agricultural lands, and wildlife habitat—either alone or in combination. The community might choose water consumption and wastewater treatment as indicators, with water consumption or treatment per household as attributes. Variables might include residential and commercial land use in already developed areas (or in "infill" parcels that lie between already developed parcels) as compared with the development of "greenfield" sites outside the edges of the existing developed area. Constraints might include local, state, and federal regulations as well as the capacity of water treatment plants. These initial val-

Summary of Large Lot (BAU)

	%	Large Lot
Total Acres		15645.36
Open Space Area	10.09%	1578.31
Commercial Area	1.30%	203
Institutional Area	0.93%	145
Residential Area	87.68%	13718
Dwelling Units (.4279)		5869
Population (2.4)		14085.6
Jobs (.75)		4402

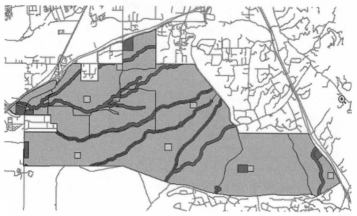

Figure 11.1: Scenario Constructor can be used to create charts that can be integrated into Microsoft Word documents or PowerPoint presentations. This chart shows particular land use characteristics under a "Business as Usual" scenario in Santa Fe, New Mexico.

ues allow a user to create the formulas employed to analyze the impact of each scenario.

Scenario Constructor facilitates community discussion by allowing participants to modify the values used in the formulas. For example, participants might decide to increase or decrease water consumption per household to evaluate the influence of assumptions regarding water supply and the capacity of wastewater treatment facilities. As users ask multiple "what if" questions, the software instantly updates any GIS layers stored in the system as well as the resulting analysis, allowing users to make real-time comparisons of alternative proposals. This analytical flexibility allows collaborators to explore both simple and complex decisions regarding managing local facilities, conserving natural resources, and enhancing community planning. Users easily can embed the resulting charts and proposals into customized reports based on Microsoft Word or PowerPoint (see Figure 11.1).

Forecasting Using the Policy Simulator Component

While Scenario Constructor compares the immediate impact of different variables and proposals, Policy Simulator allows users to explore a range of policy options, select a course of action (or nonaction), and simulate possible longer term outcomes. As users experiment with such policies as land use regulations, infrastructure, budgeting for new public buildings, and tax options, Policy Simulator displays its forecasts in graphical form as charts, tables, or maps (see Figure 11.2). By allowing users to view and analyze both anticipated possibilities and unforeseen consequences, Policy Simulator reveals the cumulative effects of decisions over time.

Policy Simulator employs a relatively new technique called "adaptive agent-based modeling" for community-based modeling. The units of such models are a community's households and businesses rather than the aggregates of demographic, economic, and land use variables commonly used in statistical modeling. Policy Simulator models the decisions, behaviors, and interactions of these actors, such as where they will shop and work. This agent-based simulation model is stochastic: the agents' behavior includes an element of randomness that produces some variability from one simulation to the next, even if the scenario inputs remain the same.

In this kind of modeling, one small change in an agent's behavior can lead to very large changes in the resulting system. For instance, in one simulation an agent may decide to purchase a tract of land, while in another the agent may decide not to purchase that tract even if the other parameters do not change. Such a small decision may lead to major changes in outcomes over the short and long term. This element of randomness allows users to explore more than one possible future for the same policy option.

Agent-based modeling is bottom up: agents adapt and learn, and the environment in which they operate changes constantly throughout mul-

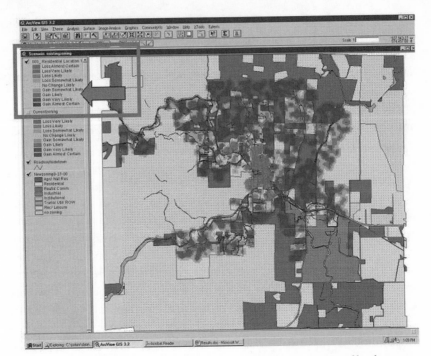

Figure 11.2: Policy Simulator graphically shows the consequences of land use regulations, infrastructure development, and budgeting and tax decisions.

tiple iterations of the simulation. This means that the actions of the agents themselves create the environment.[5] Agent-based simulations predict changes at a micro level and sum them up to produce a larger total. Analysis of multiple simulation results will reveal a "trend" in these outcomes. It is up to the user to decide the likely aggregate outcome.

Traditional simulation, in contrast, usually operates under strict assumptions. Such models also provide little explanation; because they are equation based, they express their outcomes through correlation and causation, whereas agent-based models allow users to trace the participants' decision-making process.

Visualization Using the Sitebuilder3D Component

Yet another vital aspect of effective community planning is the ability to see how changes may alter the local landscape. CommunityViz™ allows users to create three-dimensional images of their environment that they can virtually walk or fly through by simply pointing the mouse. For example, a community using Scenario Constructor or Policy Simulator to weigh two subdivision proposals might use Sitebuilder3D to evaluate the visual impact of each (see Figure 11.3).

Sitebuilder3D leverages GIS data maintained in ArcView, as well as themes that users create with Scenario Constructor and Policy Simulator, into a three-dimensional image. The software creates this image using types of buildings and vegetation chosen from a library of features; users may also add new features to the library.

The power of these software tools stems, in part, from their interactivity: changes, additions, and deletions to information in one component automatically appear in the other two components. For example, a policy choice made in Policy Simulator to limit new home development in outlying greenfield sections of a community might be reflected in a long-term reduction in sewerage and school construction capital costs in Scenario Constructor's indicator or constraint setup and analysis. Likewise, a relatively compact, clustered development pattern resulting from the Policy Simulator model may be displayed in the Sitebuilder3D visual representations of the community over time.

Prototyping and Beta Testing CommunityViz™

Between 1997 and 2000, the software development team, with the coordination of the Environmental Simulation Center (ESC), a pioneer in applying information technology to community planning, completed the initial design and testing of the CommunityViz™ system. By April 2000, Orton Family Foundation representatives were demonstrating a prototype of the CommunityViz™ software at the American Planning Association's annual meeting and in similar forums. Initial reaction to the prototype software was quite enthusiastic. For example, a working paper issued by the Centre for Advanced Spatial Analysis at the University College of London reported in October 2000 that "the CommunityViz™ project combines an agent-based simulation with their 3-D visualization model. . . . The project is being applied to a small town—Ascutney, Vermont—which has a population of roughly 1,000. The agent-based model simulates the socioeconomic evolution of every individual in the population, building up an incredibly detailed composite of the community."[6]

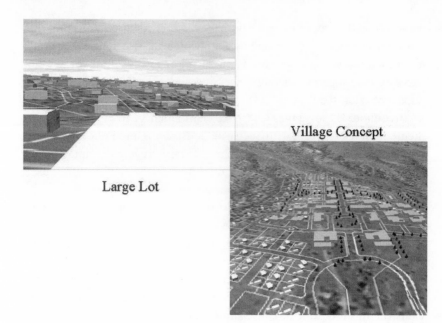

Large Lot

Village Concept

Figure 11.3: Sitebuilder3D allows participants in the planning process to compare the visual impact of alternative development proposals.

While the prototype was clearly impressive to observers, it had yet to be tested in the field by potential end users. To take the software past the prototype stage, Orton, with the help of ESC and CommunityViz™ software creators, issued a request for proposals (RFP) regarding participation in a CommunityViz™ beta (early prerelease) test in January 2000. The RFP was sent to communities and organizations that had expressed an initial interest in the software during the prototype development phase. Orton received some forty responses to the RFP, from which it selected teams in eight communities in Maine, Vermont, New York, Colorado, and New Mexico to participate. Each of the eight teams had among its members at least one representative from a regional or county planning agency, a non-profit with an interest in land use planning, or a professional consulting organization. A ninth team, representing a crew at the University of Wisconsin that had extensive experience working with planning software, was also included in the beta test.

In June 2000, the Orton Family Foundation provided the test teams with a three-day training session at a state-of-the-art GIS training lab generously made available by Middlebury College in Vermont. After completing their training at Middlebury, the test teams returned to their communities and began to develop their applications of CommunityViz™. At each of the eight community sites, interns from either Middlebury College or the University of Colorado at Denver (CU-Denver) were available to provide general project assistance and to observe software usage practices.

The Orton Family Foundation contracted with the Department of Planning and Design at CU-Denver to chronicle the adoption of CommunityViz™ at each site. Partial funding for this work was provided by the university's Program for Western Lands. CU-Denver evaluators interviewed project teams and foundation staff and also had access to discussions occurring over a password-protected Web site, conference calls, and occasional meetings. Interns at each site collected much of the raw data by keeping journals, conducting interviews, and providing biweekly reports on the adoption and use of CommunityViz™. Evaluators posted all of this material on the secure Web site.

Evaluators used conversations with participants to develop several research questions to guide the assessment of CommunityViz™:

1. Were planning organizations and advocates in the test communities experienced enough in geographical analysis to operate CommunityViz™ and make productive use of the results?
2. Were data and external resources available to support the use of CommunityViz™?
3. Is CommunityViz™ suited to the kinds of planning activities and problems that the communities think are important?
4. Did CommunityViz™ generate information or representations that shifted the scope or character of planning discussions?
5. Can communities readily integrate the test application of CommunityViz™ into their planning processes?
6. Did the use of CommunityViz™ create new interorganizational relationships or broaden participation in discussions?

The experiences of the test team located in Santa Fe, New Mexico, best illustrate the benefits and challenges experienced by the beta test group in using the CommunityViz™ software to enhance local land use decision making. As discussed toward the end of this chapter, the experience and

feedback gained during the beta tests led users to develop novel and exciting applications of the CommunityViz™ software.

The Santa Fe Case

The Santa Fe region faces an intense development struggle typical of
amenity-rich communities across the West and in parts of the northeast-
ern United States. On the one hand, both Santa Fe City and County are
growing rapidly, producing a rapid rise in housing prices and concern about
the availability of affordable housing. On the other hand, the region has
a strong growth management ethos and an interest in promoting regional
architecture and effective urban design. These competing pressures in New
Mexico's capital city, with its world-renowned reputation as a place of
extraordinary beauty, have contributed to a setting in which approvals for
new subdivisions can become contentious and controversial.

During the 1990s, the Trust for Public Land (TPL), a national conservation nonprofit organization, has worked closely with Santa Fe County to
promote long-term land use planning and open space conservation. For
example, following the New Mexico state legislature's passage of a bill in
1996 that allows counties to issue bonds for open space protection, TPL
worked closely with Santa Fe County officials in 1998 and 2000 to conduct
polling for, campaign for, and implement two successful open space bond
bills.[7] Having developed a close working relationship, TPL and the county
jointly submitted a proposal to be a CommunityViz™ beta site in the first
quarter of 2000.

In their proposal, TPL and the county expressed the idea that
CommunityViz™ could serve as a tool to help mediate divergent views
among various local interests regarding the appropriate use and development of the Community College District, a 16,000-acre area designated
for development on the south side of Santa Fe. The Community College
District includes the first major subdivisions near the city to reach serious
consideration by local planning officials in many years.

The design concept for the Community College District evolved out of
public discussions during the preparation of the county's Growth
Management Plan. These discussions led to the introduction and eventual
approval by the county of an innovative planning approach, the creation of
"new community districts" as alternatives to the more typical large-lot
development that had occurred in the rural residential areas in Santa Fe

County in recent decades. New community districts were intended to feature high-density residences as key components of mixed-use, neotraditional villages. The Growth Management Plan designated the Community College District as the focal point for such development south of the city—a focal point that would provide facilities for employment, education, and cultural activities as well as for housing.

In the late 1990s, the developers of the Community College District area obtained approval from the county for two village-style subdivisions, including the Community College District, which were intended to incorporate such mixed uses. At the same time, the developers indicated to county officials that other subdivisions had been slated for development. Once it was selected as one of the eight community-based CommunityViz™ beta sites by the Orton Family Foundation in spring 2000, the Santa Fe project team focused on modeling the impacts of the proposed developments in the Community College District on jobs, housing, traffic, and open space.

To assess the potential impact of possible development scenarios in the Community College District, TPL, the county, and others interested in the effort decided to explore three different scenarios for the Community College District. The first scenario—Business as Usual—constrains intensive development to its then current locations; developers build out the remainder of the district under current zoning regulations. The second scenario—Big Magnet—assumes that residential, commercial, and industrial development focuses on village centers, thus preserving open space, opportunities for recreation, and a community feel. The third scenario—Sprawl to the Mall—represents aggressive development given little or no regulation.

The project was strongly motivated by *a priori* planning ideas and a specific design concept: the test would evaluate a preferred alternative that had already gained significant political support. Rather than providing a foundation for a specific decision within planning agencies or a public setting, the test was designed to generate working documents that would build consensus among participating agencies and planning advocates. The tangible product of the beta test process—determined through a series of meetings—was a PowerPoint presentation composed of indicators, impact tables, maps, and images of alternative growth scenarios (see Figure 11.4). The meetings themselves also proved to be products of the CommunityViz™ beta test process.

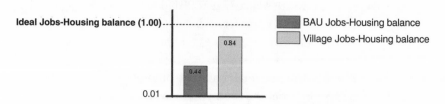

Figure 11.4: CommunityViz™ results, showing the impacts of different policies, were integrated into a PowerPoint presentation that helped to build consensus on growth management in Santa Fe, New Mexico.

The project tapped a well-equipped GIS infrastructure on the state and local level as well as extensive external expertise. Both Santa Fe City and County maintain dedicated GIS staff, while TPL's regional office includes experienced planning advocates. In addition, TPL hired the Earth Data and Analysis Center (EDAC), located at the University of New Mexico in Albuquerque, to develop data for the software and to run CommunityViz™. EDAC staff members assigned to the project had substantial training and three to five years of professional experience in employing state and federal land use data. As a state data clearinghouse and a National Spatial Data Infrastructure node, EDAC has broad access to such data. Staff from the U.S. Department of Energy and a Denver-based design firm, Design Workshop, also acted as project consultants.

With such wide-ranging information and software competence at hand, participants commented that the work entailed in building the appropriate dataset and applications (such as georeferenced information on roads, parcels, land use, and natural features) was generally straightforward, although time consuming. Indeed, in response to the research question regarding the adequacy of local GIS capabilities, respondents at Santa Fe (as well as other sites) indicated that the local technological infrastructure was able to support sophisticated GIS analysis.[8]

In light of these findings, it appears that many small and midsized communities may be capable of at least limited use of sophisticated planning support systems. We are mindful that most local planning departments have limited budgets for software and training. The tight budgets of GIS programs of small and rural governments in particular may be a significant stumbling block to the sustained use of planning support systems such as CommunityViz™. However, the benefits from using such systems may motivate communities to develop the needed budgets and staffs.

Regarding the research question that focuses on the availability of good local data, we found in Santa Fe that although most data were available to the team, there were some important gaps. For example, land parcel boundaries in Santa Fe County were not initially available in usable digital format. Limitations in GIS and other data may constrain many communities from pursuing such demanding planning support systems as agent-based modeling and 3-D representation.

Interestingly, the very process of gathering digital data for the Santa Fe project provoked considerable thought and discussion among team members. Project participants focused on the quality of available information and on how that available information could be accessed and analyzed. As reported by one project participant, thinking through these questions yielded benefits beyond the scope of the CommunityViz™ beta project itself: "Some of the data that was required—that we needed—particularly around the water issue, was also invaluable and we have it to use again. I will say that it has gotten all of us to start looking at data regionally now—and that was not there before."[9]

With regard to the research question about whether CommunityViz™ is suited to the kinds of planning activities and problems that communities think are important, the answer in Santa Fe was a qualified "yes." The county and TPL appear to have been able to use the software to create a common language among planners and other staff members in organizations with only marginally overlapping interests and expertise. As one participant commented, "Visualization for the public has been openly discussed, but I see visualization for the planners as a more important aspect of this software. As they sat and looked at the blank field they were forced to come to grips with a lot of the planning issues that were not well documented or agreed upon."[10]

In fact, Santa Fe participants described their use of the software as creating a "story" among themselves. This use parallels John Forester's conception of the "deliberative practitioner."[11] The software helped practitioners—local government officials and planning advocates alike—to jointly consider land use policy in the Community College District and develop a collective framework for discussing it.

It is important to note that TPL and other participants chose to define the planning problem associated with the use of CommunityViz™ in a relatively narrow fashion. Alternative and perhaps riskier applications of the software might have involved developing graphic, interactive presentations for public meetings, conducting a formal impact analysis, preparing parcel-level subdivision designs, or creating the Community College District master plan on the CommunityViz™ platform. In Santa Fe as well as at other sites, respondents puzzled about whether digital planning tools are better suited to tackling narrow or open-ended questions, and whether the requirements of the planning support software itself might constrain the range of questions that users can ask. Participants also wondered whether the software is more helpful in situations where community values are acknowledged and have solidified or where they are still unformed.

However, in the context in which they chose to use the software—to build a "common language" among planners and advisors to the planning project, yielding a PowerPoint presentation of indicators, graphs, and visualization that was made available to the public—the CommunityViz™ beta software appears to have been effective. The PowerPoint presentation produced by the CommunityViz™ team was used effectively to build a common vision among planners regarding the Big Magnet development scenario. Furthermore, the common vision developed by the team continued, in the words of Ted Harrison, TPL's Southwest regional director at the time, "to reinforce the values and goals of a compact development plan" that eventually led to approval by the county commissioners of the development scheme.[12]

Regarding the research questions focusing on whether CommunityViz™ shifted the scope of the planning discussions, was able to be integrated into the local planning process, and created new interorganizational relationships, we gained several important insights from the Santa Fe case. First, in using CommunityViz™ software, Santa Fe participants were forced to spend considerable time and energy agreeing on quantitative and qualitative benchmarks to be used in evaluating the performance of the proposed alternative development scenarios. A comment of a project participant is instructive: "We all treated land use differently—we knew this but here it was and we needed to talk about it. We were able to work out those different views and put them into some agreed upon indicators."[13] In Santa Fe, quantitative indicators included total population, developed acreage, water use, proximity to transit, and vehicle miles traveled for each scenario; qualitatively, the software allowed the participants to compare general land use patterns (for example, contrasting patterns of compact development

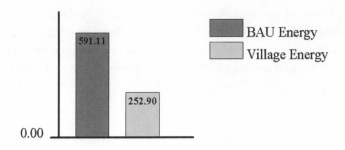

Figure 11.5: CommunityViz™ lets users see the effects of alternative development proposals on a variety of indicators, including energy consumption.

against more sprawling development patterns) as they appeared on the visualizations of each of the scenarios. The process of selecting and evaluating the simulation results associated with such indicators, in turn, appears to have focused attention on differences and points of agreement (see Figure 11.5).

Participants expressed the opinion that the process of using CommunityViz™ in the Santa Fe application did bring partners, coming from different organizations with different perspectives in the public and nonprofit sector, together around a common concept as well as around specific maps and agreed-upon metrics. In Santa Fe, this process evolved on several levels and required considerable group effort. "It took a while to get everyone going because trust had to be built. We scheduled meetings regularly to talk about what we were going to do. . . . Having them [the project participants] at the table and meeting certainly helped to create the trust."[14] The experience at Santa Fe and other sites indicates that such a process of coming together may not have occurred if the analyses had been conducted by external organizations, such as planning advocates, developers, or long-term subcontractors, who would have been less likely to meet the group's needs for flexibility and timeliness.

The Santa Fe experience also highlighted potential impediments to the "bringing together" and trust-building efforts that were integral to the successful use of this and other software-based planning support systems.[15] A mundane but telling problem concerned local timetables that guide the planning process. Santa Fe County requires developers to submit concrete plans only relatively late in the planning cycle. Assumptions regarding demographics and systems operations embedded in such plans may be difficult to test under a variety of scenarios if planners are working under a tightly constricted time schedule. Indeed, the use of tools such as CommunityViz™ by a variety of planning departments to test development proposal assumptions may prove valuable in highlighting such institutional obstacles to effective planning.

Beta Test Debriefing

At the end of the formal beta period, representatives of each of the beta teams traveled to Chicago to meet with the software developers, trainers, and analysts to participate in a two-day debriefing session. The group shared experiences regarding the striking variety of ways in which the software was used to respond to widely differing local problems and decision-making environments.[16] Beta test participants provided invaluable feedback directly to the software developers that resulted in a prioritized wish list of essential and future software enhancements and upgrades.

Debriefing participants agreed that CommunityViz™ clearly has the potential to bring more and better information to bear on land use decisions by combining maps, graphical representations, tables, and visual images into an easily understood format.

Although communities did use CommunityViz™ in unique ways, in each case the software appears to have enhanced local planning processes in three common ways. First, the tools provide a tightly structured, comprehensive vehicle for substantive research on the effects of development proposals and policies, thereby providing decision makers with more complete and timely information and enriching the analytical process. Second, CommunityViz™ expanded the visual and aesthetic experience implicit in traditional map-based analysis and comprehensive planning efforts, giving credence to Iris Murdoch's insight that "we apprehend more than we understand."[17] In the course of the beta tests, three-dimensional images did appear to extend the imaginative powers of people involved in planning processes. Third, CommunityViz™ expanded the scope of typical local land use decision making in the beta tests by allowing active community organization and public participation in the planning process.

Software Revision, Limited Release Testing, and General Release

Following the completion of the beta testing process, the Orton Family Foundation implemented a number of suggested changes to the software. For example, Santa Fe participants felt that the building types and topography in Sitebuilder3D were more suited to New England than to the Southwest. Given that feedback, the software development team came up with a more diverse library of images for use in building Sitebuilder3D three-dimensional visualizations.

After commissioning a round of software revisions, Orton proceeded with a limited release of CommunityViz™ in 2001. Twenty-one communities were selected to participate in the limited release test (see the appendix to this volume). Throughout the limited release program, users provided feedback that has been instrumental in further refining the software. Wizards have been added to increase user-friendliness, additional models have been added to the model library, and additional policy templates have been added to Policy Simulator. In addition to providing suggestions about the software itself, the Limited Release teams have provided information about their own applications of CommunityViz™ in situations ranging from urban redevelopment to farmland preservation—how they approached their applications, what worked well, how they integrated the use of the software into their local or regional planning processes, and what they learned. This information aids the foundation as it develops case studies and instructional guides for future users.

Full general release of CommunityViz™ commenced in the fourth quarter of 2001. Purchasers received the software, training, and technical support for a year and were invited to participate in the growing CommunityViz™ User Group through a Web site, phone conferences, and other mechanisms. Early purchasers included individual communities, regional planning agencies, federal government agencies, nongovernmental organizations, and consultants.

In at least one case, the use of CommunityViz™ apparently is changing how a respected conservation organization looks at the development process. Ted Harrison, now directing Conservation Ventures for the Trust for Public Land, reports that, through using CommunityViz™ on the Santa Fe case, he became convinced that scenario-based analysis was "enormously valuable." He further reports that "we are now using scenario-based GIS tools to advance our conservation practice." In October 2001, his group contracted with its first full-time GIS specialist to integrate GIS tools, including CommunityViz™, into several of its projects around the country.[18]

ACKNOWLEDGMENTS

The Orton Family Foundation is grateful to the early adopters of CommunityViz™ for their active participation in shaping future releases of the software and creation of a body of experience that will inform and help guide future users. Through this program, the participants have had access to a state-of-the-art software program. By using the software and providing detailed feedback, they have helped the Orton Family Foundation to develop a tool that builds among planning process participants a "common language" of visualization and engages citizens and planning practitioners in new ways to shape the future of their own communities.

Appendix: CommunityViz™ Limited Release Lead Organizations and Cooperating Participants

SOUTHEAST/MID-ATLANTIC REGION

1. **South Georgia Regional Development Council (RDC), Valdosta, Ga.** Participant: City of Harira. The project team is using CommunityViz™ to explore a series of projects in a number of growing rural towns in South Georgia. These projects include locating new residential subdivisions around the perimeter of town; recruiting businesses and industries to a new business park; adaptive reuse proposals for two old, abandoned industrial areas; and revitalizing an old historic southern downtown.

2. **New Jersey Office of State Planning, Trenton, N.J., and the Town of Dover, N.J.** Participants: Morris County, N.J., Dept of Transportation; N.J. Department of Environmental Protection; Rutgers University. CommunityViz™ is being used in the Town of Dover, in Morris County, as part of a Town Center design initiative that would focus on transit-oriented (re)development opportunities.

3. **Maryland Department of Planning, Town of Easton, Md.** The team is using CommunityViz™ to update the Town of Easton's Comprehensive Plan. This project will use the software to analyze and illustrate the effects of different development densities, building types, design, and access to transportation routes.

4. **Voices & Choices of the Central Carolinas, N.C.** Participants: LUTrOS, Carolinas Land Conservation Network, University of North Carolina, Land Trust for Central North Carolina, Cabarrus Regional Chamber of Commerce, Union County Chamber of Commerce. CommunityViz™ is being used to explore scenarios and build public support for the implementation of a Regional Open Space Plan.

5. **Southern Tier West Regional Planning and Development Board, N.Y.** Participants: Southern Tier West Regional Planning and Development Foundation, Town of Ellicott, N.Y. The project team is using CommunityViz™ to explore development patterns occurring along Main Street as it intersects newly designated Interstate 86. Team would like to identify and foster certain types of development patterns that are consistent with the Town's vision. They are also examining the results of possible ordinance changes.

6. **Lancaster County Planning Commission, Penn.** Participants: Millersville University, Center for Rural Pennsylvania. The team is using CommunityViz™ to help develop the Conestoga Valley Regional Strategic Comprehensive Plan. This is a multidisciplinary planning effort to develop a regional strategic comprehensive plan for three municipalities in the heart of Lancaster County. The team is focusing on evaluating alternative land use along two proposed transportation corridors.

CENTRAL REGION

7. **Liberty Prairie Foundation, Grayslake, Ill.** Participants: Liberty Prairie Conservancy, Chicago Metropolis 2020, Lake County, University of Illinois–Chicago. CommunityViz™ is being used by the project team to focus on the Liberty Prairie Reserve, an area of approximately 2,500 acres. The team will create scenarios based on zoning that existed ten or fifteen years ago to indicate how the reserve area might have been developed. The team will then develop and explore the impact of alternative development scenarios for the areas that remain to be built.

8. **Land Information Access Association, Traverse City, Mich.** Participants: Planning and Zoning Center Inc., City of Escanaba, Michigan. The team is using CommunityViz™ to assist the City of Escanaba with the development of a comprehensive highway corridor management. The purpose of this plan is to create a working template that can be applied to improve, enhance, and protect the pedestrian and driving environment of the corridor through town.

9. **Story County Planning and Zoning, Iowa.** Participants: Story County, Iowa. CommunityViz™ is being used to complete the update and revision of the County Development Plan (CDP). The CDP is the county's land use plan and serves as the basis for managing growth and development in the unincorporated areas of Story County.

10. **Alpine Township, Comstock Park, Mich.** Participants: Grand Valley State University, Grand Valley Metro Council. The project team is using CommunityViz™ to address the long-term applicability of the Township's sustainable development program. The team is using a transfer of development rights (TDR) modeling project to test the Township's sustainable urbanization policy. The team hopes to answer the question "How might Alpine Township be impacted (socially, fiscally, environmentally) if a TDR program were instituted?"

11. **1000 Friends of Minnesota, St. Paul, Minn.** Participants: Dakota County Office of Planning, Eureka Township, University of Minnesota. CommunityViz™ is being used to generate scenarios specific to Eureka Township but connected to the larger picture of Dakota County and the

Farmland and Natural Areas Project. Funded by the State of Minnesota, this project's primary goal is to develop a plan that balances resource protection and growth.

WESTERN REGION

12. **Glendale and Lyons, Colo., with Denver Regional Council of Government (DRCOG).** Participants: U.S. Environmental Protection Agency. The project team is using CommunityViz™ in two small Denver-area communities—one rural and threatened by growth all around it, the other surrounded by urban development and seeking growth within its boundaries. Lyons will use the software in an interactive public planning process to design a pedestrian environment in the downtown while addressing such significant issues as transportation, parking, transit, and increased tax revenues. Glendale will use the software for several projects, including a Comprehensive Plan Update.

13. **City of Tacoma, Wash.** Participants: Tacoma Economic Development Department. The team will use CommunityViz™ for the Tacoma Dome Area Plan, which will look at redeveloping what is now mainly a commercial/industrial area to a mixed-use district of retail and entertainment with urban housing. The team is assessing the economic impacts of the redevelopment efforts. Components of this project include rezoning, environmental analysis, transportation analysis, light-rail, and commuter rail. Infrastructure issues and visibility analysis are also being addressed.

14. **Eagle County Government, Colo.** CommunityViz™ is being used to complete a subarea master plan update for the unincorporated area of Edwards in Eagle County. The primary objective of the project is to develop an accurate future land use map based on population projections through 2020.

15. **Capital Area Planning Council, City of Lockhart, Tex.** The team is using CommunityViz™ to help determine the potential growth and developmental impacts of a new highway on the City of Lockhart. This highway, State Highway 130, is anticipated to bring a sizable amount of NAFTA-related traffic through the city. The team hopes to determine how this highway will affect growth and the economy for the city.

16. **City of Logan, Utah.** Participants: Utah State University, Cache County, Bear River Association of Governments, Cache Metro Planning Organization. The project team is using CommunityViz™ to develop a plan for residential infill within the "original" platted Townsite of Logan.

NORTHEAST REGION

17. **City of South Burlington–Chittenden County Regional Planning Commission (CCRPC), Vt.** CommunityViz™ is being used to explore a pro-

posed city center for the City of South Burlington. CCRPC and the City will use the software to help design and build consensus for the city center. The team will examine the ecological, visual, and community impacts of various site designs.

18. **Town of Hanover, N.H., Geog Data Tech Inc.** The team is using CommunityViz™ to update the Town of Hanover's Master Plan. Particularly, the team will focus on examining, visualizing, and comparing options with regard to "designing" village centers. The team is also modeling and comparing policy alternatives for the overall Master Plan concept.

19. **University of Rhode Island, Town of South Kingston, R.I.** Participants: Environmental Data Center, Rhode Island Sea Grant Advisory Services, Sustainable Coastal Communities Program, The Dunn Foundation. The project team is using CommunityViz™ to examine scenarios for a 700-bed dormitory development that could occur in two different locations on the University of Rhode Island campus. Planning the application has brought people together that normally did not meet and network on a regular basis. Most importantly, the university and the community are equal partners in exploring the impacts of the development.

20. **University of Connecticut, Cooperative Extension System; Town of Marlborough, Conn.** CommunityViz™ is being used to examine development scenarios of a ninety-five-acre area in the center of Marlborough that comprises five parcels. Currently, sixty acres are zoned residential and thirty-five acres are zoned commercial. The development scenarios include various residential layouts (conventional, cluster, open space) and changing the zoning from residential to commercial/light manufacturing.

21. **Two Rivers Ottaquechee Regional Planning Commission, Randolph, Vt.** The team is using CommunityViz™ as part of a Sustainable Challenge Grant from the U.S. Environmental Protection Agency to realize three land use planning studies. One of those studies is the Exit 4 area, a highway interchange at the center of town. The team is using the software to explore planning initiatives that would ensure that the area surrounding Exit 4 would be developed in a fashion consistent with the Town's goals.

NOTES

1. Edward Shenton, "The Happy Storekeeper of the Green Mountains," *Saturday Evening Post*, March 15, 1952, available at www.vermontcountrystore.com/Corp/artHappy_Storekeeper.asp.

2. See "About the Foundation," which describes the mission and programs of the Orton Family Foundation, available at www.ortonfamilyfoundation.org/about.htm.

3. Lyman Orton, "Bringing Citizens Back into Defining Their Town's Future," *Vermont Country Store Catalog* (fall 2001), available at www.vermontcountrystore.com/corp/editorials/bringing_citizens_back.asp.

4. See www.communityviz.com. The name CommunityViz™ has been trademarked, and various patent applications are pending.

5. Robert N. Bernard, "Using Adaptive Agent-Based Simulation Models to Assist Planners in Policy Development: The Case of Rent Control" (paper presented at the annual conference of the Association of Collegiate Schools of Planning, Pasadena, Calif., November 5–8, 1998).

6. Michael Batty et al. "Visualizing the City: Communicating Urban Design to Planners and Decision-Makers," Centre for Advanced Spatial Analysis Working Paper Series (London: University College of London, October 2000), 16.

7. The Trust for Public Land, "Santa Fe, New Mexico—Wildlife, Mountains, Trails, and Historic Places Program," in *National Programs Conservation Finance*, June 24, 2001, available at www.tpl.org/tier3_cdl.cfm?content_item_id=4527&folder_id=1365.

8. Minimum hardware requirements include a Microsoft Windows NT 4.0 operating system or Windows 2000, Intel Pentium III processor running at 500 MHz, 256 megabytes RAM, 2 gigabytes of hard disk space, and AGP OpenGL video card with 32 megabytes of onboard memory.

9. Interview with Santa Fe participant. Confidentiality agreements at each site designed to promote open discourse prevent identification of the participant.

10. Interview with Santa Fe participant.

11. John Forester, *The Deliberative Practitioner: Encouraging Participatory Planning Processes* (Cambridge: MIT Press, 1999).

12. Interview with Ted Harrison, December 2001.

13. Interview with Santa Fe participant.

14. Interview with Santa Fe participant.

15. See also Britton Harris, "Computing in Planning: Professional and Institutional Requirements," *Environment and Planning B: Planning and Design* 26 (1999): 321–31.

16. Kathy Fitzgerald provided invaluable research support by conducting the majority of the interviews. Students in the master's program at the University of Colorado and the undergraduate program at Middlebury College also assisted in background discussions for this article—most particularly Michael Stoddard and Frank Potempa. The authors would also like to thank Towny Anderson, Helen Whyte, Bruce Biggi, and Nina Seamen for their invaluable contributions to this article.

17. Iris Murdoch, *The Sovereignty of Good* (New York: Schocken, 1971). The relevant quote is, "Where virtue is concerned we often apprehend more than we clearly understand and grow by looking."
18. Interview with Ted Harrison, December 2001.

PART IV

THE INTERNET AGE AS CONTEXT FOR
CONSERVATION INNOVATION

Eras of rapid network expansion provide several types of opportunities for conservationists to make progress in efforts to protect land and biodiversity. Not only do professional conservation scientists, educators, advocates, and resource managers often find opportunities to use new technologies and communities-of-interest to advance their practice, as described in the third part of this book. In addition, policy makers, the founders of new types of nonprofit organizations, private entrepreneurs, and philanthropists find that such eras of dramatic change provide excellent context for achieving historic innovations—landmark conservation initiatives that can bring enduring change to the policy landscape as well as the physical one.

Indeed, as we begin the twenty-first century, bold conservation initiatives are being launched in the public, nonprofit, and private sectors. For example, as described by Bob Durand and Sharon McGregor in Chapter 12, the Massachusetts Executive Office of Environmental Affairs is changing the way the state approaches environmental planning and regulation. By securing passage of such measures as the Community Preservation Act, by using the watershed as the primary scale for environmental analysis, and by encouraging local communities to envision their own land use futures with the use of sophisticated geographic information systems software, they are successfully encouraging local involvement in planning for open space and biodiversity protection.

Such a local approach may seem incongruous in an age of globalism. However, as Joel Hirschhorn points out in Chapter 13, given the locational flexibility and mobility of many Internet age organizations, quality of life in highly localized areas is becoming increasingly important to the electorate, to public decision makers, and to private entrepreneurs. In Chapter 14, Peter Stein and James Levitt explain that a series of network entrepreneurs have, throughout the course of American history, played important roles as conservation philanthropists, leaders, and innovators and how they are again stepping into such roles in the Internet age. James Levitt's conclusion to the book, Chapter 15, points out the critical need for further innovation commensurate with the complex challenges facing the conservation community in the twenty-first century.

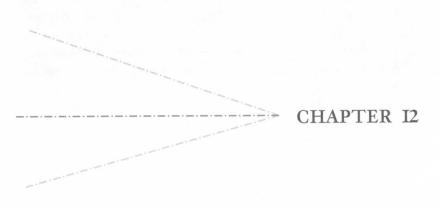

CHAPTER 12

The Watershed Approach, Biodiversity, and Community Preservation

BOLD INITIATIVES IN CONSERVATION
IN MASSACHUSETTS

Bob Durand and Sharon McGregor

In the twentieth century, the rapid expansion of residential, commercial, and industrial development irrevocably altered the natural landscape, consumed large volumes of water, and discharged significant quantities of sewage, chemical, and other waste products into our rivers, groundwater, air, and land.[1] Like the rest of the United States and other developed nations, Massachusetts responded with two decades of heavy regulation to curb such pollution. In the 1970s and 1980s, we built wastewater treatment plants, installed pollution control equipment, began to recycle wastes, and banned the use of some pesticides. Actions taken in the state and elsewhere eliminated the biggest sources of pollution, restored natural resources that had been pronounced dead—including Lake Erie, the Cuyahoga River, and Boston Harbor—and measurably improved environmental quality and human health.[2]

Although the sledgehammer approach was the right technique for the time, it also compartmentalized nature into the separate media of air, water, and land: cleaning up one medium often meant channeling pollution into another, in a kind of environmental shell game. The heavy hand of regulation also absolved the general public from responsibility for the health of the environment, as the job was assigned to government regulators and industry.

In the 1990s, Massachusetts began to see the folly of single media regulation, acknowledging that one medium should not serve as a repository for waste removed from another. The state proceeded to require industry to meet absolute pollution standards that did not allow the transfer of contaminants.[3] New regulations also required companies to submit monthly compliance reports to state and federal regulatory agencies and subjected firms to unannounced inspections. The total volume of some pollutants discharged into the environment, including mercury, lead, trichloroethylene (TCE), and dioxin, dropped noticeably.[4]

The 1990s also ushered in a new, more holistic approach to environmental protection, with the public sharing responsibility for environmental actions directed to whole environmental systems. First exemplified by the watershed approach and then woven into more recent initiatives, this new way of addressing environmental issues has, on one hand, taken us back to basics, stressing a rediscovery of our connections to nature and natural systems, and on the other hand, catapulted us forward with a new reliance on computers and the Internet.

Massachusetts helped pioneer the watershed approach, which bases environmental assessment, planning, and decision making in the watershed—the land area within which all water flows to a common point, such as a river, lake, or ocean.[5] The watershed approach is "back to basics" in its recognition of the watershed as the "circulatory system" from which humans extract life-sustaining water and dispose of the waste by-products of human activities, and in its recognition of the utility of using watershed boundaries as the primary context for managing human use and impacts on the environment.

In Massachusetts, we have moved the watershed approach to the next step in its growth: to become an *ecosystem approach* to environmental assessment, planning, and decision making. The watershed approach, with its keen recognition of the interplay of human activities and the water cycle, is being enhanced with a little more "back to basics"—emphasis on the relationship of biological diversity, vibrant natural communities, and intact ecosystems to maintaining healthy watershed systems.[6]

Through a new Biodiversity and Ecosystem Protection Initiative, we have elevated biodiversity and ecosystem protection to a top priority in Massachusetts's implementation of the watershed approach and all other environmental programs. In addition, we have created a new Community Preservation program to attack the monumental problem of sprawl—the seemingly unchecked expansion of human development into green spaces. This program, too, is employing a back to basics approach and is integrated in important ways with the Biodiversity and Ecosystem Protection Initiative. For example, the Community Preservation Initiative provides assistance to communities in evaluating where their current zoning and development buildout projection will take them, and it helps them create and implement a vision for the future that respects the ecological and environmental limits (or "carrying capacity") of the community.

By these actions, the Massachusetts Executive Office of Environmental Affairs (EOEA) is ushering in a new era in the environmental movement. The new era is about human reconnection to nature and human action to protect *the whole environment* on which animals, plants, and humans depend. The emergence of the Internet age just happens to coincide with the start of this new era in environmental protection. That is fortunate, given the enormous challenge of communicating the "why" and "how" of the watershed approach, biodiversity conservation and ecosystem protection, and community preservation. The ambitious new approach to environmental protection requires the support of a broad, informed, and empowered constituency, and the Internet and its associated tools are proving crucial in engaging a diverse public in that effort.

The Massachusetts Watershed Initiative

Rather than top-down government command and control, the Massachusetts EOEA's watershed approach emphasizes the participation of grassroots stakeholders in setting and implementing environmental priorities. The Internet and Internet-based geographic information systems (GIS) are central to this approach.

The Massachusetts Watershed Initiative operates on a five-year rolling schedule of outreach, data gathering, assessment, planning, and implementation and evaluation (see Figure 12.1). Groups organized around smaller streams and tributaries are key to this process, with technical and financial

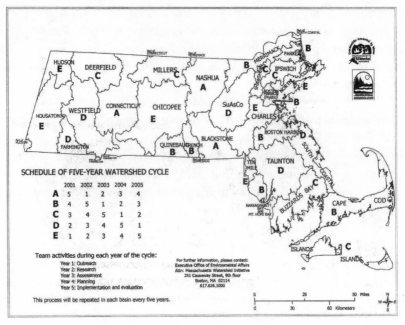

Figure 12.1: As part of the Massachusetts Watershed Initiative, watershed plans are created and implemented on a five-year cycle. (Map from Massachusetts Executive Office of Environmental Affairs, Watershed Initiative, 2001.)

assistance coming from twenty-seven cross-agency state watershed teams.[7]

During years 1 through 3, these groups convene and rally local partners to identify problems undermining the health of subwatersheds and to propose actions that will advance the protection and restoration of the subwatershed, as well as that of the full watershed, including the water resources, biodiversity, and ecosystems. In year 4, watershed-wide collaborations combine local strategies into a larger action plan. Implementation begins in year 5, when state and federal agencies, corporations, and municipalities that have participated in the process approve the final plan and begin to undertake its stipulated actions. At the conclusion of year 5, the five-year cycle begins again, allowing for ongoing implementation, expanded outreach, additional data gathering, and updating of the action plan (see Figure 12.2).[8]

The Watershed Initiative relies heavily on GIS information and the power of the Internet to invite diverse interests, including businesses, municipal governments, researchers, and citizen groups, to become active members of the local collaboration.[9] Participants use the Internet to share information and consult CD-ROMs that contain annual work plans. The five-year watershed action plans are also posted on the Internet to remind participants of the education, research, implementation, and monitoring activities they have committed to undertake. Citizens can also consult the Watershed Initiative Web site (www.state.ma.us/envir/mwi/watersheds.htm) for basic facts, the top five watershed priorities identified by participants, recent success stories, contact information for the watershed team leader, and links to participating organizations.

At present, the EOEA is in the process of prioritizing watersheds and subwatersheds for total maximum daily load (TMDL) analyses, which will produce "pollution budgets" based on natural background levels of contaminants, the assimilative capacity of soils and vegetation, the sensitivity of species and ecosystems, and current pollutant loads. The pollution budgets will describe the maximum levels of different pollutants the watershed or subwatershed can absorb and will assign discharge allowances to point and nonpoint sources to keep the watershed or subwatershed within the pollution cap. Watersheds or subwatersheds found to be exceeding the cap will require a mix of pollution control measures, including wetlands restoration, stormwater management practices, and upgrades or retrofits of municipal and industrial wastewater facilities.[10]

We plan to use Internet tools to inform dischargers and citizens about TMDL analyses and pollution budgets. The Internet will also help us disseminate information on how dischargers can cut their net pollution loads and trade pollution allowances within the pollution cap.

Using the Internet to Protect Biodiversity and Ecosystems

The Internet and Internet-based tools are central to the task of addressing a root cause of environmental problems in Massachusetts, across the United States, and around the globe: people's lost connection to the natural world. Sprawling development patterns are removing nature from our backyards and communities, making the "great outdoors" distant and difficult to access. As in tropical rain forests, we are losing habitat in Massachusetts every day, even before we know what lives in these habitats and exactly what we are losing. By building public awareness and under-

Action	Proposed lead parties	Possible funding source	Calendar Years 02	03	04	05	06	Subwatershed
Goal #1: Restore and Promote the Water Quality of Rivers and Ponds								
Objective 1.1: Identify and minimize point sources of pollution throughout the watershed								
Proposed actions for the next five years:								
1.1.1 Attleboro and North Attleborough priority areas sewering	City of Attleboro and Town of North Attleborough	SRF						
1.1.2 DEP 1997 Water Quality Assessment monitoring recommendations								Lower Ten Mile, Bungay
- Dissolved oxygen monitoring	Volunteers and DEP	EOEA Volunteer Monitoring Grant and technical assistance						
- Biomonitoring stations	Volunteers and DEP	EOEA Volunteer Monitoring Grant and technical assistance						
- Bungay River shoreline survey	Volunteers and Riverways	Internal						
- Seven Mile River shoreline survey	Volunteers and Riverways	Internal						
- Sampling plan for Bungay River	DEP DWM	604(b)						
1.1.3 Reissue minor NPDES permits	EPA and DEP	Internal						Lower Ten Mile, Bungay
Actions already planned or underway:								
1.1.4 Headwaters industry enforcement and remediation	Town of Plainville, DEP, EPA	Internal						
1.1.5 Major NPDES permits	EPA and DEP	Internal						

Figure 12.2: Watershed plans include a matrix that lists goals for the next five years as well as strategies for achieving them. (Graph from Massachusetts Executive Office of Environmental Affairs, Ten Mile River Watershed Team Action Planning Subcommittee, Five-year Watershed Action Plan, 2001.)

standing of the wealth of living things that reside in our own home state, the EOEA is inspiring action to inventory and protect this diversity of animals and plants and the ecological processes on which they and humans depend. Modern Internet technology is our primary tool for promoting this understanding and awareness. In the back to basics vein, our premise is that unless we protect biodiversity and ecosystems, we cannot achieve environmental and economic goals, because living things, together with the nonliving elements of air, water, and soil, comprise the life-support system that sustains environmental and economic systems.

The Biodiversity and Ecosystem Protection Initiative has three main objectives. First, we are pursuing the difficult task of building a constituency for protecting biodiversity and ecosystems through public outreach. Our programs target all ages and sectors of society, including teachers, students, corporate leaders and employees, municipal officials, and other residents. Second, we are pursuing projects that protect and restore ecosystems through land protection and ecological restoration. Third, we are promoting consideration of biodiversity and ecosystem health by private citizens in their daily lives, corporate leaders as they establish and expand their businesses, developers and land use officials as they make land use decisions, and government officials in the full range of policy making and program administration.

To build a constituency for protecting biodiversity and ecosystems, we are pursuing education projects that get a broad range of citizens of the state, children and adults alike, back into the outdoors. Through our School Visits Program, EOEA representatives, including this chapter's authors, spend time in the classroom and outdoors teaching kids about the natural world, about how human well-being depends on the well-being of plants and animals, about threats to biodiversity, and about how each individual can help to protect this diversity. A slideshow used in these visits informs students about their "watershed address" and about specific local biodiversity restoration projects, such as our fish stocking program. The message is reinforced by additional classroom tools developed by the EOEA, including our *Exploring Biodiversity Workbook*; *Guide to the Critters of Massachusetts*; the *Vernal Pool Guide*; guidance on the use of GIS environmental data layers in the classroom; and an *Education Resource Guide*, which contains a full range of environmental information sources, including Web-based sources.

When we are with the students, we ask them the question, "In the last week, how many of you have spent three or more hours watching TV?" Almost every hand goes up. Then we ask, "How many of you have spent three or more hours surfing the Internet?" Almost every hand goes up. Finally we ask, "How many of you have spent three or more hours outdoors, exploring a park, woodland, meadow, pond, or stream?" Invariably, only a few hands go up. This demonstrates how apart from nature our kids—and, it follows, we their parents, guardians, and teachers who supervise how they spend their time—have become. It seems appropriate to point to technology as a primary cause of our becoming distanced from nature. We would rather seize the opportunity of technology as a tool for bringing our kids and us *back* to nature.

Our Outdoor Classroom Grant Program complements the school visits program by awarding funds to elementary, middle, and high schools for establishing wooded school yards and nearby ponds, streams, and estuaries as permanent settings for teaching schoolchildren about science, ecology, and the environment. The funds can also be used to design curricula, build trails, and purchase reference materials, computer software, and field equipment for teacher and student use. Along with the science teachers, math teachers are using woods, ponds, and streams as settings to teach how geometry and algebra apply to environmental inquiry and problem solving. English teachers use these settings to inspire students to learn about natural history authors and to write about nature. And history and social science teachers use outdoor settings to show how the environment has shaped—and has been shaped by—the evolution of human society. Studies indicate that students master relevant subject matter much more easily when such outdoor classrooms are integrated into the curriculum.[11]

With the Massachusetts Department of Education, the EOEA has just completed a statewide Environmental Education Plan. Posted on the Internet at www.state.ma.us/envir/education/meep.pdf, this plan is intended to guide a network of environmental education leaders in special projects to advance environmental literacy in the state. Learning about biodiversity and ecosystems is a primary objective of this plan.[12] Initial success with the plan includes the incorporation of biodiversity and ecosystem concepts in the state's learning standards for science and technology.[13] Regional environmental education centers, through training programs, on-site school visits, conferences, and the Internet, will promote education on the biodiversity and ecosystems of each region.[14]

The crown jewel of our constituency-building programs, for which the Internet and Internet tools are central, is the state's Biodiversity Days effort. We held our first Biodiversity Days event, which focused on the southeastern part of the state, on June 9–11, 2000. Over that three-day period, some 15,000 people—including 10,000 students from 150 schools and 5,000 adults representing 100 cities and towns—inventoried animals and plants in their backyards, school yards, neighborhoods, and communities. Experts in flora and fauna—from wildflowers, mushrooms, and lichens, to birds, mammals, fish, marine invertebrates, and insects—helped participants explore and document their discoveries. Participants recorded 2,810 species—a world record for a cooperative event held in one region.

The success of this event inspired us to plan what was the nation's first statewide citizens' Biodiversity Days, on June 8–10, 2001. With the help of nonprofit partners, including the Massachusetts Audubon Society, watershed associations, bird and nature clubs, and science and nature centers selected through competitive bidding, we expanded our reach to clusters of cities and towns across the state. Some 30,000 people, representing 260 of the state's 351 communities, participated in Biodiversity Days 2001.

We believe these participants were drawn by their natural curiosity about the living things with which we share this planet. But we are learn-

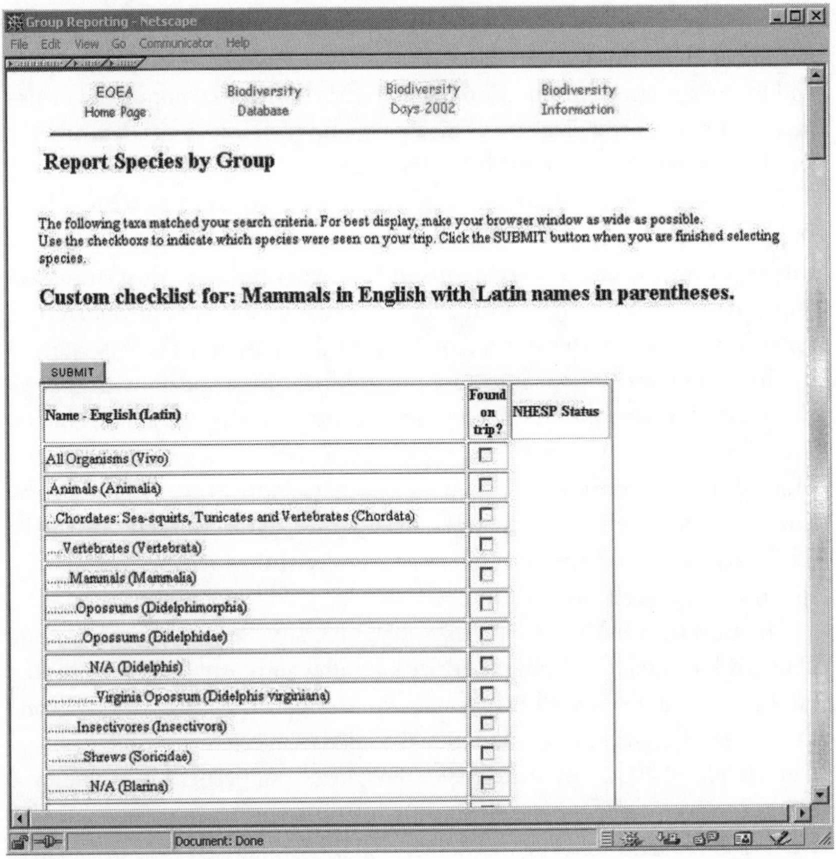

Figure 12.3: Citizens participating in Massachusetts's Biodiversity Days recorded their observations of flora and fauna on a checklist that could be submitted to EOEA's central database via the Internet or on paper.

ing that technology, specifically the Internet, which has been pointed to as one cause of people becoming out of touch with nature, may well be what hooks people into developing a long-term relationship with nature. We are actively facilitating this connection. For example, participants in Biodiversity Days 2000 submitted their lists of species to the EOEA, and naturalist and author Peter Alden worked with the Office of Geographic and Environmental Information Systems (MassGIS) to compile the information into a central database. This information complements data on rare species compiled by state wildlife officials.

For Biodiversity Days 2001, we made the database, which lists species identified town by town and by specific location, accessible to Internet users. Citizen participants in Biodiversity Days, with the assistance of expert naturalists, sent their observational data into the EOEA for incorporation into the database, either directly over the Internet or by filling out and forwarding to a central location a specially prepared checklist of Massachusetts flora and fauna (see Figure 12.3). The checklist—which incorporates everything from the vascular plants of Massachusetts compiled by state biologists Paul Somers and Bruce Sorrie to the Web-based species list of the Massachusetts Butterfly Club—is the state's first compilation of all visible species (that is, any species larger than one millimeter).

Dr. Edward O. Wilson, Pellegrino University Research Professor at Harvard University and Pulitzer Prize–winning author of several works on biodiversity, and Russell Mittermeier, president of the Washington, D.C.–based Conservation International, joined us in celebrating Biodiversity Days 2000 and 2001. Dr. Mittermeier recognized our event as "the first government-sponsored, citizen-directed biodiversity inventory in the world." Dr. Wilson said he believes Massachusetts's Biodiversity Days to be "an historic moment" and predicted that we will soon see events and activities based on this model across the United States and the globe.[15]

We plan to employ the biodiversity database developed in conjunction with Biodiversity Days as the basis for an online field guide to Massachusetts's flora and fauna. Building this database through Internet-based interaction with biodiversity experts and interested citizen scientists will be an exciting challenge. With a couple clicks of a mouse, users will be able to call up their city's or town's data, select a specific location, and call up a picture of and descriptive information about any species listed for that location.

As of this writing, we are about to celebrate Biodiversity Days 2002, with most of the state's 351 cities and towns and an expected 50,000 citizens participating. Biodiversity Days and the biodiversity database are initiating what we hope will be year-round inquiry and data collection by citizens. The database is part of a larger biodiversity Web site (www.state.ma.us/envir/biodiversity.htm) that provides general information, including our popular publication *Exploring Biodiversity*. The online copy of this publication, which duplicates the text and illustrations of the hard-copy form but adds hyperlinks to additional information sources, explains biodiversity, the threats to it, and the relationship of biodiversity and ecosystem health to human well-being. It emphasizes where and how to observe and inventory species and habitats, how to record data for personal use, and how to submit information and tap into the central database.[16]

Clearly, we view the Internet as a primary tool in our efforts to build a constituency for protecting biodiversity and ecosystems. The enormous potential to reach large numbers of people of all ages—to teach them, learn from them, and rally them in support of environmental protection—is unprecedented in the history of the environmental movement. The Internet and the Internet tools we create enable us to establish an army of citizen scientists to help gather important information on changes in wildlife or ecosystem health for use by state and local officials. We encourage citizens to put this information to use by working with their local conservation commission or environmental nonprofit group to create a Local Species and Habitat Registry to guide local and watershed conservation planning.

Preserving Biologically Important Land and Water

To fulfill the second objective of the Biodiversity and Ecosystem Protection Initiative—to protect and restore ecosystems—the EOEA is implementing an aggressive land protection program that uses biodiversity as a primary criterion. The agency has selected six geographic focus areas, which contain important biological and water resources, working forests, and agricultural lands for priority protection. Massachusetts governor Paul Cellucci, succeeded by Governor Jane Swift, set the goal of protecting by the year 2010 an additional 200,000 acres in Massachusetts, above and

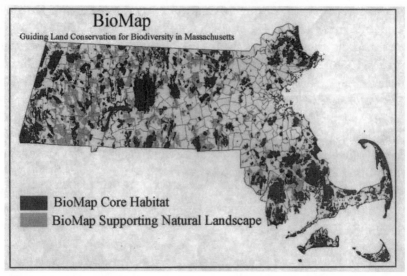

Figure 12.4: A statewide BioMap of important habitat will enhance efforts to protect and conserve biodiversity in Massachusetts. (Map from Massachusetts Executive Office of Environmental Affairs Division of Fisheries and Wildlife, Natural Heritage and Endangered Species Program.)

beyond the substantial amount of land already protected by 1998. With 120,000 acres protected between 1998 and June 2002, we are well on our way to meeting and even exceeding that goal.

The centerpiece of the land acquisition program is the creation of bioreserves, which are designed to be large, unfragmented parcels of biologically important lands open to the public for passive recreation. In partnership with a city government and the Trustees of Reservations, a nonprofit land trust, we created the state's first bioreserve in southeastern Massachusetts, in the city of Fall River (a community of 70,000 people) and the suburban towns of Freetown and Dartmouth. The 14,000-acre Southeastern Massachusetts Bioreserve—distinctive in abutting a highly urbanized area, with 50 percent of the city of Fall River's limits within the bioreserve—protects and provides for the cooperative management of entire ecosystems. We are working to create our second bioreserve in a region in the north-central part of the state, which includes a vast undeveloped interior forest, and a third bioreserve in the Berkshire Highlands of western Massachusetts, which connects five wildlife management areas.

To guide these efforts, the EOEA's Division of Fisheries and Wildlife

Natural Heritage and Endangered Species Program has completed statewide field inventories and mapping of terrestrial rare species habitat, exemplary natural communities, and associated ecological lands that support and buffer the rare species habitat and natural communities (see Figure 12.4). The next phase of this BioMap project, in progress, is to inventory and map aquatic species and habitats. Aerial photos of vernal pools have already been completed for the aquatic habitats analysis. The map products of these inventories are statewide or regional GIS maps of biodiversity "hot spots" and priority areas for biological conservation.

We have compiled the vernal pool data on CD-ROMs for distribution to state and local officials. BioMap information is also available—for the entire state, or by town or watershed—via a GIS data layer on the MassGIS Web site (www.state.ma.us/mgis/massgis.htm). We have plans to create an interactive Web site and CD products for BioMap as well. Both the interactive Web site and the CD will allow users to click on a polygon and view pictures and descriptions of the species and natural communities within that area, bringing BioMap alive.[17]

Mapping activities in several regional watersheds, supplemented with data collected by citizens throughout the year, will help refine BioMap. This rich data source will also help provide a framework for conserving large core areas linked via riparian and upland corridors to smaller local habitats. The plan is for state, regional, and local analyses to form an interconnected web of green, or a "greenprint" for conservation, and to direct human infrastructure outside this ecological infrastructure. With BioMap, state and municipal officials, nonprofit land trusts, and other environmental decision makers will have at their disposal a powerful tool for identifying land with important ecological values and for prioritizing their conservation efforts.[18]

Second only to development pressures as a cause of biodiversity habitat disruption is the growing presence of invasive species. Conservationists agree that an army of citizens will be needed to combat the unwanted spread of these nonindigenous species, ranging from Purple Loosestrife and Common Reed in wetlands, to Oriental Bittersweet in fields, Hemlock Woolly Adelgids in woodlands, Eurasian Watermilfoil in lakes and ponds, and Green Crabs in our coastal zone. Responding to this threat, we are establishing an invasive species SWAT team—a network of volunteers who will scout out and report pioneer invasions of invasive species. The volunteers will do this with the help of an invasive species page on our biodiversity Web site, which will contain pictures, descriptions, and habitat information on the most egregious invaders. By clicking on a species, users can report a siting and call for help in removing the invader before it overwhelms an established community of native species.

The EOEA has also launched, as part of its efforts to protect and restore the state's ecosystems, a series of ecological restoration programs, including the Wetlands Restoration Program, with the initial goal of restoring 3,000 acres of wetlands in Massachusetts by 2010; River Restore, which prioritizes dams for removal in the interest of restoring anadromous fish habitat (that is, habitat for fish such as the Atlantic Salmon that migrate up rivers from the sea to breed in freshwater); the Upland Habitat Management Program, designed to restore grassland ecosystems and other early successional habitats critical to a wide variety of birds and plants; and the Lakes and Ponds Restoration Program, which aims to restore ecologically significant inland bodies of water in the state (an example of Web-based information on such programs is available at www.mass.gov/envir/mwrp/).

Creating Better Decision Making

In setting the third goal of the Biodiversity and Ecosystem Protection Initiative—to promote consideration of biodiversity and ecosystem health by private citizens in their daily lives, by corporate leaders as they establish and expand their businesses, by developers and officials as they make land use decisions, and by government officials in policy, regulatory, and management decisions—we acknowledge that if we are to protect the integrity of valuable ecosystems, a wide variety of decision makers must consider biodiversity and ecosystem health. This objective, perhaps the most difficult to achieve of the initiative's three goals, will require both time and innovative strategies that employ the Internet as a key tool.

Guidelines on sustainable practices teach citizens how to consider biological resources in purchases, household maintenance, backyard landscaping, and transportation decisions (see www.state.ma.us/envir/ biodiversity.htm). For example, information provided on backyard landscaping describes the use of native plants as an alternative to green lawns and ornamental shrubs and provides links to other sites that contain detailed information on providing food, cover, shelter, and breeding habitat for native wildlife. Business guidelines, currently under development, will build on the Coalition for Environmentally Responsible Economies (CERES) Principles and The Natural Step (TNS) to teach companies how to emphasize protection of biological resources in their decision making.[19] Developer and municipal official guidelines, also under development, will explain site selection and design practices that protect biological resources in subdivisions and commercial and industrial developments.

Our Forest Vision program (see www.state.ma.us/envir/forestvision.htm) is the EOEA's commitment to biodiversity-based decision making in forest management. This program promotes the protection of large contiguous forest tracts—interspersed with harvested open areas—to maintain healthy forest ecosystems and preserve their wildlife, recreation, water supply, and economic values. Forest Vision is being implemented through interagency coordination of state forest and wildlife management programs and practices as well as by an outreach and technical assistance program for private landowners, who own 80 percent of our forests.

Outreach to private landowners on biodiversity-based forest management techniques will rely heavily on the use of Internet tools. In partnership with the University of Massachusetts, we are developing an Internet-based knowledge management system. This system will include an information clearinghouse, references, contact information for experts who develop biodiversity-based forest-cutting plans, a description of the process and requirements for earning "green" (or sustainable) forest management certification, and links to other sources—all available through the Web.

Preserving Communities

As we protect and restore the natural environment through our Watershed Initiative and Biodiversity and Ecosystem Protection Initiative, we are also helping shape the built environment through our Community Preservation Initiative. Cited recently by the Trust for Public Land (TPL) as one of seven exemplary greenprint initiatives (as defined by TPL, a greenprint is "a smart growth strategy that emphasizes land conservation to ensure quality of life, clean water and air, recreation and economic health"), the Community Preservation Initiative, like the Watershed and Biodiversity and Ecosystem Protection initiatives, relies on GIS and the Internet for its success.[20]

The campaign for community preservation began in 1984, when as a freshman state legislator Bob Durand was a leading cosponsor of statewide land bank legislation.[21] Based on the successful Nantucket and Martha's Vineyard land bank laws, this bill would have authorized—after local approval—a real estate transfer fee of up to 2 percent to fund land protection and affordable housing. At the height of the 1980s building boom, environmental and housing communities viewed the bill as a much-needed tool for mitigating the negative impacts of economic prosperity—the loss of open space and skyrocketing housing prices.

The real estate lobby opposed the statewide land bank bill and prevented a vote by the full House and Senate for five years. In 1989, although a poll of legislators indicated widespread support for the bill, our hopes were dashed when, in the waning hours of the session, the House rejected the bill, preempting any vote in the Senate.[22] After that difficult loss, we convened multiple interests to reshape the land bank bill into Community Preservation legislation (still framed as a local option), to include historic preservation among its purposes. In ensuing years, the bill enjoyed growing support from a strong coalition of open space, affordable housing, and historic preservation advocates.

In 1997, the Massachusetts Legislature passed land bank legislation for Cape Cod, authorizing local land conservation funded through a 3 percent surcharge on property taxes.[23] In 1998, all fifteen Cape Cod communities approved the idea in local elections. In 1999, when Bob Durand became secretary of environmental affairs, he joined Governor Cellucci in pushing for passage of the Community Preservation Act with a property tax surcharge as the funding mechanism. On September 14, 2000, after more than fifteen years of effort by a broad coalition of interests, Governor Cellucci signed the bill into law.

Despite the many years required to pass this bill, the effort to help communities protect their quality of life has just begun. At no other time in Massachusetts history has our quality of life been as threatened. The monster is sprawl—not development per se, but development in the wrong places. As in many states, inappropriate development is causing unprecedented strain on communities' natural, historic, and cultural resources. Each resident has experienced the shock and disappointment of passing a favorite woodland, meadow, or park and seeing bulldozers transforming the tranquil scene into a bustling mall, office park, subdivision, or parking lot. For many of us, the unique structures and landscapes that define our communities—and our sense of who we are and where we live—exist only as childhood memories.[24]

Massachusetts Growth Trends since 1980

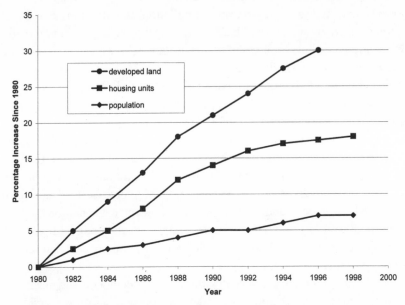

Figure 12.5: In Massachusetts, the growth in developed land has far outpaced population growth in the past twenty years. (Data from Massachusetts Executive Office of Environmental Affairs, *The State of Our Environment*, 2001.)

Population growth is not to blame: between 1950 and 1990, the amount of developed land in Massachusetts grew by 188 percent while population grew by only 28 percent (see Figure 12.5).[25] The built environment that was once concentrated in downtowns has spread over acres and acres of land, connected by miles and miles of roads, water and sewer pipes, and electric utility lines. Each person now consumes more than twice as much developed land: since 1950, population density on developed land has dropped from 11.19 persons/acre to 4.97 persons/acre.[26]

As noted previously, development is the leading cause of biodiversity loss, as it destroys great swaths of habitat and isolates remaining habitat. Development in the wrong places and of the wrong design also pollutes the environment. Whereas point-source discharges from industrial facilities and municipal wastewater treatment plants were once the primary sources of contamination, nonpoint runoff from roadways, parking lots, farmland, and lawns is now the number one cause of pollution of our rivers, streams, lakes, ponds, and groundwater.[27]

Sprawl also imposes many other costs, including deterioration of city and town centers, traffic congestion, and air pollution from the rising number of miles we drive in our cars and trucks. Municipal services—including roads, water and sewer lines, police and fire protection, schools, and libraries—are much more costly when development spreads out.[28] Local property taxes skyrocket as communities try to keep up with the costs of these services.

The Community Preservation Initiative is designed to empower citizens and local officials to combat sprawl and protect their quality of life by directing growth to appropriate areas and conserving important natural resources. The EOEA has teamed up with the Executive Office of Transportation and Construction, the Department of Housing and Community Development, and the Department of Economic Development to give each city and town a detailed GIS-based buildout map and analysis of current conditions and the consequences of maximizing potential future growth, along with $30,000 in funding and technical assistance to develop a community plan that reflects the community's vision for itself.

Each buildout map, produced by the EOEA, documents the maximum potential future growth allowable under current state regulations and local zoning. An accompanying analysis shows the results of this maximum growth in terms of population, housing units, school enrollments, gallons of water consumed, sewage generated, miles of roads built, and other key indicators. The maps and analyses are available via our community preservation Web site (see www.mass.gov/envir/cpa/communitypreservation.htm). Included are digital versions of the buildout maps (in Adobe Acrobat PDF format), companion analyses (in ESRI ArcView 3.2 format, and a spreadsheet in Microsoft Excel format), our growing inventory of community photographs (in JPEG format), and statistical reports derived from the analyses (in Microsoft Word format).[29]

In addition, we are offering communities two software tools to help them test alternative buildout futures dynamically. Alternative Futures allows localities to analyze the impact of changes in zoning, bylaws, and other regulations. Fiscal Impact enables local financial officers to examine the effects of different development scenarios on tax revenues and the cost of new schools, police and fire services, and infrastructure upgrades or expansions.[30] Communities can also use the state GIS to evaluate the impacts of different scenarios, including pollution from impervious surface area, the effects on the water supply of different population and land use scenarios, the effects on schools of different housing types, and infrastructure needs stemming from different land use configurations.

The EOEA also intends to inventory all brownfields and other sites with redevelopment potential and compile this information on a CD-ROM. This tool will help redevelopment advocates channel economic revitalization to urban neighborhoods and away from parks and other green spaces that serve these populations.

We are encouraging every community to develop, as a final product, *one map* that represents existing downtown and outlying developed areas, future areas of concentrated village-like development accessible to downtown, and conservation lands linked via riparian and upland corridors. This overlay of the community's human and ecological infrastructures will provide a powerful blueprint for future land use decisions.

Never before has so much information been available to our communities, and the MassGIS office is maximizing the accessibility of that information using the Internet. But convincing municipal officials to make the most of the new community-planning information and GIS and Internet tools will also require creative outreach. In partnership with the University of Massachusetts (UMass), the EOEA has established a Community Preservation Institute at the Westborough, Dartmouth, and Lowell campuses. The institute enrolls classes of twenty-five Massachusetts residents who are employed or volunteer in municipal government, environmental advocacy, environmental and urban planning, historic preservation, housing advocacy, and transportation planning. The seven-week course includes land use planning to protect water quantity and quality, biological diversity, and ecosystem health; redevelopment to encourage affordable housing and preserve historic structures and sites; transportation planning to minimize dependence on the automobile; and restoration and economic use of brownfields. Teachers include professional educators, government officials, and citizen group leaders.

We plan to offer the course at five of six UMass campuses as well as an expanded curriculum for academic credit.[31] But to extend the reach of the institute, the EOEA may also offer the community preservation course online through both live and delayed feed. The online version would introduce users directly to GIS and Internet tools for watershed management, community preservation, and biodiversity and ecosystem protection.

With the EOEA's focus shifting from public awareness of sprawl as the problem and community preservation as the solution to a focus on practical community preservation techniques, we also intend to use the Internet to disseminate tools to individuals who may not enroll in the institute. The EOEA's Sustainable Practices Guidelines for land use decision makers and developers, posted on the community preservation Web site, will help decision makers steer development away from rare species habitats, exemplary natural communities and associated ecological lands documented on BioMap, vernal pools and other wetlands, working farm and forest lands, aquifer recharge areas, historic landscapes, cold water streams, "high-quality antidegradation waters" designated under the federal Clean Water Act, and state-designated Areas of Critical Environmental Concern (so designated because of their concentration of sensitive natural features). In exchange for choosing less environmentally sensitive sites or—even better—redeveloping existing structures and brownfield sites served by existing water, sewer, and other infrastructure, developers will enjoy lower costs, fewer regulatory requirements, and possibly expedited environmental review by state and local agencies.

The Sustainable Practices Guidelines will also show how to develop a site in the most environmentally sustainable way. These guidelines are based on a "green neighborhood" methodology developed by Randall Arendt and applied by a coalition of municipal governments, businesses, nonprofit groups, and developers on Massachusetts's North Shore.[32] Through coordinated planning and zoning, 50 to 70 percent of developable land area in a subdivision is set aside as protected open space, providing a return to the landowner from the sale of clustered housing and a profit to the developer, who saves through lower road and utility costs.[33] Cities and towns can also apply this technique on a community-wide or regional scale by delineating areas desired for protection and directing growth to areas where infrastructure can best support it.[34] The Internet offers tremendous opportunity to promote and apply this design technique across Massachusetts.

Information on other land-saving techniques—including zoning and nonzoning bylaws, the creation of water resource and historic districts, and the use of transferable development rights to shift development from one location to another—is distributed electronically to municipal planners and land use decision makers through our community preservation Web site and community preservation e-letter. With the recent passage of the Community Preservation Act, people are asking technical questions about aspects of the act and what other communities are doing to pass it locally. The community preservation Web site is set up to provide detailed information about the act, the amount of money each community can raise, an updated list of communities that have adopted the act, and state and nonprofit resources (see www.mass.gov/envir/cpa/communitypreservation.htm). The site includes links to the Web sites of nonprofits that help tailor implementation to local needs, as well as to a PowerPoint presentation on the act and a question-and-answer bulletin.[35]

Summary

Together, the Watershed Initiative, the Biodiversity and Ecosystem Protection Initiative, and the Community Preservation Initiative are important contributors to the creation of a rich and valuable source of information on our state's pattern of growth; its biological, water, and other natural resources; threats to such resources and to historic sites; and the effects of various development scenarios on infrastructure and fiscal health. Our work is all about connections—connecting conservation decision makers and citizens in an Internet-based information and communications network, concentrating and connecting development where development and infrastructure already exist, and preserving connections in the undeveloped landscape to achieve ecological and environmental sustainability. With ready access to good information, we expect citizens—including businesspeople, public officials, community groups, and individuals—to make short-term and long-term decisions that take biodiversity and other natural resources into account. By making this information available over the Web and through related technologies, we aim to free up staff to provide direct assistance on fully integrating watershed, biodiversity, and community preservation objectives into local goals.[36] Our hope is that other states and countries can benefit from our experience in using Internet-based information and communications networks to forge connections and take charge of our common environmental future.

NOTES

1. Massachusetts Executive Office of Environmental Affairs (EOEA), *The State of Our Environment* (Wellesley, Mass.: Susan Nappi Creative, 2000), 15–17.
2. Clive Ponting, *A Green History of the World: The Environment and the Collapse*

of Great Civilizations (New York: Penguin, 1991), 375; Massachusetts EOEA, *The State of Our Environment*, 63–81.

3. Roy Manik and Lee A. Dillard, "Toxics Use Reduction in Massachusetts: The Blackstone Project," *Journal of Air and Waste Management Association* 40 (October 1990): 1368–71.

4. Mark Smith, Massachusetts Department of Environmental Protection, personal communication, June 22, 2001.

5. Stephen M. Born and Kenneth D. Genskow, *Exploring the Watershed Approach: Critical Dimensions of State-Local Partnerships*, final report of the Four Corners Watershed Innovators Initiative (Madison: University of Wisconsin Extension, 1999), 6.

6. "Biological diversity," or "biodiversity," is "the variety of living things." Reed F. Noss and Allen Y. Cooperrider define biodiversity to include "the variety of living organisms, the genetic differences among them, the natural communities and ecosystems in which they occur, and the ecological and evolutionary processes that keep them functioning, yet ever changing and adapting" (*Saving Nature's Legacy* [Washington, D.C.: Island Press, 1994], 5; modified from "Final Consensus Report of the Keystone Policy Dialogue on Biological Diversity on Federal Lands" [Keystone, Colo.: The Keystone Center, 1991]). The term is generally used to refer to native species in their proper places and in proper balance with one another.

7. Paula M. Jewell et al., *The Massachusetts Watershed Initiative: Opportunities and Challenges in Reshaping Government* (Boston: Massachusetts Executive Office of Environmental Affairs, 1996), 3–4.

8. Jewell et al., *The Massachusetts Watershed Initiative*, 5.

9. Massachusetts Watershed Initiative Steering Committee, "The Massachusetts Watershed Approach and Its Implementation: Status Report" (1995), 6.

10. Massachusetts Department of Environmental Protection, "DEP's Proposed Total Maximum Daily Loads (TMDLs) Strategy to Improve the Water Quality of Massachusetts Rivers and Lakes" (1998), 3.

11. Gerald A. Lieberman and Linda L. Hoody, *Closing the Achievement Gap: Using the Environment as an Integrating Context for Learning* (Poway, Calif.: Science Wizards, 1998), 19.

12. Alan Lee Hankin and the Secretary's Advisory Committee on Environmental Education, "Massachusetts Environmental Education Plan: Education to Protect, Restore, and Preserve Our Commonwealth" (2000), 1–4.

13. Massachusetts Department of Education, "Science and Technology/Engineering Curriculum Framework" (2001), 37–51.

14. Hankin and the Secretary's Advisory Committee on Environmental

Education, "Massachusetts Environmental Education Plan" (2000), 8–11.

15. Edward O. Wilson and Russell Mittermeier, remarks at Biodiversity Days Celebratory Event, June 10, 2000, Dartmouth, Mass.

16. Massachusetts Executive Office of Environmental Affairs (EOEA), *Exploring Biodiversity: A Workbook* (Braintree, Mass.: MacDonald and Evans, 2000).

17. Henry Woolsey, coordinator, Massachusetts Division of Fisheries and Wildlife, Natural Heritage and Endangered Species Program, personal communication, June 22, 2001.

18. Massachusetts Executive Office of Environmental Affairs (EOEA), Division of Fisheries and Wildlife, Natural Heritage and Endangered Species Program, "BioMap: Guiding Land Conservation for Biodiversity in Massachusetts" (2001).

19. The CERES Principles are simple standards of environmental conduct adopted by Exxon Corp. and promoted to businesses around the world in the aftermath of the 1989 oil spill in Valdez, Alaska. The Natural Step is a comprehensive set of environmental principles and guidelines developed in 1989 by Dr. Karl-Henrik Robèrt and promoted to governments and businesses worldwide.

20. Trust for Public Land, *Greenprint Gallery 2000* (San Francisco: Trust for Public Land, 2000), available at www.tpl.org.

21. As a historical note, this was well before the State House became computerized; PCs and the Internet were some years away. We used an electric typewriter to prepare the first draft of the land bank bill. Needless to say, this first draft made heavy use of "white-out."

22. Massachusetts Legislature, *Legislative Record* (1989), 106S.

23. Massachusetts General Laws, chap. 144 of the Acts of 1997.

24. Massachusetts EOEA, *The State of Our Environment*, 22–40.

25. Massachusetts Executive Office of Environmental Affairs (EOEA), "Planning for Growth: Implementation of Executive Order 385 within Agencies of the Commonwealth" (1996), B-1.

26. Massachusetts EOEA, *The State of Our Environment*, 24.

27. Massachusetts Department of Environmental Protection, Division of Watershed Management, "Summary of Water Quality 2000" (2000), 24.

28. Massachusetts EOEA, *The State of Our Environment*, 27–28.

29. Massachusetts Executive Office of Environmental Affairs (EOEA), "Community Preservation Vision" (2001), 13.

30. Massachusetts EOEA, "Community Preservation Vision," 8.

31. Massachusetts EOEA, "Community Preservation Vision," 5.

32. Randall Arendt, *Conservation Design for Subdivisions: A Practical Guide to*

Creating Open Space Networks (Washington, D.C.: Island Press, 1996).

33. Arendt, *Conservation Design for Subdivisions,* 7.

34. Arendt, *Conservation Design for Subdivisions,* 6.

35. Massachusetts EOEA, "Community Preservation Vision," 12.

36. Massachusetts EOEA, "Community Preservation Vision," 13.

CHAPTER 13

Natural Amenities and Locational Choice in the New Economy

Joel S. Hirschhorn

Despite predictions to the contrary, the importance of place is not declining in the Internet age. To the contrary, enhanced mobility, unlimited electronic access to images and information about the natural world, and rising prosperity are expanding people's desire to enjoy environmental amenities. The natural world is no longer just a source of raw materials to be extracted and exploited, but a place of multiple pleasures for people who have transcended their survival needs.

The Internet age has given rise to a "new economy," a transformation that, as economist Lawrence H. Summers has said, is "both palpable and amorphous—more often declared than defined." Summers continued: "But if there is one fundamental change at its heart, it must be the move from an economy based on the production of physical goods to an economy based on the production and application of knowledge."[1] In the old economy, natural resources were valued chiefly as raw materials to be extracted and exploited to sustain society and propel growth. Workers chased jobs, and companies made location decisions without stressing an area's environmental attributes. In the new economy, companies chase knowledge workers and base their location decisions on the amenities, including natural amenities, that will attract those workers. These amenities have therefore become essential to successful economic development.

Doug Henton and Kim Walesh provide an important insight into the relationship between place and economic change:

> The New Economy values quality of life more than the old economy, because it values people more than the old economy. . . . In the New Economy, quality of life has become a community's most valuable economic asset. . . . The New Economy values the natural environment as an important quality-of-life asset. . . . Increasingly, the New Economy recognizes that protecting the natural environment is in its long-term self-interest. . . . The New Economy brings the potential for a type of qualitative growth that is more compatible with environmental preservation and conservation. . . . In the old economy, growth typically was associated with degradation of quality of life. In the New Economy, growth will change—but must not degrade—quality of life.[2]

This paradigm shift signifies a potentially powerful expansion of public support for conserving natural resources. For decades, such support has been based primarily on a moral responsibility to protect and preserve irreplaceable resources for the good of people today and tomorrow—or "doing the right thing." Now, conservation combined with smart growth principles offers an effective strategy for stimulating economic development while also combating the loss of natural amenities and quality of life in high-growth communities.

The Scope and Importance of Natural Amenities

Clean air and water are clearly threshold environmental amenities of central importance to human health. However, natural amenities also include scenic landscapes, natural habitats, wilderness areas, and wetlands that people do not necessarily encounter directly. As economist Thomas Michael Power has said: "People take pleasure in simply knowing that certain natural treasures have been protected or put value on keeping their options open or feel that it is important that a bit of our past natural heritage be preserved for future generations."[3] Offering credence to Power's argument, voters approved some $17.5 billion to preserve open space from 1998 to

2000, and many state governors have created additional state-funded land conservation programs in recent years.

At the more active end of the spectrum of people who appreciate natural amenities are individuals, families, and organizations who use (or who think they may use) local "outdoor" resources, such as waterways, beaches, parks, nature reserves, and forests. In many locations, of course, such activities constitute a major business sector. Outdoor activities based on natural amenities generate more than $40 billion annually nationwide and account for nearly 800,000 full-time jobs.[4] Activities in national parks alone contribute more than $10 billion annually to local economies.

New economy companies are rapidly migrating to both urban and rural areas rich in such natural amenities as parks, biking and walking trails, and greenways. High-tech meccas have been lauded for their environmental amenities. Seattle, Boston, and San Francisco made *Bicycling* magazine's list of the top ten cities for cyclists, and Austin, Raleigh, and Seattle made *Walking* magazine's list of America's best walking towns.

Considerable research has verified the importance of natural amenities in companies' location decisions. A survey of 500 business owners and managers in three rural Montana counties found that scenic beauty and a high-quality environment were more important than traditional financial incentives in luring new firms and retaining existing ones.[5] A survey of executives at 118 foreign-owned companies operating in North Carolina found that the quality and availability of labor and transportation, the overall quality of life, and the general business climate were the most important factors in the firms' choice of location.[6] Tax incentives, location assistance from government, government financing, and state marketing assistance ranked at the bottom. Another survey confirmed that high-tech firms ranked environmental quality ahead of housing costs, cost of living, commuting factors, schools, climate, government services, and public safety.[7]

One study of nonmetropolitan counties between the Mississippi River and the Rockies and north of Texas found that those with recognizable environmental amenities recorded job growth, while those with the highest concentrations of resource-extractive industries (agriculture, mining, timber) and presumably fewer amenities did not.[8] In separate research, Richard Florida, professor of regional economic development at Carnegie Mellon University, found that "environmental quality was the top-rated factor" when new economy firms decided where to open offices. He concluded that environmental amenities are a "critical component of the total package required to attract talent and, in doing so, generate economic growth."[9]

Studies have similarly supported the importance of place in individuals' and families' location decisions. A survey conducted in 2000 for *Money* magazine's "The Best Places to Live" asked people to rank the importance of thirty-seven quality-of-life attributes in their choice of where to live.[10] Clean water ranked number one and clean air number three (low crime ranked number two). Good public schools, low property taxes, and low cost of living scored lower than clean water and air. Another study found that 60 percent of newcomers to western wilderness counties cited wildlands as playing a significant role in their location decisions, while 45 percent of existing residents said that proximity to such lands was an important reason for remaining in the area.[11]

A 1999 survey of residents in six high-tech cities found that the "environment is the attribute with the most powerful impact on overall quality of life satisfaction."[12] The most important factor determining respondents' future outlook was their region's ability to control urban sprawl. In an *Information Week* salary survey in 2000, nearly a fifth of information technology workers said geographic location was a key feature of their job, and about the same number ranked job commute as a key quality-of-life factor.

Richard Florida has used focus groups to confirm that workers are more interested in where they live than ever before. Young technology workers told Florida that they had "a strong preference for high-amenity locations, with high levels of environmental quality, and a range of outdoor recreational options."[13]

Natural Amenities as Part of a Complex Mix of Location Decision Factors

Of course, even in the new economy, people make choices on where to live based on a complex set of criteria. For any given location, potential newcomers may ask a wide variety of questions, including the following:

- Is the community relatively safe? Is it "walkable"? Does it offer racial harmony and diversity?
- How good are the schools? How good are local public services?
- Is there convenient road and air travel? How good are local utilities? Is the local infrastructure in good shape?

- Are cultural, shopping, and entertainment amenities readily available?
- Are the local recreational and outdoor "natural" attractions—such as rural landscapes, streams, rivers, and farms—accessible?
- Are efforts to revitalize distressed urban cores and older suburban neighborhoods under way? Do local and state policies steer development and check unrestrained growth?

Florida has spoken eloquently of this importance of a complex quality of place:

> In this digital world of the Internet and the world wide web and globalization and bits and digits flying around all over the place, we were taught that place was supposed to no longer matter, that place would be annihilated, that people could do things anywhere, any time, any place. Boy, were they wrong. In fact, place has become the defining feature of the new economy and the quality of place is really the critical factor. . . . We need to refocus our attention away from companies and toward people—away from business attraction to talent attraction and quality of place. . . . By quality of place I mean three or four things: lifestyle, environmental quality, a vibrant music and art scene, and natural and outdoor amenities. . . . Place replaces job and career as the central source of identity. . . . Place is where the over-connected, hyper-connected, always-on individual can get some downtime and escape from the tremendous pace of work in the new economy. It is where the participants in the new economy unplug. . . . [Knowledge workers] want environmental quality. They will not go to a dirty, fouled, polluted city. . . . They want a city that is easy to get around, not too congested, where you can walk to work, where you can cycle, use mass transit, with smart growth and sustainability.[14]

The importance of proximity in this complex mix is rarely discussed; the fact is that new economy workers with high-intensity work habits want easy access to a host of social, cultural, and natural amenities. The land use patterns associated with suburban developments built in the second half of the twentieth century do not often offer such easy access. The new economy has rekindled interest in land use configurations that again attempt to join home, work, and play into single communities, with the natural environment an important and compatible element. Both knowledge workers and older, affluent baby boomers preparing for retirement are increasingly choosing to live in pedestrian- and transit-friendly, mixed-use communities that offer housing, offices, shopping, schools, and recreational areas with plentiful environmental amenities—whether in urban, suburban, or exurban locations.

Amenity-Driven Growth Strategies

Explicit recognition that environmental amenities constitute "natural capital" by which a location can gain competitive advantage is driving markedly new approaches to promoting economic growth. For local governments, this means shaping an attractive "people climate" rather than the right business climate, and creating parks and bike trails rather than beltways and business parks. Green spaces may need revitalization, and local rivers may need cleanup. Infrastructure investments can focus on improving local transportation without resorting to cars—surveys show strong public support for promoting walking, biking, and public transportation and not for building new roads.[15] Such investments make more sense as an economic development strategy than do traditional tax concessions and other corporate subsidies.

Concern about the loss of urban green infrastructure has been growing. In January 2001, American Forests released a report documenting dramatic tree loss in Houston and linked it to greater stormwater runoff, energy use, and air pollution over a twenty-seven-year period.[16] Although the report's authors viewed such losses as an environmental rather than a development issue, commentators did not fail to note the report's economic implications. One newspaper story on the importance of quality of life noted: "Known for its laissez-faire, pro-business attitude, Houston is now intent on improving air quality, expanding green space, sprucing up neglected parks, and expanding bike lanes. The approach . . . not only works for longtime residents, but can help the city develop economically by making it more appealing to the new waves of 'knowledge workers.'"[17]

Another analysis concluded: "Houston's challenge to attract talent presents a uniquely aligned situation for the environmental, business, and government communities. Environmental quality and amenities have now become a bottom line issue in a new way. . . . Our local governmental agencies need to see the connection between environmental improvement and stimulating the new economy."[18]

Of course, communities cannot completely overlook traditional factors related to business climate, such as basic government services and the local tax structure, and excellent Internet capability is essential for success. However, a quality-of-place development strategy that promotes a "conservation economy" represents the synergistic convergence of human and natural capital and benefits everyone.

Many high-tech meccas have promoted economic development by investing in green infrastructure. One of the more impressive public-private projects occurred in Boston, which demolished a 500,000-square-foot concrete parking garage in the heart of the financial district and replaced it with an underground garage covered with a two-acre park. The Park at Post Office Square, which features a 143-foot formal garden, a walk-through sculpture fountain, and a café, has boosted economic values and activity in the surrounding area.[19]

Austin, a new economy winner, has linked its economic development strategy explicitly to environmental amenities, including recreational water resources; local business leaders and government officials alike recognize that nature is the city's competitive advantage. In the 1990s, Austin voters approved more than $130 million in local bonds to create parks and greenways and protect critical watershed lands. This strategy has worked: more than 800 high-tech companies moved to the area, and the population has swelled from 465,000 to some 630,000 people from 1990 to 1999. Austin is projected to grow by another 170,000 people, to a total of 800,000, by 2010.[20]

Proof that Austin is moving in the right direction comes from a 1999 survey of seventeen quality-of-life attributes in six high-tech cities: Atlanta, Austin, Boston, Raleigh-Durham, San Jose, and Seattle.[21] Austin residents ranked eight such factors as high quality, residents of four other cities assigned the highest ranking to only two factors, and Boston residents cited only one. However, a large proportion of Austin residents were pessimistic about the future because they thought the city was not controlling sprawl and traffic congestion. Surveys have shown that the more educated Austin residents are, the more likely they are to perceive a decline in the quality of life, and that such people are more likely to plan to leave Austin in the next five years.[22]

Oregon deserves special recognition for promoting sustainable economic development and a conservation economy. The state's plan, known as "Oregon Shines: An Economic Strategy for the Pacific Century," emphasizes a three-pronged strategy: (1) transforming the state's population into a world-class, twenty-first-century workforce; (2) creating an "international frame of mind" that positions Oregon as the gateway to the Pacific Rim instead of as the end of the Oregon Trail; and (3) pitching the state's extraordinary environmental amenities as a comparative advantage. A "circle of prosperity" underlies this strategy: a talented workforce, environmental amenities, and high-quality public services attract good jobs, which in turn provide a tax base that enables the state to maintain its public services and protect its environmental amenities.

Other states are getting on board. In 2000, Michigan launched a $5 million ad campaign aimed at attracting high-tech workers from other states, which emphasized the state's quality of life, particularly its recreational opportunities. Kathy McMahon, of the Michigan Economic Development Corporation, said, "We shifted from recruiting companies to recruiting people because skilled workers are the No.1 thing we hear our businesses need."[23] And in his 2001 state-of-the-state address, Governor John Engler proclaimed:

> The New Economy is transforming the old, and a new Michigan is emerging—the Next Michigan. . . . The Next Michigan protects her unique natural treasures. . . . The Next Michigan means an economy even more prosperous, a government even more responsive and a quality of life even more inviting. . . . In addition to our work protecting the Great Lakes, its harbors and fisheries, and our inland lakes and streams, conserving Michigan's land resources has also been a focus. In fact, during the past decade, more than 46,000 acres of land were acquired by the Department of Natural Resources and local governments for public use. . . . Our conservation strategy envisions development that would balance the desire for open space with the need for more housing. How? By creating incentives to preserve open space with existing natural features like wetlands and woods while more intensively using less acreage to develop family-friendly neighborhoods.[24]

Meanwhile, in his 2001 state-of-the-state address, South Carolina Governor Jim Hodges said: "South Carolina needs to grow. But imagine a South Carolina with no woods to hunt in and no parks to play in. If we are going to keep South Carolina livable, profitable and enviable, we must protect our natural resources. Now is the time to plan for the next decade of explosive growth. We must promote and preserve open spaces."[25]

In a similar fashion, a recent analysis of the importance of quality of life to the health of Louisiana's economy offers this observation: "The interaction between environmental amenities and residential location decisions has important implications for economic-development policy in Louisiana and elsewhere. As both households and firms become more footloose, the natural resources of a place increasingly will contribute to the structure and growth of the local economy through their influence on household location. In relative, if not absolute terms, the use of natural resources as a source of raw material for industrial production will become less important."[26]

A regional council presented a similarly enlightened view for the Pacific Northwest: "Environmental amenities, in general, and healthy watersheds, water quality and native fish runs, in specific, are key environmental amenities. The competition to attract new people and investments will be fierce. Those communities that take the initiative to protect, and where necessary, restore their environmental amenities will be the winners. Those that hesitate or refuse to acknowledge the inevitable are the sure losers."[27]

Natural Amenities in the New West

Nowhere are environmental amenities more significant than in western states, which have seen explosive growth in the past several decades. We can learn much about the value people place on conservation by investigating the complex transition from the natural resource–based economy of the Old West to the amenities-based economy of the New West.

Support for an amenity-based strategy is widespread in the West. According to Pete Geddes, of the Montana-based Foundation for Research on Economics and the Environment: "The West's attractive environment has tremendous economic value. Roadless lands, wilderness, free-flowing rivers, national parks and forests, and healthy wildlife habitat stimulate much of its new economic activity. These amenities attract entrepreneurs. For example, Bozeman, Montana has 40 high-tech firms in a town of 30,000. Freed by FedEx and the Internet, 'modern cowboys' (and cowgirls) arrive seeking high environmental quality."[28]

Supporting Geddes' contention regarding the value of western amenities are the recent statements of politicians and policy analysts across the region. In his 2001 state-of-the-state address, Utah Governor Michael O. Leavitt said: "In the New Economy, quality of life is an economic development tool. . . . In a world where most jobs can be located anywhere, now, more than ever, preserving our quality of life is an economic imperative. The natural beauty of Utah and opportunities for recreation are a major draw. For this reason I ask you to join me in a major drive to spruce up, clean up and keep up our state parks and monuments."[29]

Since 1970, Montana added over 150,000 new jobs, and not one of the new net jobs was in mining, oil and gas, farming, ranching, or the wood products industry. Though not traditional western jobs, these many occupations depend in their own way on open space and natural resources, because of the desire for natural amenities. Economist Ray Rasker has written about environmental amenities in the West: "These amenities are economic assets in very much the same way timber and minerals resources are. They serve an important function, to retain existing people and business, and to attract potential entrepreneurs."[30] According to Rasker: "Ralph Hutchinson and his Scientific Materials Corporation relocated from Oregon to the Gallatin Valley of Montana in 1989, in part because of the valley's quality of life. His company, which produces special crystals used in lasers, satellites, and other high-tech equipment, is a prime example of the new knowledge-based industries."[31] Rasker and Jerry Johnson reiterate: "Rural communities can design a community development strategy based on preservation of environmental amenities rather than one based on recruitment by traditional methods such as tax incentives, waivers of protective environmental regulations, or increased exploitation of natural resources. Such traditional development policy may, in fact, foreclose future community development options if it results in losing community amenities."[32]

As timber production has slowed, public recreational use of the forests has mushroomed to thirteen times its 1950 rate. The largest recreational use (35.8 percent) is driving for pleasure on the more than 80,000 miles of Forest Service roads maintained for passenger vehicles. And despite this primary use, polls conducted by the Idaho Conservation League found that majorities across the West support policies to protect and preserve roadless areas on public lands. Thomas Power concluded: "The primary economic role of these federal lands is not timber but the valuable amenities that these lands provide to residents: recreation opportunities, scenic beauty, water quality, wildlife, fisheries, etc. These non-commodity values of Forest Service land are of growing economic significance."[33]

Despite such support, debate over extracting natural resources on public lands versus safeguarding their environmental amenities remains contentious. New West advocates generally promote curtailing all forms of extraction, or at least allowing extractive industries to decline, making an explicit connection between conservation and shifting away from resource-based industries to an amenity-based economy. The editor of a Colorado newspaper proclaimed: "The war between extractive interests and the environmental movement for control of the Interior West's public lands is drawing to a close. The timber era, the cattle era, the mainstream big-dam era, the wise use era are ending. An immense landscape is going from one set of uses to another set of uses, from one way of life to another, in an astoundingly short time."[34]

But western traditionalists are not ready to abandon extractive industries. Part of their agenda may be a desire to limit population growth resulting from people seeking locations with a high quality of life. Traditionalists may also see such amenity-driven growth as just as damaging to the West as extractive industries, especially in loss of open space. However, natural amenities and extractive industrial activities do not need to be viewed as either-or propositions. New technologies and effective regulatory enforcement can reduce the negative impacts of extractive industries while those industries continue to contribute to local economies.

What's more, states and regions that attract talent may build mini-clusters, clusters, and finally regional concentrations of new economy companies thriving synergistically. However, rural areas may need to reach only a mini-cluster stage to help revitalize the local economy.

Preserving Amenities through Smart Growth

One of the more perverse aspects of the growing focus on natural amenities is that success too often kills the goose that lays the golden egg. Many early new economy regions—including Atlanta, Silicon Valley, and the Northern Virginia suburbs of Washington, D.C.—have experienced the kind of rapid, haphazard growth that undermines quality of life and threatens future growth. Knowledge workers can react quickly to such declining quality of life, causing retention problems for companies, low worker morale, and lower productivity.

Richard Florida confirms:

Sprawl poses a particularly vexing problem for rapidly growing high-technology regions. Part of their appeal in the first place came from their manageable size and high quality of life. Growth generates pressures that threaten these qualities. . . . Deteriorating air quality, traffic congestion, and damage to natural amenities are some of the negative outcomes that challenge prospering high-technology regions. In extreme cases, unmanaged growth may eventually destroy the appeal of a region and create an impediment to growth and make other regions relatively more attractive location choices.[35]

High rates of growth undermine natural amenities in several ways:

- Growing population and economic prosperity boost demand for development, most of which entails converting farmland, forests, and other open space into sprawling subdivisions as well as commercial and retail projects. In coastal areas, the concentration of both primary and vacation homes and supporting infrastructure has proved particularly destructive to natural amenities.
- Lower air and surface water quality may inhibit outdoor activities, while loss of natural habitats and biodiversity also detracts from some natural amenities.
- Growth in amenity-rich areas brings more people attempting to gain access to those amenities. Many may be seasonal tourists less concerned than longer term residents in preserving natural amenities. Greater use of natural settings interferes with important habitats, litters scenic places, pollutes surface waters, and creates forest fires.

Of rising importance, therefore, is not merely attracting knowledge companies and workers but also retaining them through smart growth. Such an approach includes a host of strategies that enable both the public and the private sector to preserve quality of life.[36] These strategies include conserving land and using the remaining land more effectively, and applying transportation planning that recognizes the importance of public transportation and walking and biking options. Indeed, the rapid consumption of land has propelled the smart growth movement nationwide.

Baltimore is an important example of an older city that is capitalizing on the state's leading-edge smart growth policies and converting older industrial buildings relatively close to Chesapeake Bay harbor and residential areas into the "cool space" desired by high-tech companies. One of the most visionary and successful developers is Struever Brothers, Eccles & Rouse, which has developed the long-abandoned American Can Company manufacturing facility into a remarkable mixed-use facility, which has invigorated the local community. The site's four buildings, which total

300,000 square feet, are now home to 700 jobs and forty separate businesses, including high-tech companies, restaurants, cafés, and bookstores. Many employees walk to work. Meanwhile, the Maryland Economic Development Corporation, in partnership with the University of Maryland, Johns Hopkins University, and Morgan State University, is designing an associated Emerging Technology Center to fulfill the needs of high-growth information technology companies.

In Chattanooga, Tennessee, civic and corporate leaders have made environmental amenities the centerpiece of a successful economic revitalization strategy that has included redeveloping the riverfront as a recreation area. Chattanooga's freshwater aquarium and downtown green spaces give it a distinctive character that has been lauded by observers from across the nation.[37]

By applying smart growth principles early, Portland, Oregon, has experienced substantial growth based largely on the activities of some 1,200 high-tech companies while preserving its quality of life. Bureau of Planning director Gil Kelley explained: "We're reinventing ourselves as a very urban place by incorporating the natural environment, transportation, parks and neighborhoods."[38] *Money* magazine's article "Best Places to Live"—which "focused on economically vibrant cities that are also successfully managing their growth and providing the highest quality of life around"—rated Portland the country's top big city.[39]

In Austin, public-spirited organizations understood that the city's rapid growth could undermine its success, particularly in attracting talent. City leaders developed a smart growth strategy based on the identification of Desired Development Zones and Drinking Water Protection Zones. A 1998 bond initiative raised $65 million to buy 15,000 acres in southwest Austin to protect water quality and to offer incentive packages—mostly breaks on development fees—for firms such as Dell Computer to move into zones designated for development. The smart growth initiative also targets improving the quality of life, including protecting environmental quality, and enhancing the tax base. The initiative's Web page outlines these principles.[40]

In Atlanta, the Georgia Conservancy has played a major role in advancing smart growth as a strategy for reconciling economic and environmental goals. While local and state government has focused on traffic congestion, the Conservancy has stressed the loss of some 1,500 acres of forest and farmland a month resulting from development designed to accommodate more than 3 million additional residents in the metro Atlanta region over the past twenty years.

While smart growth principles can extend a location's carrying capacity, limiting growth—particularly land consumption—to preserve quality of life may be inevitable. In some regions, nature provides obvious limits to growth, such as the mountains surrounding Los Angeles and the major waterways abutting coastal cities such as New York. In unconstrained locations, suburban sprawl and the loss of open spaces and other natural amenities are likely to stimulate more calls for smarter growth.

Quality of Life Supports Pragmatic Conservation

The environmental movement has been driven for decades by core principles that call for doing the right thing for the good of the planet and future generations. The movement has also tried to combat the view that environmental goals—particularly as implemented through government regulation—threaten growth and prosperity by raising costs and constraining economic activities.

In the new economy, a fortunate confluence has emerged among the interests of companies, workers, and government. Natural amenities support a high quality of life even for people—now the vast majority of Americans—who do not use natural resources directly as a source of income. Conservation has broad practical value as an economic development strategy when combined with smart growth principles. Time will tell whether this more pragmatic and personal—even selfish—valuing of natural amenities is more consequential than a moral valuing, but it would seem to promise greater popular support for conservation. Ironically, the fast pace and rapid loss of land to development induced by the Internet age make this need urgent but also put the possibility of fulfilling it within reach.

NOTES

1. Jonathan Rauch, "The New Old Economy: Oil, Computers, and the Reinvention of the Earth," *Atlantic*, January 2001, available at www.theatlantic.com/issues/2001/01/rauch.htm.
2. Doug Henton and Kim Walesh, "Linking the New Economy to the Livable Community" (San Francisco: James Irvine Foundation, April 1998), available at www.irvine.org/frameset3.htm.
3. Thomas Michael Power, "The Economics of Wilderness Preservation in Utah," testimony before the U.S. House of Representatives Committee on

Resources and the Environment (June 29, 1995), available at www.suwa.org /newsletters/1995/winter/insert1.html.

4. Outdoor Recreation Coalition of America, "Economic Benefits of Outdoor Recreation," *State of the Industry Report, 1997*, available at www.out-doorindustry.org/.

5. Jerry D. Johnson and Ray Rasker, "The Role of Economic and Quality of Life Values in Rural Business Location," *Journal of Rural Studies* 11:4 (1995): 405–16.

6. Dennis A. Rondinelli and William J. Burpitt, "Do Government Incentives Attract and Retain International Investment? A Study of Foreign-Owned Firms in North Carolina," Kenan Institute of Private Enterprise, University of North Carolina at Chapel Hill, 1999. The report concluded that such incentives can divert public expenditures from investments in human resources, quality of life, infrastructure, and services that companies consider more important.

7. Paul Gottlieb, "Amenities as an Economic Development Tool: Is There Enough Evidence?" *Economic Development Quarterly* (August 1994): 270–85.

8. Mark Drabenstott and Tim R. Smith, "The Changing Economy of the Rural Heartland," in *Economic Forces Shaping the Rural Heartland* (Kansas City, Mo.: Federal Reserve Bank, 1996).

9. Richard Florida, "Competing in the Age of Talent: Environment, Amenities, and the New Economy," report prepared for the R. K. Mellon Foundation, Heinz Endowments, and Sustainable Pittsburgh (Pittsburgh: Carnegie Mellon University, January 2000), available at www2.heinz.cmu.edu/~florida/talent.pdf.

10. Nick Pachetti and Alan Mirabella, "The Best Places to Live," *Money* (2000), available at http://money.cnn.com/best/bplive/.

11. Power, "The Economics of Wilderness Preservation in Utah."

12. IntelliQuest.com, "Environment Is of Greatest Concern to Residents of High Technology Communities" (June 20, 1998), available at www.intelliquest. com/press/archive/release84.asp.

13. Florida, "Competing in the Age of Talent," 44.

14. Richard Florida, "Place and the New Economy," Champions of Sustainability lecture broadcast on WQED (Pittsburgh), August 27, 2000, transcript available at www.heinz.cmu.edu/~florida/.

15. "In the Fast Lane: Delivering More Transportation Choices to Break Gridlock" (Washington, D.C.: National Governors Association, November 2000).

16. "Urban Forests May Solve Houston's Problems," *The Forestry Source*, February 2001, available at www.safnet.org/archive/houston201.htm.

17. Paul Van Slambrouck, "Lifestyle Drives Today's Workers," *Christian Science Monitor*, September 11, 2000.

18. Justus Baird, "New Economy Is Environmentalists' New Friend: Why City Hall and the Business Community Need Environmentalists in the New

Economy," *CEC Newsletter* (December 2000), available at www.cechous-ton.org/newsletter/nl_12-00/presletter.html.

19. Steve Lerner and William Poole, "Open Space Investments Pay Big Returns" (June 23, 1999), available at www.tpl.org/tier3_cd.cfm?content_item_id=964&folder_id=765.

20. City of Austin, "Smart Growth Initiative," available at www.ci.austin.tx.us/smartgrowth/.

21. IntelliQuest.com, "Environment Is of Greatest Concern to Residents of High Technology Communities."

22. Henton and Walesh, "Linking the New Economy to the Livable Community."

23. Mary Deibel, "States Raise Stakes in Jobs Hunt," Scripps Howard News Service, available at www.jrnl.net/news/00/May/jrn106220500.html.

24. John Engler, "Governor's Address" (January 31, 2001), available at http://www.nga.org/governors/1,1169,C_SPEECH^D_1203,00.html.

25. Jim Hodges, "Governor's Address" (January 17, 2001), available at www.nga.org/governors/1,1169,C_SPEECH^D_1113,00.html.

26. ECONorthwest, "Quality of Life and Its Impact on Louisiana's Economy," available at www.riversentinel.net/first.htm.

27. "The Northwest Faces Critical Choices," *Freeflow: Journal of the Pacific Rivers Council* (winter 1996).

28. Pete Geddes, "Economy and Ecology in the Next West," *Journal of Forestry* 96:8 (August 1998). availabe at www.free-eco.org/pub/980800pg.html.

29. Michael O. Leavitt, "Governor's Address" (January 16, 2001), available at www.nga.org/governors/1,1169,C_SPEECH^D_1075,00.html.

30. Raymond Rasker, "A New Look at Old Vistas: The Economic Role of Environmental Quality in Western Public Lands," *University of Colorado Law Review* 65 (1994): 365–99.

31. Ray Rasker, "Entrepreneurs of the New West," available in 2001 at www.yel-lowstonescience.com/views/rasker/entrepreneurs.html.

32. Jerry D. Johnson and Ray Rasker, "Local Government: Local Business Climate and Quality of Life," *Montana Policy Review* 3:2 (1993): 11–19.

33. Thomas M. Power, "Do Jobs Follow People or Do People Follow Jobs?" KUFM commentary (March 30, 1998), available in 2001 at www.cas.umt.edu/econ/Power/kufm/1998/033098.htm.

34. Ed Marston, "Beyond the Revolution, *High Country News*, April 10, 2000.

35. Florida, "Competing in the Age of Talent," 23.

36. Joel Hirschhorn, *Growing Pains: Quality of Life in the New Economy* (Washington, D.C.: National Governors Association, June 2000).

37. Chattanooga News Bureau, "Chattanooga: America's Talking," available at

www.chattanooga-chamber.com/newsandinfo/factsheets/americatalking.htm.

38. Nick Pachetti and Alan Mirabella, "The Best Places to Live: Best Big City," *Money* (2000), available at www.money.com/money/depts/real_estate/bplive/portland.html.

39. Pachetti and Mirabella, "The Best Places to Live: Best Big City."

40. City of Austin, "Smart Growth Initiative," available at www.ci.austin.tx.us/smartgrowth/.

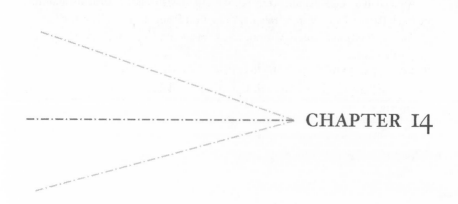

CHAPTER 14

Conservation Philanthropy and Leadership

THE ROLE OF NETWORK ENTREPRENEURS

Peter R. Stein and James N. Levitt

The long-term protection of land and open space requires dedicated people. In the United States and throughout the world, many land protection projects also require substantial financial capital—capital supplied by governments, by private individuals and corporations, and by the foundations and not-for-profit groups funded by public and private entities. For many generations, Americans, acting as private conservationists and philanthropists, have used their personal and family wealth to protect and conserve natural areas in urban, suburban, and rural settings, on small parcels and expansive landscapes. The total fiscal magnitude of this public-spirited conservation philanthropy is unknown. The return on this investment in the landscape belongs to us all.

Conservation philanthropy is generated by Americans of all descriptions and levels of affluence, including individuals of very modest means who contribute long hours and the proceeds of bake sales to save a cherished local park; schoolchildren who donate their nickels and dimes to help save the Amazonian rain forest; and families of great wealth who work over decades to protect entire island ecosystems or vast forest tracts. Particularly with regard to large-scale, big-ticket land acquisition and conservation easement transactions, the philanthropy, initiative, and leadership of affluent entrepreneurs, industrial leaders, and their families has played an essential role in the successful completion of multifaceted conservation fundraising campaigns. From Big Sur to Jackson Hole to the coast of Maine, large leadership gifts and inventive efforts, often made with little fanfare or recognition, have made it possible for many ambitious conservation initiatives to be successfully consummated.

The sources of private sector wealth that provide such large, leadership funding are as varied as the modern economy. Look over the donor lists of leading nonprofits that focus on the conservation of land and biodiversity, and you will find substantial philanthropy from individuals, families, and foundations associated with a broad range of enterprises and industries, from clothing manufacturing and investment banking to fast food, pharmaceuticals, and forestry.

Among the most prominent conservation philanthropists in American history, from the early 1800s to our day, are individuals and families who have been closely identified with the growth of the nation's transportation and communications networks. The fact that such network entrepreneurs and their kin have been in the forefront of conservation philanthropy and related private conservation efforts is consistent with the prominent role of network-related enterprises in the U.S. economy at large. Such enterprises have long played an essential role in our national development, from the advent of large railroad and telegraph enterprises in the last half of the nineteenth century, to the rise of automobile, petroleum, telephone, and electric power interests beginning around the turn of the twentieth century, to the remarkable growth of broadcast networks, networked information technology firms, and advanced logistics enterprises in more recent decades. Remarkably enough, as networked information technology companies play a critically important role in the economy of the early twenty-first century, entrepreneurs and families associated with those firms have stepped up to take their places among the leading conservation philanthropists of our day.[1]

There is no single motivational reason why network entrepreneurs, their families, and their foundations have become so deeply involved in conservation philanthropy and initiatives over the course of American history. As the profiles that follow illustrate, the motivations of such individuals and organizations range from enlightened self-interest, to a desire to set a good example and leave a good mark on the world, to a genuine affection for nature, to a heightened sense of social responsibility. It is notable, however, that the network entrepreneurs profiled here— including such remarkable personalities as Thomas Jefferson, J. Sterling Morton, Frederick Billings, Gilbert H. Grosvenor, Laurance Rockefeller, David Packard, Bill Ford, and Gordon Moore—have in their lives exhibited an exceptional range and acuity of technological and social vision, with the ability to provide important insight into a wide spectrum of disciplines, as well as the proclivity to discern small and critical details in their chosen fields of endeavor. With vision, the wherewithal to act, and a range of deeply felt personal motivations, the network entrepreneurs profiled here, along with their families, enterprises, and foundations, have been able to make historic contributions to the conservation of land and biodiversity.

Thomas Jefferson and Natural Bridge

Thomas Jefferson played a pivotal public role in the opening of the Louisiana Territory. Following in the footsteps of his father, Peter, as noted in Chapter 2, he acted as a historic network entrepreneur, sending Lewis and Clark on a mission to find "the most direct & practicable water communication across this continent, for the purposes of commerce."[2] In laying out his dreams for the unexplored West, Jefferson planted the idea that, while the waterways and adjacent lands would necessarily be used for the commercial purposes of the United States, large portions of the vast western territories should be reserved for Indians.

In addition to his interest in setting aside land owned by the government, Jefferson acted as a private citizen to protect and enhance remarkable landscapes. His role as a thoughtful planner and planter at Monticello is familiar to many Americans. Less familiar is his role as the owner and protector of what is now known as the Natural Bridge of Virginia, near the Blue Ridge Mountains.

Jefferson visited and acquired the property relatively early in his career. In 1774, just two years before he penned the Declaration of American Independence, Jefferson purchased Natural Bridge and the surrounding acreage from King George III's government. The bridge, a massive natural stone arch standing some 215 feet above the bed of Cedar Creek, inspired his awe. He described it fervently in his book *Notes on the State of Virginia:*

> The Natural Bridge, the most sublime of nature's works. . . . Looking down from [the top] of the arch about a minute gave me a violent head ache. If the view from the top be painful and intolerable, that from below is delightful in an equal extreme. It is impossible for the emotions arising from the sublime, to be felt beyond what they are here; so beautiful an arch, so elevated, so light, and springing as it were up to the heaven, the rapture of the spectator is really indescribable.[3]

Jefferson held on to Natural Bridge all of his life, despite his declining financial fortunes. While he once considered its sale, and generated modest income from the property through activities such as the manufacture of gunshot, by 1815 the Sage of Monticello had "no idea of selling the land." Importantly, he reported, "I view it in some degree as a public trust, and would on no consideration permit the bridge to be injured, defaced or masked from public view."[4]

As Natural Bridge's owner, Jefferson tirelessly promoted "this singular landscape."[5] He urged several accomplished artists, including John Trumbull and Charles Willson Peale, to paint it.[6] Reportedly, many of the era's most distinguished Americans, including John Marshall, James Monroe, Henry Clay, Sam Houston, Daniel Boone, Andrew Jackson, Martin Van Buren, and Thomas Hart Benton, made the pilgrimage to the celebrated site.[7]

Why did Jefferson promote Natural Bridge so persistently? For several reasons, we believe. First, as his testimony in *Notes on the State of Virginia* indicates, he was genuinely in awe of its natural beauty. Among the nation's founding fathers, Jefferson offered perhaps the most ardent expressions of what is now known as *biophilia*, defined by Harvard biologist E. O. Wilson as "a human dependence on nature that extends far beyond the simple issues of material and physical sustenance to encompass as well the human craving for aesthetic, intellectual, cognitive, and even spiritual meaning and satisfaction."[8]

Jefferson also viewed his natural wonder as a powerful attraction to potential European visitors, and association with Natural Bridge as a way to be long remembered. For example, Jefferson wrote to Maria Cosway, the Frenchwoman who had captured his heart, in 1786: "The Falling

Spring, the Cascade of Niagara, the Passage of the Potowmac [*sic*] through the Blue Mountains, the Natural bridge. It is worth a voyage across the Atlantic to see these objects; much more to paint, and make them, & thereby ourselves, known to all ages."[9]

In addition, in the words of Jefferson scholar Charles Miller, Jefferson saw that "the portrayal of American nature served a nationalist as well as an aesthetic end," demonstrating to Europeans that the New World was fully the equal of the Old World.[10] Furthermore, this man who never traveled west of the Blue Ridge Mountains was likely to have seen Natural Bridge, in the opinion of Gerald Gewalt, manuscript curator at the Library of Congress, as a symbolic "gateway to the West."[11] By sending distinguished visitors, essayists, and painters there, he likely hoped to inspire in others the passion he felt for the frontier, an area that he saw as having vast potential.[12] Caleb Boyle, an artist who apparently never visited Natural Bridge or drew Jefferson from life, evidently got the message: in a portrait of Jefferson painted circa 1801, Boyle pictured Jefferson, then U.S. president, standing with his hat off before Natural Bridge, with verdant hills and an illuminated sky forming the background to the great man and his natural wonder.[13]

After his retirement from public office, Jefferson carried on, encouraging his fellow Americans to expand to the west, in the interest of establishing an "empire for liberty."[14] He strongly endorsed John Jacob Astor's efforts to establish an American presence at the mouth of the Columbia River—the end point of the trail that he commissioned Lewis and Clark to blaze: "I considered as a great public acquisition the commencement of a settlement on that point of the Western coast of America, & looked forward with gratification to the time when its descendants should have spread themselves thro' the whole length of that coast, covering it with free and independent Americans."[15]

Since Jefferson's death, Natural Bridge has several times been bought and sold by various private interests. It remains in private hands, with designation as a National Historic Landmark, to this day. For a modest fee, the bridge is available for public enjoyment.

Although the view from on top of Natural Bridge has been largely blocked by wooden fencing, the view from the valley floor continues to inspire in visitors reverent appreciation. Perhaps as significantly, Jefferson's idea that the natural wonder on his private property was to be protected, to some degree, in the public interest was an important precursor to the conservation practices of generations of Americans who have followed.

J. Sterling Morton and Arbor Lodge

By later in the nineteenth century, railroads had surpassed water routes as the primary means of transporting goods across the continent. During the 1850s and 1860s, the corporations that formed to build and operate the railroads (and the telegraph networks essential to railroad operations) became, in the words of eminent business historian Alfred Chandler, "the first modern business enterprises."[16] While many of them were never far from controversy, the individuals who led the railroads exhibited corporate imagination and vision on a grand scale. Among the powerful railroad interests of the late nineteenth century were several individuals who, along with their families, showed similar foresight through their private investments in conservation.

Consider, for example, J. Sterling Morton, who as a young man in 1854 moved to the Nebraska Territory to achieve "some great undertaking."[17] Morton rose rapidly to become a prominent newspaper editor, a Democratic politician (he served a term as territorial governor), and a highly effective publicist, lobbyist, and negotiator for several railroads, including the Burlington & Missouri (B. & M.). Amid all his other activities, Morton, serving on the state agricultural board in 1872, promoted the creation of a tree-planting holiday that he suggested be called Arbor Day. With help from the state, the first Arbor Day was an enormous success; by one reckoning, Nebraskans planted more than 1 million seedlings that day.[18]

Morton continued to invest his personal funds, alongside in-kind and financial contributions made by a number of public and private interests, including several railroads, in the promotion of tree planting, arboriculture, and fruit grown in Nebraska. By the 1880s, Arbor Day had been instituted by states across the country. The messages delivered at Arbor Day celebrations took on a rhetorical style familiar to our own ears. In 1885, for example, Morton urged his audience to consider their legacy: "Each generation takes the earth as trustees. . . . We ought to bequeath to posterity as many forests and orchards as we have exhausted and consumed."[19]

Morton's homesite in Nebraska City, a place that he and his family came to call Arbor Lodge, had extensive orchards and groves. Visitors came to the site from far and wide to see how a plot of land in what was once known as the Great American Desert could be transformed into a forested show-case. In the 1890s, Morton went on to become secretary of agriculture in Democrat Grover Cleveland's second administration as well as president of the American Forestry Association. As he had hoped as a young man, he had achieved more than one "great undertaking" and was widely recognized at a national level for doing so.

Morton's apparent motivation to invest his own money in conservation was not dissimilar to Jefferson's. From early on, Morton sensed that the idea would earn him a place in history. On April 8, 1874, he wrote in his diary: "*Arbor Day*, an invention of mine, now become a public holiday, destined to become a blessing to posterity as well as to ourselves. . . . On the Morton Place, today, two Hundred Elmns [*sic*] Ash & Linden trees are set out on East Line and East Avenue."[20] Furthermore, as a landowner and politician, Morton appears to have been motivated by the idea that extensive tree planting on private lands and in public spaces would enhance Nebraska's image to easterners and Europeans. In a letter to the *Omaha Herald* on the occasion of the first Arbor Day, Morton wrote: "What infinite beauty and loveliness we can add to the pleasant plains of Nebraska by planting forest and fruit trees upon every swell of their voluptuous undulations, and in another short decade, make her the Orchard of the Union, the Sylvan queen of the Republic."[21] Importantly, the railroads that employed Morton concurred. For example, as early as 1873 the B. & M. laid plans to send a shipment of Nebraska apples and other fruits to the Smithfield Fair in England, with the apparent aim of demonstrating the fertility of the prairie land, advertising the property that the railroad owned and hoped to sell, and increasing foreign immigration to Nebraska.[22]

Morton passed away in 1902, after successfully passing his conservation ethic on to his children. After using Arbor Lodge as a summer residence for two decades, the family donated the site to the State of Nebraska. Today, Arbor Lodge, recently renovated, is a Nebraska state park.[23] It sits across a valley from the impressive headquarters of the National Arbor Day Foundation, sponsor of Tree City USA and other widely applied conservation programs.

Morton's oldest son, Joy, went on to lead the Morton Salt Company in Lisle, Illinois. He built there a highly regarded arboretum, which has been expanded over the decades to encompass some 1,700 acres. The philanthropic tradition of the Morton family has made it possible for tens of thousands of annual visitors to the Nebraska and Illinois sites to enjoy the outdoors, participate in conservation education programs, and learn about the frontier politician and railroad promoter who successfully urged his fellow citizens to "Plant Trees."

Frederick Billings and His Investment in Conservation

Five years before J. Sterling Morton headed for Nebraska, a young man from Vermont named Frederick Billings accompanied his sister Laura on a steamship to California. Billings, a graduate of the University of Vermont, soon became an important player in California legal circles. As noted in Chapter 2, Billings, as an attorney for John C. Fremont and Jessie Benton Fremont and an active supporter of the still-young Republican Party in California, was a member of the team that helped to persuade the Lincoln administration to set aside Yosemite as a state reservation in 1864. His broadly ranging interests also led him to become a founder of the institution that was to become the University of California at Berkeley. When he returned east to become an officer and board member of the Northern Pacific Railroad (the NP), Billings was integral to the 1872 effort to create at Yellowstone the world's first national park. And as president of the Northern Pacific in the late 1870s, he championed several corporate efforts related to conservation, including tree planting along the route of the NP.

It was, however, at his estate in Woodstock, Vermont—the former home of legendary conservationist George Perkins Marsh—that Billings made his greatest personal investments in conservation. Under Billings's watchful eye, his Woodstock farm became a model known throughout Vermont and beyond for its progressive forestry, farming, and livestock operations. Billings worked for years to ensure that the system of roads and tree plantings that he bestowed upon Woodstock's Mount Tom—the majority of which he owned but left open to the public—would make a contribution to the community that would long be remembered. Billings thought that the wide roads, which would accommodate two carriages, "were to be my monument."[24] In another act of philanthropy, Billings gave a reported $250,000—a very large sum of money in the 1880s—to the University of Vermont (UVM) for the construction of a grand library to house the books, and preserve the memory, of George Perkins Marsh. The library, which is perhaps UVM's most famous building, was converted into a student center in the 1960s. Still, in addition to highly regarded degree programs related to natural resource management and conservation leadership, scholarship regarding George Perkins Marsh is ongoing at UVM: the university hosts an extensive George Perkins Marsh Online Research Center.[25]

Like Jefferson and Morton, Billings's private conservation efforts and philanthropy appear to have been informed by his experience in establishing a new transportation network. His biographer, Robin Winks, notes that Billings was asked to join the Vermont Forestry Commission in the early 1880s, as "Billings already had a reputation for reforestation, since he was introducing Austrian larch and Norwegian pine into his estate, and he was a public figure who, by virtue of his intimate relationship with the railroads, could be counted on to view realistically the problems railways caused."[26] Indeed, Billings was also keenly aware of how the railroad interests could be a positive force in the re-greening of Vermont, as the railroads were closely aligned with the state's increasingly important tourism and hospitality industries.

While Billings, who died in 1890, did not live to see the widespread recovery of Vermont's forests, his leadership gift to UVM and the reforestation example that he set on his own estate were significant milestones in the building of Vermont's reputation as an environmental leader among the states of the Union. Furthermore, as will be discussed in further detail later in this chapter, the Billings family tradition of conservation leadership continues to this day, in part through the efforts of Billings's granddaughter Mary. Mary and her husband, Laurance Rockefeller, have made very significant contributions to the conservation movement throughout much of the twentieth century.

Gilbert H. Grosvenor, Tenderfoot

In the last several decades of the nineteenth century, the circle of prominent Americans concerned with the careful management of land, wildlife, and other natural resources widened considerably. That growing circle included a number of individuals and families who were associated with burgeoning communications networks. One such individual was Gilbert H. Grosvenor, son-in-law of Alexander Graham Bell.

Bell, of course, was the inventor of the telephone and a founder of what was to become AT&T, the world's largest communications company through most of the twentieth century. He had a deep interest in natural history and earth science and served as one of three regents of the Smithsonian Institution from 1898 to 1916. At his mother-in-law's urging, Bell in 1898 took over the presidency of a small organization founded by his recently deceased father-in-law and business partner, Gardiner H. Hubbard. With Bell's encouragement and financial assistance, that organization, the National Geographic Society, was to become a great American institution.

Most of the credit for the Society's growth is given to "Bert" Grosvenor, who married Bell's daughter Elsie in 1900. Grosvenor worked tirelessly, in concert with Bell, to bring new members to the Society and vivid photography to its famous yellow-bordered magazine. As a pioneer of photojournalism, Grosvenor helped to define modern mass media communications, delivering striking images of the wide world into the homes of millions of Americans.

As editor of *National Geographic* for fifty-five years, Grosvenor had a firm policy about staying away from stories of a controversial nature.[27] Still, he did allow the magazine to run features promoting the cause of conservation. *National Geographic* published articles supporting Teddy Roosevelt's and Gifford Pinchot's conservation initiatives in the first decade of the 1900s as well as articles supporting the creation of a National Park Service in the next decade. Demonstrating his commitment to conservation, according to Warren Bielenberg, "Grosvenor provided $20,000 of National Geographic Society funds to supplement a $50,000 congressional appropriation to buy Giant Forest and add it to Sequoia National Park."[28] It was on a trip to see the redwoods that Grosvenor, the well-mannered son of an Amherst College professor, earned the affectionate nickname "Tenderfoot." He remained a strong supporter of national parks for many years to come.

Almost a century later, the Grosvenor family continues to provide leadership to the National Geographic Society's board of directors. The Society's Exploration Committee continues to fund critical research in the field of conservation biology, and the editorial voice of the Society in support of environmental protection and conservation, through its magazines, books, television programming, and Internet presence, is stronger today than ever. Alexander Graham Bell, who wrote to a friend in 1872 that "the study of Nature is undoubtedly one of the most interesting of all pursuits," would very likely approve.[29]

The Rockefeller Tradition of Conservation Philanthropy

One family name is perhaps most readily associated with conservation philanthropy in the twentieth century—Rockefeller. The family's tradition has its roots in the philanthropic practices of John D. Rockefeller Sr., founder of the Standard Oil Company, in its many forms and affiliations. The senior Rockefeller built his fortune in the several decades before and after 1900 as automobiles first became a prominent feature of the American transportation scene. The petroleum products brought to market by Standard Oil, its successors, and the other large integrated oil companies have literally fueled the ongoing growth in American automobility.[30]

The Rockefeller family's conservation vision has expanded throughout the course of four generations, spreading its impact from coast to coast in the United States, and beyond to the far reaches of the globe. Generation after generation of family members has continued to devote time, energy, and dollars to wide-ranging conservation projects. With the diversification of the family's portfolio into a wide range of investments, including some notable venture capital successes in high technology, there is good reason to believe that this intergenerational commitment to conservation will continue well into the future.

John D. Rockefeller Sr. (JDR Sr.) made an early and important conservation philanthropy gift in Cleveland, where he had built the foundations of Standard Oil and had many of its first oil refining facilities erected. During the 1890s, Cleveland experienced rapid—and sometimes alarming—industrial growth. In 1893, the city council passed into law a park act calling for the creation of a series of connected green spaces along the eastern edge of the city, so that the city's burgeoning population would have a place to escape to the "country."[31] William Gordon, an industrialist, donated 122 acres of waterfront along Lake Erie, on the city's east side, complementing an inland parcel that had been donated by Jeptha Wade, founder of the Western Union Telegraph Company. In 1896, on the occasion of the city's centennial, Rockefeller donated 270 acres of his summer home estate at Forest Hills to the system, helping to complete "the jigsaw puzzle, allowing Gordon Park and the newly created Wade Park to be linked."[32]

Rockefeller's donation came with three conditions: one, that the city could not authorize the sale of alcoholic beverages in the park; two, that the land was to be used only for the purposes of public recreation and drives; and three, that the city must contribute $600,000 within two years to improve the park facilities. Rockefeller Park has since become an important link in the Cleveland metropolitan area's green infrastructure. Efforts to enhance its value are ongoing: in 2001, the Cuyahoga County Planning

Commission proposed that the park be integrated into the conceptual planning for a countywide trail system, as part of the overall Cuyahoga County Greenspace Plan.

Born in 1874, John D. Rockefeller Jr. (JDR Jr.), the founder's son, grew up in Cleveland and New York, attended Brown University, and played an important role in the management of Standard Oil in the early part of his adult life. Even then, he devoted considerable energy to philanthropic efforts of his own. Following his decision in 1910 to "redirect his life's work from being manager of his father's vast business interests to being a philanthropist,"[33] JDR Jr. "made the giving away of his money his chief business," devoting his attention to the fields of "medicine, education, conservation, and improved race relations," as he "cast the Rockefeller conservation net worldwide."[34] His commitment to the protection of places of great natural beauty, which he sustained until his death in 1960, was both impassioned and unwavering. To name only two examples relevant to the expansion of the National Park System (NPS), Rockefeller devoted twenty-seven years to the construction of an elaborate system of carriage roads, which he donated to Acadia National Park in Maine; he also made major land acquisitions in the northwestern corner of Wyoming, spending a total of $17.5 million and engaging in complex negotiations with the federal government to make the Grand Teton National Park one of the system's "crown jewels."[35]

In both of these projects, as in many others, JDR Jr. appears to have been motivated by a strong sense of social responsibility in the context of the dramatic economic, technological, and demographic changes of his day. As explained by his granddaughter Ann Rockefeller Roberts: "My grandfather grew up amid all this change, watching the agricultural ideal give way before the industrial maelstrom. . . . In fact, nearly every invention born of modern technology [by the 1920s] served to insulate people more and more from the land. . . . To JDR, Jr., road building was much more than a passion indulged in for its own sake. It was a consummate skill used in service of a large mission: to unfold for the public the wonders of nature in Acadia National Park and offer the chance to reestablish bonds with the earth." Roberts's sense of JDR Jr.'s motivation is echoed by the journal entry of Horace Albright, an early NPS director and Rockefeller's close friend: "Mr. Rockefeller . . . took every opportunity he felt possible to step in and save his fellow humans from the onslaught of the crippling effects of an industrial society."[36]

Much like their father, each member of the family's third generation, including John D. III, David, Nelson, Winthrop, Abby, and Laurance Rockefeller, as well as their spouses, followed through with an interest in and generosity toward conservation and other areas of philanthropy. For example, Peggy Rockefeller, David's wife, was a key supporter of early efforts to conserve scenic beauty on the Maine coast as well as farmland throughout the nation.

Of particular interest for the purposes of this essay is the profound interest and passion for the protection of nature shown by one of the Rockefeller brothers, Laurance, and his wife, Mary, granddaughter of Frederick Billings. Throughout his life, Laurance S. Rockefeller (LSR) has been intimately involved with a remarkably wide range of conservation initiatives: he has chaired presidential commissions, including the Outdoor Recreation Resources Review Commission, which made historic contributions to the nation's conservation policy; he has provided key gifts for important research regarding conservation policy, including sponsorship of a series of meetings examining the role of national parks in the twenty-first century that resulted in the "Vail Agenda"; he played an instrumental role in the creation, expansion, or improvement of existing state and national parks in places as far flung as the U.S. Virgin Islands, Hawaii, California, and New York; along with his four brothers, he was instrumental in the private effort to preserve the family's estate at Pocantico and allow limited public access there; and with Mary, he made possible the creation of the Marsh-Billings-Rockefeller National Park and the NPS Conservation Study Institute in Woodstock, Vermont.

In his two separate biographies of Frederick Billings and Laurance Rockefeller, Robin Winks explains that each man saw a complementarity between "commerce *and* conservation." Billings, for example, worked diligently to open within Yellowstone National Park a set of amenities, including a grand hotel, owned by interests associated with the Northern Pacific Railroad. In a similar fashion, Laurance Rockefeller, who was also a key investor in early passenger airline enterprises, including Eastern Airlines and McDonnell Aircraft, "combined his venture capitalism with the promotion of national parks through his Rockresorts. These resorts, built between 1955 and 1969, were precursors to eco-tourism."[37] At far-flung locations, including Hawaii, the Virgin Islands, and Woodstock, Vermont, the Rockresorts, as well as other Rockefeller-affiliated lodges in the Grand Tetons, offered LSR an opportunity to do well and do good: namely, to build a portfolio of profitable hospitality-related businesses at the same time as he was working to protect the nation's natural heritage.

It is important to note that LSR's investments have not been limited to air transportation and hospitality. He also acted as an early and prescient venture capitalist. Both on his own and through the Rockefeller-affiliated Venrock Associates, LSR has made early-stage investments in such enterprises as Intel, Apple Computer, and StrataCom (subsequently acquired by Cisco Systems), which have produced key pieces of what is now a global electronic communications network.[38]

The women and men of the fourth generation of Rockefellers, known as the cousins, as well as their children, are carrying on the family's tradition of conservation philanthropy and investment. Among a broad spectrum of endeavors, they are providing resources and guidance to critical land protection efforts; working diligently and effectively on natural resource advocacy efforts; and investing in innovative initiatives, in both the for-profit and not-for-profit sectors, that should help to advance conservation for many years to come.

The Initiative of the Packard Family

As noted in earlier chapters, as the end of the twentieth century approached, a fresh set of new networks—including information technology–related networks such as the Internet, cable television systems, and advanced logistics networks—came to play an increasingly critical role in the national economy. For many of the same reasons that earlier generations of network-related conservation philanthropists came forward—out of a sense of genuine affection for the natural world, a keen appreciation for social responsibility, and the opportunity to take on new and complex challenges—a new generation of conservation leadership has emerged from the ranks of contemporary network entrepreneurs and their families. The family of David and Lucile Packard provides one outstanding example.

The Packard family, first through direct philanthropic action by David and Lucile Packard and later by the David and Lucile Packard Foundation, raised the bar for magnitude and scope of charitable giving in several conservation-related fields. David Packard, who with William Hewlett in 1939 founded the Hewlett-Packard Corporation (HP), was, in the words of one remembrance, "a giant of a man, with a 'boarding house reach' so broad that it could take in the high technology revolution, ocean conservation, and the overpopulation crisis all at once."[39] The company he cofounded grew up in the shadow of Stanford University to become a critical catalyst in the phenomenal post–World War II growth of Silicon Valley economic activity. For the past several decades, HP has been at the forefront of the development of both proprietary computer networks (for example, networks owned and operated within a single company) and open, non-proprietary electronic communications networks such as the Internet.

David Packard's initial interest in marine technology was sparked when, as deputy to Defense Secretary Melvin Laird in 1969, he oversaw the secret recovery of a sunken Soviet submarine, an operation that relied on the sophisticated technology onboard the Glomar Explorer. His daughters Nancy Burnett and Julie Packard, who both studied marine biology in graduate school in the 1970s, were instrumental in getting David and his wife, Lucile, more deeply interested in marine science and conservation. By 1984, the family funded the construction and operations of a novel educational institution, the Monterey Bay Aquarium (now headed by Julie Packard), which was designed to "inspire conservation of the oceans."[40] In 1987, the family extended its commitment, funding the creation of a sister institution, the Monterey Bay Aquarium Research Institute, to conduct deep-sea scientific investigations. Today, with the continued leadership and support of the Packard family and its foundation, the two organizations are among the most inventive and effective of their kind in the world.[41]

The conservation philanthropy and vision of the Packard family touch the land as well as the sea. For example, in March 1998, the David and Lucile Packard Foundation initiated a five-year, $175 million program, the Conserving California Landscapes Initiative (the CCLI). It is designed to protect large expanses of open space, farmlands, and wildlife habitat in three California regions—the Central Coast, the Central Valley, and the Sierra Nevada. The program is further intended to develop organizations and policies that will support the protected landscapes over time. In late 2000, the board of the Packard Foundation voted to extend the effort by two years and to double the targeted CCLI financial commitment to $350 million.

The Packard Foundation CCLI is notable not only for its scale but also for its three-pronged design. It is intended to provide (1) capital support for specific transactions on a leveraged (matching) basis; (2) capacity building of land conservation nongovernmental organizations with a specific emphasis on landscape scale conservation planning and land use policy improvements; and (3) interim financing of land deals through the use of program-related invest-

Box 14.1

Saving the Peninsular Coast

The 7,000 acres owned by the Coast Dairies and Land Company was one of the largest privately owned coastal properties between San Francisco and the Mexican border. It is also among the most beautiful. The property covers more than six miles of coastline and beaches, has seven distinct watersheds, contains rich agricultural lands, and is host to several endangered species and rare plant communities. The unbroken views of the coast evoke earlier times.

In 1997 a Nevada development company held an option on the land, situated north of Santa Cruz near the tiny town of Davenport, and was poised to split the land into 139 lots in an attempt to develop luxury homes. With a grant from the Conserving California Landscapes Initiative (CCLI), Save-the-Redwoods League temporarily halted the threat by purchasing the development company's option on the property. The Trust for Public Land (TPL) then secured and exercised the development rights, again with the help of CCLI, taking ownership of the property and turning it into a 7,000-acre preserve. Along with other nongovernmental organizations, private donors, and foundations, including the William and Flora Hewlett Foundation, TPL is now crafting an innovative beach and upland management plan.

The Coast Dairies and Land Company property is an important parcel on the peninsular coast, running from Santa Cruz to San Francisco. The coast offers denizens of burgeoning Silicon Valley a nearby place to escape. On places such as the Coast Dairies property, they find relatively undeveloped, highly scenic landscapes that provide critical habitat to the coast's unique biodiversity. According to Peninsula Open Space Trust (POST) director Audrey Rust, "We have the only rural accessible coast in a major metropolitan area in the whole world."* Such landscapes are of tremendous value to people like Gordon Moore, who with Robert Noyce founded Intel Corporation in the Silicon Valley community of Santa Clara. The

Box 14.1 continued

Gordon and Betty Moore Foundation and the David and Lucile Packard Foundation each made contributions of $50 million in April 2001 to POST's Saving the Endangered Coast Campaign, designed to protect additional parcels of the peninsular coast. In Moore's words: "This is probably the last opportunity we have to save something like this. If it becomes a concrete jungle, L.A. moved north, a lot of us would consider that a disaster."*

* Michael McCabe, "$100 Million Pledged for Coastal Open Space," *San Francisco Chronicle*, April 19, 2001, A18.

ments (PRIs), loans made by foundations to land trust organizations to allow transactions to close in advance of the permanent funding availability.

Arguably, the Packard Foundation was the first large-scale funder to use these three linked strategies in specific geographic regions. One of the stellar success stories of this effort was the protection of the Coast Dairies ranch in northern Santa Cruz County, protecting seven miles of Pacific Coast shoreline (see Box 14.1).

And the Work Goes On . . .

Given the complex, considerable challenges facing the conservation community in the twenty-first century, the work of the conservation community, and the need for philanthropic support of such activities, is greater than ever. The encouraging news is that, along with colleagues from every sector in the economy and in concert with more established network-related philanthropic sources, fresh ranks of network-related philanthropists continue to step forward to participate in critically important ways.

These conservation philanthropists are widely varied in description and intent and include among their ranks Paul Brainerd, Paul Allen, the Seattle-based funders of the Wilburforce Foundation, and many, many others. In both prominent and discrete ways, they are playing key roles in the advancement of conservation practice and achievement. While they cannot all be named in this essay, several network-related entrepreneurs have made recent commitments of personal energy and financial resources that merit particular mention. Indeed, such fresh commitments indicate that we have begun a new round of conservation philanthropy that may rival, or even surpass in significance, the extraordinary philanthropy of earlier generations of American network entrepreneurs.

Ted Turner

Ted Turner, the legendary cable television entrepreneur and key AOL Time Warner stockholder, is undertaking conservation philanthropy and direct action on a grand scale. For example, the conservation philanthropy of the Turner Foundation amounts to about $25 million per year.[42] The related conservation investments of the Turner-sponsored UN Foundation increase the family's level of financial commitment to conservation and sustainable development severalfold. Furthermore, like Frederick Billings and J. Sterling Morton in the nineteenth century, Turner is practicing conservation on much of the scenic and working landscapes that he owns outright. As the largest private landowner in America (as of 1999), Turner is striving to restore a large portion of his 1.35-million-acre holdings, mostly in western ranch land, to a less developed condition. He explains that his methods are relatively straightforward: "All we're doing is allowing the ecosystem to be as natural as possible. We're trying to replace as many missing pieces to the environment as we can. . . . We don't kill rattlesnakes. We've reintroduced prairie dogs. I got the bison because they were native originally."[43]

Several environmental organizations have praised Turner's efforts. As reported by Donovan Webster in *Audubon* magazine, "the depth of conservation initiatives Turner has built in the last decade is on a par with that of The Nature Conservancy, the National Audubon Society, or other conservation organizations in the past 50 years or more." Turner's stated desire, like many conservation leaders before him, is both to set a good example and to leave a positive mark: "Everything I'm doing these days, from my ranches to the U.N. money to what I'm trying to do with Time Warner, I'm trying to leave the world a better place."[44]

The Libra Foundation

The Libra Foundation, with funds left by Elizabeth B. Noyce (former wife of Intel cofounder Robert Noyce), played a critical role in a landmark conservation effort in Maine, two-thirds of the way across the continent from

Ted Turner's western ranches. Libra made a multimillion-dollar gift to the New England Forestry Foundation (NEFF) campaign to complete the largest conservation easement deal in American history.[45] With funding from Libra and many other sources, NEFF was able to acquire, for a total of $30 million, an easement that will prevent development on more than 760,000 acres (an area larger than Rhode Island) of working forestry property owned by the Pingree family in northern Maine.

Bill Ford

Bill Ford, chairman and chief operating officer of Ford Motor Company, the enterprise founded by his great-grandfather at the turn of the twentieth century, is outspoken among his colleagues in the auto industry with regard to his commitment to sustainability and conservation. Ford speaks openly about the dual challenges of being socially responsible and running a profitable business operation—in effect, combining commerce and conservation: "In the twentieth century, the automobile provided tremendous benefits, and raised the standard of living in much of the world. But it also had a major negative impact on the environment. I struggled to reconcile my environmental belief with working for an industrial company. . . . When I became Chairman of Ford Motor Company last year, we began to form a very different vision. We have an opportunity to have a major positive impact on society. We cannot afford to miss this opportunity."[46]

Ford has personally, with his siblings, his larger family, and the philanthropic resources of Ford Motor Company, made substantial contributions to the causes of conservation and environmental responsibility. He has, for example, publicly committed his company to substantial improvements in the fuel efficiency of the vehicles it produces; actively promoted the advent of fuel cells; committed Ford Motor to join with Conservation International to set up a Center for Environmental Responsibility in Business; and helped to underwrite the efforts of conservationists from the Middle East to Michigan. Reminiscent of John D. Rockefeller Jr., Bill Ford has a very personal and deep commitment to his cause, as reflected in his closing comments as the keynote speaker at the 2000 meeting of the Coalition for Environmentally Responsible Economies: "If some of the challenges appear insurmountable to us, we must remember that our predecessors faced challenges that appeared equally insurmountable to them. They found success because they chose to cast aside the received wisdom, cut through old boundaries, and chart a new course into the future. . . . This is our time. For the sake of our children and our grand-children, we must prevail."[47]

Gordon Moore

Gordon Moore, cofounder of Intel Corporation, along with his wife, Betty, is making in the first decade of the new century a series of landmark gifts to the conservation community that are remarkable in scope and estimated to be unprecedented in size. Moore provided leadership at Intel over several decades, helping to build an industrial giant that provides the key components for computing and communications equipment now deployed in every corner of the world. In high-technology circles, Moore is probably best known for his technological vision: in the late 1960s, he made a prediction, now codified as "Moore's Law," that the number of transistors packed onto a microprocessor would double every approximately eighteen months over a period of years. His prediction, which has proven to be quite accurate over a period of decades, pointed to a key factor in the success of the information revolution of the late twentieth century.

Now retired from Intel but still involved in the high-tech world, Moore has set his sights on several new horizons. He and his wife have set up the Moore Foundation, to which they plan to give about half of their Intel stock—worth several billion dollars—to fund higher education, scientific research, and environmental initiatives. In addition to providing funding for conservation efforts close to home, such as the Saving the Endangered Coast Campaign noted in Box 14.1, the environmental focus of the Moores spans the globe. As chairman of the board of Conservation International (CI), Gordon Moore has for years championed efforts to save global bio-diversity hot spots. He collaborated with E. O. Wilson in August 2000 to sponsor the conference "Defying Nature's End" at the California Institute of Technology. That conference came up with a highly ambitious blueprint to protect twenty-five specified biodiversity hot spots in more than forty countries. And in December 2001, the Moore Foundation announced that it will give CI a series of grants totaling $261 million over the next ten years to launch the initiative. CI estimates that the Moore Foundation gift is the largest gift ever given to a private conservation group.[48]

Conservationists such as Moore's friend E. O. Wilson and others argue that "Moore has contributed very substantially and usefully to the world—

supporting just the kind of research and field programs that are most effective in global conservation."[49] As noted in Chapter 2, Alexander Graham Bell, almost a century ago, had the vision to warn that global industrial growth might result in an atmospheric "greenhouse effect." At the beginning of the twenty-first century, Gordon Moore, another historic leader in information technology, has warned that "the rate of species loss and habitat destruction demands immediate action."[50] Perhaps future generations will note that Moore's warning and his willingness to take action were as significant as Bell's.

Conclusion

Conservation philanthropy and direct conservation action by network-related entrepreneurs and their families in America is not a new phenomenon. As discussed here and elsewhere, U.S. network entrepreneurs since the time of Thomas Jefferson have protected their own land "in some degree as a public trust." And, of course, Jefferson's ideas had roots in the even more distant past. It is likely that Jefferson's thinking on the subject was influenced by the scientific work of European naturalists such as Buffon as well as by the land management practices of Jefferson's eighteenth-century contemporaries and predecessors who held expansive estates in England and France.

Furthermore, network entrepreneurs are not in any way alone in their support for conservation. To offer one of many relevant examples, Percival Baxter, whose donation of Baxter State Park to the people of Maine was a signal event in early-twentieth-century conservation history, was the youngest son of James Phinney Baxter, the six-time mayor of Portland, Maine. Percival inherited his fortune from his father, who made his money by discovering a new technique for canning corn. The Baxters were not prominent network entrepreneurs of their day.

Nevertheless, as we have argued here, a long line of entrepreneurs who have pioneered the growth of network-related enterprises in the United States, along with their families, heirs, and foundations, have repeatedly played key philanthropic roles in critical conservation initiatives. Such network-related philanthropists have acted out of a variety of motivations, including but clearly not limited to enlightened self-interest in the pursuit of conservation and commerce; a heightened sense of social responsibility, related in several cases to the disruptive impacts of the enterprises and technologies that they themselves advanced; a desire to leave behind a good example and a good mark on the world; a desire to conceive and articulate a vision of what the future may hold; and, perhaps most importantly, a genuine affection for the natural world.

If the past is in any way a prologue here, two realities are likely to confront the conservation community as it strives to make continued progress in the twenty-first century. First, new and increasingly powerful networks may well play an important part, both constructively and disruptively, in the land and biodiversity conservation challenges that will face us. Second, entrepreneurs and others associated with the emergence of such networks may well have the opportunity, the vision, the motivation, the resources, and the skills to make outstanding contributions to land and biodiversity conservation efforts that are responsive to those challenges in the decades to come.

NOTES

1. In addition to the brief case histories offered in this chapter, the history of conservation is replete with stories of other entrepreneurs closely associated with the growth of communications and transportation networks who were vitally interested in conservation issues. These include, among others, the stories of Henry Ford, who pioneered materials recycling in the automotive industry; Andrew Carnegie, who forcefully addressed the first White House Conference in 1908 in support of proposed forestry conservation measures; and George Washington Vanderbilt, son of railroad titan Cornelius Vanderbilt, who underwrote at his Biltmore estate in North Carolina the first forestry school in America. Recounting all of their histories, however, is beyond the scope of this brief chapter.

2. Thomas Jefferson, letter of instruction to Meriwether Lewis, June 20, 1803, available at www.library.csi.cuny.edu/dept/history/lavender/jefflett.html.

3. Thomas Jefferson, *Notes on the State of Virginia* (book online), available at http://xroads.virginia.edu/~hyper/jefferson/ch05.html.

4. Thomas Jefferson to William Caruthers, March 15, 1815, cited in "Jefferson and the Natural Bridge," available at www.monticello.org/resources/interests/natural_bridge.html.

5. Thomas Jefferson to John Trumbull, February 20, 1791, in Julian Boyd, ed., *The Papers of Thomas Jefferson* (Princeton, N.J.: Princeton University Press, 1950), 298.

6. S. Allen Chambers, "Thomas Jefferson Takes His Granddaughters to Natural Bridge: A Trip Attended with Disasters and Accidents," *Lynch's Ferry* 1 (fall 1998): 7.

7. Robert J. Smith, "CPC Case Study: Natural Bridge of Virginia," Center for Private Conservation (posted June 1, 1998), 5, available at www.cei.org/gencon/025,01355.cfm.

8. Edward O. Wilson, in Stephen Kellert and E. O. Wilson, eds., *The Biophilia Hypothesis* (Washington, D.C.: Island Press/Shearwater, 1993), 42.

9. Thomas Jefferson to Maria Cosway, October 12, 1786, in Thomas Jefferson, *Thomas Jefferson: Writings*, ed. Merrill D. Peterson (New York: Library of America, 1984), 870.

10. Charles A. Miller, *Jefferson and Nature: An Interpretation* (Baltimore: Johns Hopkins University Press, 1988), 107.

11. Library of Congress, "The West," in *Thomas Jefferson*, exhibition at the Library of Congress, Washington, D.C., 2000, available at www.loc.gov /exhibits/jefferson/jeffwest.html. In the Web site that accompanies the exhibition, Manuscript Division curator Gerald Gewalt provides the following commentary: "Viewed by Jefferson as the symbolic gateway to the West, the Natural Bridge was about as far west as Jefferson personally ventured."

12. Note that, despite his enthusiasm for the West, Jefferson did not make investments in western land. Declining an invitation in 1783 to invest in a western land company, Jefferson wrote Francis Eppes on February 15, 1783: "I have never ventured in this way in my own country because being concerned in the public business I was ever determined to keep my hands clear of every concern which might at any time produce an interference between private interests and public duties" (in Donald Jackson, *Thomas Jefferson and the Stony Mountains: Exploring the West from Monticello* [Champaign: University of Illinois Press, 1981], 16).

13. Charles A. Miller, "The Rich Fields of Nature," in Garry Wills et al., *Thomas Jefferson: Genius of Liberty* (New York: Viking Studio with Library of Congress, 2000), 60.

14. Thomas Jefferson to James Monroe, April 27, 1809, cited in Garry Wills et al., *Thomas Jefferson: Genius of Liberty* (New York: Viking Studio with Library of Congress, 2000), 133.

15. Thomas Jefferson to John Jacob Astor, May 24, 1812, cited in Garry Wills et al., *Thomas Jefferson: Genius of Liberty* (New York: Viking Studio with Library of Congress, 2000), 115.

16. Alfred Chandler, *The Visible Hand* (Cambridge: Belknap Press of Harvard University Press, 1977), 81.

17. Carolyn Morton to Emma Morton, April 7, 1855, cited in James C. Olson, *J. Sterling Morton* (Lincoln: University of Nebraska Press, 1942), 49.

18. James C. Olson, *J. Sterling Morton* (Lincoln: University of Nebraska Press, 1942), 166.

19. J. Sterling Morton, "An Address by J. Sterling Morton on Arbor Day 1885," as reported in the *Nebraska City News*, available on the National Arbor Day Foundation Web site at www.arborday.org/arborday/morton1887.html. For further discussion of Morton's contribution, see James N. Levitt, "Innovating on the Land: Conservation on the Working Landscape in American History" (paper prepared for Private Lands, Public Benefits: A Policy Summit on Working Lands Conservation, National Governors Association, March 16, 2001).

20. Olson, *J. Sterling Morton*, 166.

21. Morton's letter to the *Omaha Herald* is quoted in Olson, *J. Sterling Morton*, 165.

22. Olson, *J. Sterling Morton*, 199.

23. For additional information on Arbor Lodge, see www.ngpc.state.ne.us/parks/parkinfo/historical/arbor.html; additional information on the National Arbor Day Foundation is available at www.arborday.org. The Morton Arboretum in Lisle, Illinois, hosts a Web site at www.mortonarb.org.

24. Robin Winks, *Frederick Billings* (Oxford: Oxford University Press, 1991), 300. Winks quotes from a document in the Billings Mansion Archives in Woodstock, Vermont, box 25, that details Billings's supplementary instructions on the administration of his estate. The document was prepared in 1890, the last year of Billings's life.

25. George Perkins Marsh Online Research Center, available at http://bailey.uvm.edu/specialcollections/gpmorc.html.

26. Winks, *Frederick Billings*, 296.

27. C.D.B. Bryan, *The National Geographic Society: 100 Years of Adventure and Discovery* (New York: Abrams, 1987), 405.

28. Warren Bielenberg, "Biographical Vignettes: Gilbert H. Grosvenor, 1875–1966," in *National Park Service: The First 75 Years*, available at www.cr.nps.gov/history/online_books/sontag/grosvenor.htm.

29. Edwin S. Grosvenor and Morgan Wesson, *Alexander Graham Bell: The Life and Times of the Man Who Invented the Telephone* (New York: Abrams, 1997), 269.

30. See statistics on vehicle miles traveled and petroleum consumed by Americans in the past century provided by the U.S. Department of Transportation and by the U.S. Department of Energy (DOE), respectively. For example, the DOE Energy Information Administration, in its *Annual Energy Review 2000*, reports that U.S. petroleum consumption grew from 20 quadrillion BTUs in 1960 to 39 quadrillion BTUs in 2000. Cited in *The World Almanac and Book of Facts 2002* (New York: World Almanac Books, 2002), 157.

31. The Nature Center at Shaker Lakes, "History of the Nature Center," available at www.shakerlakes.org/history.htm.

32. Mary Mihaly, "All About University Circle: How the Circle Grew," available as of November 2001 at www.universitycircle.org/go5.htm.

33. Ann Rockefeller Roberts, *Mr. Rockefeller's Roads* (Camden, Maine: Down East, 1990), 47, 48.

34. Robin Winks, *Laurance S. Rockefeller: Catalyst for Conservation* (Washington, D.C.: Island Press, 1997), 12–13.

35. Winks, *Laurance S. Rockefeller,* 57. Winks explains that "JDR, Jr. spent a million and a half dollars [on land preservation in the area] between 1926 and 1933." He also details how, in 1943, Rockefeller convinced the federal government to take his land for incorporation into the park, threatening a sale to private buyers if the government refused to accept.

36. Roberts, *Mr. Rockefeller's Roads,* 6. In her book detailing the construction of the Acadia carriage road system, Roberts offers her own point of view and quotes from Albright's diaries, which are in the personal collection of Marion Schenk, Studio City, Calif.

37. Winks, *Laurance S. Rockefeller,* 54.

38. Winks, *Laurance S. Rockefeller,* 55. See also recent information on Venrock Associates at www.venrock.com.

39. Marcia McNutt, "How One Man Made a Difference: David Packard" (paper presented at the symposium Oceanography: The Making of a Science, Scripps Institution of Oceanography, La Jolla, Calif., February 8, 2000), available at www.mbari.org/about/howonemanmadeadifference.pdf.

40. For further information on the Monterey Bay Aquarium, see www.montereybayaquarium.org; information on its sister institution, the Monterey Bay Aquarium Research Institute, is available at www.mbari.org.

41. See, for example, National Audubon Society, "Audubon Society Awards Julie Packard '98 Audubon Medal," available at www.audubon.org/news/release98/jpackard.html. In giving Julie Packard the 1998 Audubon Medal, John Flicker, Audubon president, noted that "Julie Packard's vision and leadership have helped create a unique facility that sets the standard for environmental education."

42. Donovan Webster, "Welcome to Turner Country," *Audubon,* January/February 1999, 56.

43. Webster, "Welcome to Turner Country," 48–56.

44. Ibid.

45. Owen Wells, "Message from the President," The Libra Foundation, Portland, Maine, 2001, p. 2, available at www.librafoundation.org/messagepresident/welcome.html.

46. Bill Ford, speech to the Coalition for Environmentally Responsible Economies (CERES) 2000 conference, April 14, 2000, available at www.ford.com/en/ourcompany/environmentalinitiatives/environmentalactions/billfordspeaksattheceres2000conference.htm, pp. 1–10.

47. Ibid.

48. Conservation International, "Conservation International Unveils Solution to Prevent Global Species Extinctions," press release, December 9, 2001, available at www.conservation.org/xp/CIWEB/newsroom/press_releases/2001/120901.xml.

49. Kris Christen, "Biodiversity at the Crossroads," *Environmental Science and Technology* (March 1, 2000): 123A–28A.

50. Conservation International, "Conservation International Unveils Solution to Prevent Global Species Extinctions."

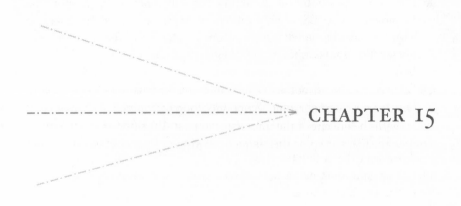

CHAPTER 15

Conclusion

A CALL FOR CONSERVATION INNOVATION

James N. Levitt

Today, in the early twenty-first century, North Americans can look back on the past century with some measure of amazement at the technological, economic, and social progress they have made. In 1900, an average voter—remember that only male citizens could vote at that time—might have been able to reach his state capital in a day's journey by train to voice his opinion on a pressing matter. Today, his great-granddaughter can journey by air across the continent in five hours to take her seat in the U.S. Senate, or span the globe within twenty-four hours to make a business meeting in Beijing, Buenos Aires, or Brussels. In 1900, a very brief cross-country telegram was something most families sent only on special or urgent occasions; today, college students living on tight budgets routinely open dozens of e-mails from places far and wide on a daily basis.

Only half a century ago, President Eisenhower proposed an interstate highway system for the United States that, as explained in Chapter 2, has actually brought many of the constructive benefits that expert witnesses forecast before members of the U.S. Senate in 1954. The interstates have also been associated with a range of disruptive impacts on the American landscape and environment, as foreseen by only a few prescient observers, such as Howard Zahniser in the 1950s. A host of observers—from the Sierra Club to Bill Ford, chairman of Ford Motor Company—now acknowledge a broad range of problems associated with modern highways and cars, including air pollution and landscape fragmentation.

The dramatic demographic, economic, and environmental change associated with the advent of the interstates in post–World War II America— along with such factors as the eloquent writing of Rachel Carson, the civil rights movement, and protests against the Vietnam War—provided an exceptional context for numerous conservation and environmental initiatives. The Wilderness Act that Zahniser campaigned for over the last decade of his life passed in 1964, as did legislation creating the Land and Water Conservation Fund advocated by Laurance Rockefeller. The Endangered Species Act, the Clean Air Act, the Clean Water Act, and the creation of the Environmental Protection Agency followed within a decade.

The debate over the appropriate role of highways, automobiles, and oil exploration continues today, as we argue over measures to control the gases that contribute to global climate change, whether to allow oil drilling on Alaska's North Slope, and how to implement smart growth policies intended to get us out of our cars and onto community sidewalks. While these debates are far from resolved, we have intensively considered the impacts of highways and cars. In contrast, we are only beginning to understand the constructive potential and possible disruptive impact on biodiversity, land use, and the environment of the next generation of networks, exemplified by the Internet and sophisticated logistics systems. As noted by Iowa Governor Tom Vilsack in the Foreword, it is high time that such considerations begin in earnest.

The Challenges at Hand

William Mitchell, in his exceptionally lucid style, has given us some idea in Chapter 3 of the "fragmentation and recombination" of the built and natural environments catalyzed by major new communications and transportation systems. Mitchell's outlook reminds us that dramatic change in the physical structures of our cities and lives can be wonderfully stimulating, opening broad opportunities and horizons that had previously been impractical, even unimaginable.

Clearly, however, the first decade of social change in the Internet age has been somewhat different from the scenario many conservationists might have imagined or hoped for. Despite the inspired urgings of new urbanists and smart growth advocates, we saw in the 1990s—the decade in which the Internet first widely penetrated our businesses and homes—a continued deconcentration of the U.S. population. As described by demographer Kenneth Johnson in Chapter 4, we witnessed a net positive migration from the city to the country—a movement attributable in part to a new generation of technology that allows knowledge workers and others to settle outside metropolitan areas and stay well connected to work, friends, and family. As explained by Ralph Grossi in Chapter 5, the impact of such demographic deconcentration on the working landscape, as well as on open spaces reserved for recreation and biodiversity habitat, is not encouraging. Indeed, the growing, increasingly mobile, and affluent American population consumed open space in the mid-1990s at a rate 79% greater than that recorded a decade earlier.

As John Pitkin and James Levitt showed in Chapter 6, the counties around Bend, Oregon, reveal interesting and statistically significant correlations between migration to physically attractive nonmetropolitan areas and the use of the Internet and other new networks. Recent in-migrants who cited natural or recreational amenities as important reasons for choosing Bend were significantly more likely to use the Internet at home than were longer term residents or other in-migrants. While this preliminary investigation does not somehow prove that the Internet "causes" people to move to places with beautiful scenery, it does give credence to the hypothesis that the new networks and information technologies enable people, as telecom entrepreneur Craig McCaw says, to "live where they like."

In short, historical precedents indicate that new communications and transportation networks are typically associated with broad changes in North American demographic and land use patterns, and there is growing empirical evidence that our own era is no different. To the extent that deconcentration into formerly remote and environmentally sensitive areas continues well into the new century, the impacts on wildlife habitat may well be quite disruptive. As Andrew Hansen and Jay Rotella detail in Chapter 7, expanding human settlement of the Greater Yellowstone ecosystem, particularly in relatively lush riparian habitats, appears to exert a negative effect on the health of bird, mammal, and other biological populations throughout the region.

Innovation, But of What Kind?

Fortunately, rapid social and technological change associated with new communications and transportation networks also appears to offer fertile ground for important innovations in conservation. As discussed by Governor Vilsack and others throughout the book, a considerable number of promising innovations are occurring in the field.

But not all innovations are good innovations, if history is any guide; most new enterprises and practices fail to catch on and instead proceed to die out. The pregnant question facing conservationists in the early twenty-first century is, "What advanced practice or revolutionary innovation can we implement that is both commensurate with the complex challenges at hand and likely to leave a lasting impact on the landscape?"

While we may not be able to accurately predict exactly which conservation innovations will stick, we can, with some confidence, think about the *characteristics* that are likely to distinguish truly outstanding conservation innovations. The criteria developed by the John F. Kennedy School of Government's Innovations in American Government program are particularly helpful in weighing historic and contemporary innovations.[1]

According to those criteria, outstanding conservation innovations are likely to be *novel*, to the degree that they demonstrate a leap in creativity. For example, when in the 1930s Hugh Hammond Bennett and his associates invented the idea of a Soil and Water Conservation District that would bring farmers together with government planners, irrigation and electric utility officials, wildlife biologists, and local policy makers to address critical soil erosion problems, the idea was untested and contrary to traditional rural decision-making practices. At demonstration sites in Wisconsin and elsewhere, however, the experimental districts made good progress in resolving such seemingly intractable problems as widespread soil loss and were soon replicated throughout the nation. In the early twenty-first century, the vast majority of counties have Soil and Water Conservation Districts or their equivalents, working closely with the U.S. Department of Agriculture and various state agencies, to tackle such problems as nonpoint sources of water pollution.

Similarly novel, the CommunityViz™ program developed by the Orton Family Foundation (see Chapter 11) uses three-dimensional software based on geographic information systems to make community land use planning transparent and comprehensible to nonspecialists. Its developers hope that CommunityViz™, in its current form or perhaps in a future iteration, will become a standard part of land use planning and decision making in the United States and abroad.

Outstanding conservation innovations are likely to be *effective*, to the degree that they achieve tangible or measurable results. When Charles Eliot created the world's first regional land trust in Massachusetts in 1891, the organization went directly to work preserving landscapes of special beauty as well as historic and recreational value, including Virginia Woods in Stoneham and other sites throughout the state. Eliot's organization, now known as the Trustees of Reservations, has protected some 45,000 acres throughout Massachusetts and has set an example followed around the globe.[2] A 2000 census indicates that 1,263 local and regional land trusts in the United States, following the Trustees of Reservations' lead more than a century ago, have protected more than 6.2 million acres of open space.[3]

Comparably, the BirdSource program jointly sponsored by the Cornell Lab of Ornithology and the National Audubon Society (see Chapter 9) has, in less than a decade, had notable success in mobilizing more than 50,000 citizen scientists across North America to watch the birds in their back-yards, school yards, and favorite recreational places and to report their observations to a central database. The results, available to the growing community on the www.birdsource.org Web site, are remarkable. On time-series maps composed from the data that BirdSource participants provide, schoolchildren and seniors, teachers, and field scientists can track the migration of avian species throughout spring, summer, winter, and fall and compare behavior from one year to the next, in the process becoming active stewards of the natural world.

Outstanding conservation innovations are likely to be *significant*, to the degree that they address an important problem of public concern. Theodore Roosevelt and Gifford Pinchot pursued a multipronged campaign over the course of several presidential administrations to advance irrigation and conservation. Their efforts eventually resulted in a number of highly significant achievements, including the creation of the U.S. Forest Service within the Department of Agriculture and the expansion of the National Forest system to the eastern states.

In a similar vein in the 1990s, the advocacy campaign engineered by the Natural Resources Defense Council (NRDC) and its partners to protect whale habitat at Laguna San Ignacio on Mexico's Baja peninsula (see Chapter 10) used electronic media as well as more traditional methods to attract international attention. Through persistent and well-coordinated effort, the coalition mobilized a broad public constituency in the United States, Europe, and Asia, in the process capturing the attention of the presidents of Mexico and Mitsubishi Corporation. The campaign to halt the construction of a Mitsubishi salt plant was ultimately successful, and efforts to further protect the whale nursery and promote sustainable economic development in the region are under way.

Outstanding conservation innovations are likely to be *transferable*, to the degree that they have inspired, or show promise of inspiring, successful replication. The world's first national park, at Yellowstone in 1872, propelled by the promotional efforts of Jay Cooke and Frederick Billings of the Northern Pacific Railroad (see Chapter 2), did not initially inspire widespread replication. As explained by Richard West Sellars, except for the creation in 1875 of a national park of about 1,000 acres on Mackinac Island, Michigan, "after Yellowstone nearly two decades passed before the national park idea spread to any significant degree."[4] It was not until 1916, the year the National Park Service was created as an administrative body, that the number of parks grew to more than a dozen. Since that day, however, admirers have replicated the national park model at home and abroad. As of 2001, the U.S. National Park System includes 385 units, encompassing more than 83 million acres.[5] The idea of nationally protected areas has spread to nations on every continent, from Kenya to Japan and from Argentina to Iceland.

The Internet, with the benefit of network effects, can replicate modern conservation innovations globally in a relatively short time frame. The Species Analyst project, launched in the late 1990s at the University of Kansas (see Chapter 8), allows researchers to share and analyze data from natural history collections around the world. As of early 2002, dozens of institutions in the United States as well as Latin America, Europe, Africa, and Australia are using Species Analyst, and many more partners have signed agreements to join the network. Some fifty-three natural history institutions also rely on a sister effort called Specify, which gives them advanced ability to manage their collections.[6] By providing remote access to information on biodiversity around the world, these two tools should markedly advance scientists' ability to model and predict the behavior of complex ecosystems.

When considering innovations in conservation—a field that often needs to maintain its efforts in perpetuity—we need to add a fifth criterion to *novelty, effectiveness, significance,* and *transferability.* Outstanding conservation innovations also should be *enduring,* to the degree that they have proven to have, or show promise of having, a durable impact on the challenges facing the conservation community.

Soil and Water Conservation Districts, regional land trusts, the U.S. Forest Service, a coast-to-coast system of national forests administered by the U.S. Forest Service, and the world's first national park fulfill all five criteria. Each innovation was novel in conception, measurably effective in implementation, significant with regard to national concerns, and transferable to other sectors and jurisdictions within and outside the United States. And each innovation has proved to have enduring value, playing an important role in conservation efforts in the twenty-first century. Only time will tell if CommunityViz™, BirdSource, Species Analyst, and NRDC's conservation campaign techniques will have comparably enduring impacts.

Political Will and Personal Determination

It is the political will and personal determination that we, in the early twenty-first century, bring to the table that will shape our children's judgment regarding the enduring value of such contemporary innovations, as well as our ultimate success in protecting open space and biodiversity from further dramatic loss.

As detailed in Chapters 12 and 13, people working at the state and local levels are marshaling considerable creativity and political will to pursue multifaceted, multiyear campaigns to advance a conservation agenda in the Internet age. For example, the Community Preservation Act—which allows cities and towns to tax real estate transactions to fund open space protection, historic preservation, and affordable housing—required fifteen years of concerted effort to become law in Massachusetts. The next generation may well cite that act, as well as similar creative efforts across the country, as examples of landmark conservation legislation. The law appears to have the hallmark characteristics of novelty, effectiveness, significance, transferability, and the promise of making an enduring difference.

Who will provide the leadership that will bring the most promising initiatives to successful realization and ensure their sustainability? Innovation, by its nature, is unpredictable. The inventiveness, determination, and resources required to devise and sustain conservation innovations commensurate with the grand challenges of the twenty-first century are likely to come from people of a wide diversity of backgrounds and beliefs. Some individuals and groups are as yet unknown, while others are already established in the public, private, nonprofit, and academic sectors.

Leadership may come from relatively remote rural communities, such as the Nez Perce Tribe based in central Idaho. When the state governments of Idaho, Montana, and Wyoming declined to participate in a U.S. Fish and Wildlife initiative to reintroduce the gray wolf to the Northern Rockies, the Nez Perce—relying in part on their own wildlife specialists—stepped in to manage the program. The bold effort, celebrated as a tribal governance success story by the Honoring Nations program, has proved to be a striking biological and intergovernmental success.[7]

Leadership in conservation innovation may come from schoolboys in Alabama who, fascinated with life on earth from ants to snakes, become preeminent university biologists. Such is the story of Edward O. Wilson.[8] Wilson's scientific findings and eloquent prose have helped convince national governments from the United States to Surinam, and international bodies including the World Bank and the United Nations, to make substantial investments to protect biodiversity hot spots around the world.

Leadership may come from women such as Julie Packard, executive director of the Monterey Bay Aquarium, whose institution is establishing a new standard for interactive environmental education. The aquarium provides hands-on displays of marine life and close connections to the cutting-edge research performed by the institution itself as well as by its sister institution, the Monterey Bay Aquarium Research Institute. And leadership may come from industrialists and philanthropists such as Bill Ford and Gordon Moore, who recognize that conservation of open space and biodiversity are essential to economic development and technological progress (see Chapter 14).

Such innovators are likely to apply effective new practices at a variety of scales (from a single critical acre to millions of acres), for a spectrum of purposes (for example, to protect wilderness areas, the working landscape, and recreational venues), and at a continuum of sites (ranging from rural to exurban, suburban, and urban locations). From wherever they come, initiatives that hope to succeed in their early lives and endure over the course of decades will require a well-articulated organizational infrastructure; consistent supplies of human and financial resources; a keenly developed sense of the competitive and cooperative environment; and a willingness to adapt to changing conditions.[9]

When should we launch bold conservation initiatives? Perhaps Americans would have been well advised to more strongly emphasize the conservation of open space and biodiversity in 1804, when Thomas Jefferson sent Lewis and Clark on their journey of discovery into the trans-Mississippi West. It would be wonderful if we could send spokesmen and women back in time, to the year 1954, to tell the assembled members of Congress about the disruptive impacts that the federal highway system would have on the North American landscape. But even enabled with the seemingly magical powers of the Internet, we cannot yet turn back the years.

What we can do is act today, in the early twenty-first century, to promote bold initiative and innovation. We have at our disposal the most powerful communications and transportation systems in the history of humankind. Conservation biologists can reach to the bottom of the oceans and into outer space with sensitive tools to gather information on forms of life—and places where life could plausibly exist—that we could not even imagine as recently as the 1990s. Our ability to devote financial resources to comprehensive conservation campaigns is far greater today than in the early 1900s, when Teddy Roosevelt and Gifford Pinchot beat the drum for imaginative conservation programs. And broad social awareness of environmental issues among schoolchildren, working parents, and seniors far exceeds that of the decades just after World War II.

In considering opportunities for conservation innovation in our day, we are well advised by the words of W. H. Murray, a Scottish mountaineer active in the 1950s:

> Concerning all acts of initiative (and creation), there is one elementary truth . . . that the moment one definitely commits oneself, . . . providence moves too. A whole stream of events issues from the decision, raising in one's favor all manner of unforeseen incidents, meetings and material assistance, which no man could have dreamt would have come his way. I learned a deep respect for one of Goethe's couplets: "Whatever you can do or dream you can, begin it. Boldness has genius, power and magic in it."[10]

Whether you believe in magic or simply the power of hard work and creativity, the impact of this message is clear: "Begin it now!"[11]

NOTES

1. For further information on the Innovations in American Government awards competition and the selection criteria it employs, see www.innovations.harvard.edu. A more in-depth treatment of the subject of innovations in the public sector is available in Alan A. Altshuler and Robert D. Behn, eds., *Innovations in American Government: Challenges, Opportunities, and Dilemmas* (Washington, D.C.: Brookings Institution Press, 1997).

2. For further information on the Trustees of Reservations, see www.thetrustees.org.

3. For further information on the 2000 land trust census, see www.lta.org/aboutlta/census.shtml.

4. Richard West Sellars, *Preserving Nature in the National Parks: A History* (New Haven, Conn.: Yale University Press, 1997), 11.

5. For a full listing of all national park units in the United States, see the National Park, Conservation Association, "About NPCA: A Primer on Federal Lands," available at www.npca.org/about_npca/park_system/default.asp.

6. For information on Specify, see http"//usobi.org/specify/.

7. "Idaho Gray Wolf Recovery: Initiative Leads to Increased Tribal Sovereignty and Cultural Rejuvenation," in *Honoring Nations: Tribal Governance Success Stories, 1999,* Honoring Contributions in the Governance of American Indian Nations, administered by the Harvard Project on American Indian Economic Development, Cambridge, Mass., 2000, p. 4.

8. Edward O. Wilson, *Naturalist* (Washington, D.C.: Island Press, 1994).

9. Charles H. W. Foster and James N. Levitt, "Beginner's Mind: Innovation and Environmental Practice," BCSIA Discussion Paper 2001-7, Environment and Natural Resources Program, John F. Kennedy School of Government, Harvard University, June 2001.

10. W. H. Murray, *The Scottish Himalaya Expedition* (1951), cited in Meredith Lee, "Goethe Quotation Identified," *Goethe News and Notes* 19:1 (spring 1998), available at www.humanities.uci.edu/tpsaine/Commitment.htm. The author is indebted to John Bennett, former mayor of Aspen, Colorado, for sharing one of the many paraphrasings of Goethe's poetry. Anuradha Oza, a master's student at the Kennedy School of Government, was kind enough to point out the connection to W. H. Murray.

11. The word *magic* has appeared several times in this book. In Chapter 2, I quote George Gilder as citing "the magic of the inter-connections." In the same chapter, and in this essay, I write about the "seemingly magical power of the Internet." *Webster's Revised Unabridged Dictionary*, quoted at www.dictionary.com, offers as one of several definitions of the word: "Producing

effects which seem supernatural or very extraordinary; having extraordinary properties." As explained by Arthur C. Clarke in his "Third Law": "Any sufficiently advanced technology is indistinguishable from magic" (*Profiles of the Future: An Inquiry into the Limits of the Possible* [New York: Bantam, 1958]).

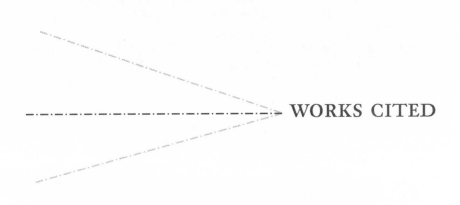

WORKS CITED

1. INTRODUCTION

Ambrose, Stephen E. *Nothing Like It in the World*. New York: Simon and Schuster, 2000.

Ambrose, Stephen E. *Undaunted Courage*. New York: Simon and Schuster, 1996.

Jackson, Donald. *Voyages of the Steamboat Yellow Stone*. Norman: University of Oklahoma Press, 1985.

Sellars, Richard West. *Preserving Nature in the National Parks: A History*. New Haven, Conn.: Yale University Press, 1997.

Uchitelle, Louis. "He Didn't Say It. But He Knew It." *New York Times*, April 30, 2000, Money and Business section.

Winks, Robin. *Laurance S. Rockefeller: Catalyst for Conservation*. Washington, D.C.: Island Press, 1997.

2. NETWORKS AND NATURE IN THE AMERICAN EXPERIENCE

"The 400 Richest People in America, 2001 Edition." *Forbes*, September 2001.

Abbott, Gordon. *Saving Special Places*. Ipswich, Mass.: Ipswich Press, 1993.

Alm, Alvin L. "NEPA: Past, Present, and Future." *EPA Journal* (January/February 1988). Available at www.epa.gov/history/topics/nepa/01.htm.

Altshuler, Alan, and Robert Behn, eds. *Innovation in American Government: Challenges, Opportunities and Dilemmas*. Washington, D.C.: Brookings Institution Press, 1997.

Altshuler, Alan, et al., eds., National Research Council. *Governance and Opportunity in Metropolitan America*. Washington, D.C.: National Academy Press, 1999.

Ambrose, Stephen E. *Nothing Like It in the World*. New York: Simon and Schuster, 2000.

Ambrose, Stephen E. *Undaunted Courage.* New York: Simon and Schuster, 1996.

American Forests. "Timeline of American Forests." Available at www.american-forests.org/about_us/history_timeline.php.

Benfield, F. Kaid, Matthew D. Raimi, and Donald D. T. Chen. *Once There Were Greenfields: How Urban Sprawl Is Undermining America's Environment, Economy and Social Fabric.* New York: Natural Resources Defense Council, 1999.

Benson, Guy Meriwether, William Irwin, and Heather Moore. "Loyal Company Grant, July 12, 1749." Chap. 16 in *Exploring the West from Monticello: A Perspective in Maps from Columbus to Lewis and Clark* (1995). Alderman Library, University of Virginia. Available at www.lib.virginia.edu/exhibits/lewis_clark/ch3-16.html.

Benson, Guy Meriwether, William Irwin, and Heather Moore. "To the Western Ocean: Planning the Lewis and Clark Expedition." Chap. 4 in *Exploring the West from Monticello: A Perspective in Maps from Columbus to Lewis and Clark* (1995). Alderman Library, University of Virginia. Available at www.lib.virginia.edu/exhibits/lewis_clark/ch4.html.

Bryan, C.D.B. *The National Geographic Society: 100 Years of Adventure and Discovery.* New York: Abrams, 1987.

Cairncross, Frances. *The Death of Distance: How the Communications Revolution Will Change Our Lives.* Cambridge: Harvard Business School Press, 1997.

Chandler, Alfred, Jr. *The Visible Hand.* Cambridge: Belknap Press of Harvard University Press, 1977.

Clinton, William J. Remarks on economic growth. December 3, 1999. Available as of October 2000 at www.whitehouse.gov/WH/EOP/OSTP/html/0014_5.html.

Cromie, William. "Roads Scholar Visits Most Remote Spots." *Harvard University Gazette,* June 14, 2001, 1. Available at www.news.harvard.edu/gazette/2001/06.14/01-roadsscholar.html.

Cronin, William. *Nature's Metropolis.* New York: Norton, 1991.

Davis, Judy, for Parsons Brinckerhoff Quade & Douglas, "Consequences of the Development of the Interstate Highway System for Transit." Transportation Research Board, National Research Council, Transit Cooperative Research Program, *Research Results Digest* 21 (August 1997). Available at http://nationalacademies.org/trb/publications/tcrp/tcrp_rrd_21.pdf.

Dertouzos, Michael. *What Will Be: How the New World of Information Will Change Our Lives.* New York: HarperCollins, 1997.

"Discounting Dynamo: Sam Walton." *Time 100* (c. 2000). Available at www.time.com/time/time100/builder/profile/walton.html.

Florida Department of Environmental Protection. External affairs communication. "Governor Bush Announces Full-Funding Commitment to Everglades Restoration." January 18, 2000. Available as of October 2001 at www.dep.state.fl.us/secretary/comm/2000/00-009.htm.

Forman, Richard T. T. "Estimate of the Area Affected Ecologically by the Road System in the United States." *Conservation Biology* 14 (2000): 31–35.

Gilder, George. "Metcalfe's Law and Legacy." *Forbes ASAP* (September 13, 1993). Available at www.gildertech.com/public/telecosm_series/metcalf.html.

Griffiths, R. T. History of the Internet: "Internet for Historians." Leiden, Netherlands: Universiteit Leiden, 1999. Available at www.let.leidenuniv.nl/history/ivh/frame_theorie.html.

Grosvenor, Edwin S., and Morgan Wesson. *Alexander Graham Bell: The Life and Times of the Man Who Invented the Telephone.* New York: Abrams, 1997.

Hardin, Garrett. "The Tragedy of the Commons." *Science* 162 (1968): 1243–48.

Infoplease.com. "Motor Vehicle Fuel Consumption and Travel in the U.S., 1960–1999." Available at www.infoplease.com/ipa/A0004727.html.

Inter-university Consortium for Political and Social Research (ICPSR). "United States Historical Census Data Browser." Available at http://fisher.lib.virginia.edu/census.

Jamieson, Paul. *The Adirondack Reader.* Lake George, N.Y.: Adirondack Mountain Club, 1982.

Jefferson, Thomas. Thomas Jefferson Papers Series 1, General Correspondence, 1651–1827. Library of Congress. Available at http://memory.loc.gov/ammen/mtjh.mtjser1.html.

Jefferson, Thomas. *Thomas Jefferson: Writings.* Ed. Merrill D. Peterson. New York: Library of America, 1984.

Johns, Joshua Scott. "All Aboard: The Role of the Railroads in Protecting, Promoting, and Selling Yellowstone and Yosemite National Parks." Master's thesis, University of Virginia, 1996. Available at http://xroads.virginia.edu/~MA96/RAILROAD/home.html.

Johnson, Kirk. "Hunter Mountain Paintings Spurred Recovery of Land." *New York Times,* June 7, 2001. Available at www.pinchot.org/gt/news/nytimes_huntermtn.htm.

Johnson, Lady Bird, and Laurance Rockefeller. "A New Conservation Century." *Washington Post,* September 14, 2000. Available at www.ilparks.org/ppupdate-cara9-15-00.htm.

Land Trust Alliance. "84% of Referenda Passed: More Than $7.4 Billion Committed to Open Space Protection." (December 1, 2000). Available in October 2001 at www.lta.org/publicpolicy/referenda2000.htm.

Lappin, Todd. "The Airline of the Internet." *Wired,* December 1966. Available at www.wired.com/wired/archive/4.12/ffedex_pr.html.

Levitt, James N. "Innovating on the Land: Conservation on the Working Landscape in American History." In *Private Lands, Public Benefits: A Policy Summit on Working Lands Conservation.* Washington, D.C.: National Governors Association, March 2001.

Lewis, Jack. "The Birth of EPA." *EPA Journal* (November 1985). Available at www.epa.gov/history/topics/epa/15c.htm.

Lewis, Tom. *Divided Highways.* New York: Penguin, 1997.

Limerick, Patricia Nelson. *The Legacy of Conquest.* New York: Norton, 1987.

Ludlow, Fitz Hugh. "Seven Weeks in the Great Yo-Semite." *Atlantic*, June 1864.

Mandel, Michael J. "The Internet Age." *Business Week*, October 4, 1999.

Marsh, George Perkins. *Man and Nature*. 1864. Reprint, Cambridge: Belknap Press of Harvard University Press, 1965.

Meigs, William. *The Life of Thomas Hart Benton*. New York: Lippincott, 1904.

Mokhtarian, Patricia. "Now That Travel Can Be Virtual, Will Congestion Virtually Disappear?" *Scientific American*, October 1997.

Nickel, Douglas R. *Carleton Watkins: The Art of Perception*. San Francisco: San Francisco Museum of Modern Art, 1999.

OECD Environment. "Sectoral Studies: Rethinking Paper Consumption." In *Sustainable Consumption*. Available at www.oecd.org/env/consumption/scp24c.htm.

Pastore, Michael. "The World's Online Populations." *Cyberatlas: The Big Picture*. Available at http://cyberatlas.internet.com/big_picture/geographics/article/0,,5911_151151,00.html.

Pinchot, Gifford. *Breaking New Ground*. 1947. Reprint, Washington, D.C.: Island Press, 1998.

Population Estimates Program, Population Division, U.S. Census Bureau. "Annual Projections of the Total Resident Population as of July 1 (NP-T1)." Available at www.census.gov/population/projections/nation/summary/np-t1.pdf.

Proceedings of a Conference of Governors: In the White House, May 13–15, 1908. Washington, D.C.: U.S. Government Printing Office, 1908.

Public Broadcasting Service (PBS). *The American Experience*. Transcript of the episode "The Telephone." Narrated by David McCullough. Available at www.pbs.org/wgbh/amex/telephone/filmmore/transcript/index.html.

Riis, Jacob. *How the Other Half Lives*. New York: Scribner's, 1890. Available at www.yale.edu/amstud/inforev/riis/title.html.

Roosevelt, Theodore. "Seventh Annual Message to Congress" (December 3, 1907). Available at www.pbs.org/weta/thewest/resources/archives/eight/trconserv.htm.

Roosevelt, Theodore. *Thomas H. Benton*. Cambridge, Mass.: Riverside Press of Houghton Mifflin, 1887.

Runte, Alfred. *Trains of Discovery: Western Railroads and the National Parks*. Boulder, Colo.: Rinehart, 1990.

Schneider, Paul. *The Adirondacks: A History of America's First Wilderness*. New York: Holt, 1997.

Sellars, Richard West. *Preserving Nature in the National Parks: A History*. New Haven, Conn.: Yale University Press, 1997.

Standage, Tom. *The Victorian Internet*. New York: Walker, 1998.

Trust for Public Land. "Case Statement for Tribal Lands Initiative." Available at www.tpl.org/tier3_cdl.cfm?content_item_id=1183&folder_id=217.

Turner, Frederick Jackson. "The Problem of the West." *Atlantic*, September 1896.

Available at www.theatlantic.com/issues/95sep/ets/turn.htm.

U.S. House of Representatives Committee on Public Works. *National Highway Program: Hearings before the Committee on Public Works.* Washington, D.C.: U.S. Government Printing Office, 1955.

U.S. Senate Committee on Public Works. *Federal-Aid Highway Act of 1954: Hearings before a Subcommittee of the Committee on Public Works.* Washington, D.C.: U.S. Government Printing Office, 1954.

van Ravenssway, Charles. *St. Louis: An Informal History of the City and Its People, 1764–1865.* St. Louis: Missouri Historical Society, 1991.

The Wilderness Society. "A Summons to Save the Wilderness." In *The Living Wilderness* (September 1, 1935). Cited in Paul Shiver Sutter, "Driven Wild: The Intellectual and Cultural Origins of Wilderness." Ph.D. thesis, University of Kansas, 1997. Available as UMI microform 9827540 from University of Michigan, Ann Arbor.

Williss, G. Frank. *Do Things Right the First Time: The National Park Service and the Alaska National Interest Lands Conservation Act of 1980.* National Park Service, September 1985. Available at www.cr.nps.gov/history/online_books/williss/adhi.htm.

Wills, Gary, et al. *Thomas Jefferson: Genius of Liberty.* Washington, D.C.: Viking Studio with Library of Congress, 2000.

Winks, Robin. *Frederick Billings.* New York: Oxford University Press, 1991.

Winks, Robin. *Laurance S. Rockefeller: Catalyst for Conservation.* Washington, D.C.: Island Press, 1997.

World Resources Institute. "No End to Paperwork." *Global Trends.* Last modified September 18, 2000. Available at www.wri.org/trends/paperwk.html.

3. THE INTERNET, NEW URBAN PATTERNS, AND CONSERVATION

Calthorpe, Peter, and William Fulton. *The Regional City: Planning for the End of Sprawl.* Washington, D.C.: Island Press, 2001.

Duany, Andreas, Elizabeth Plater-Zyberg, and Jeff Speck. *Suburban Nation: The Rise of Sprawl and the Decline of the American Dream.* San Francisco: North Point, 2000.

Graham, Stephen, and Simon Marvin. *Splintering Urbanism: Network Infrastructures, Technological Mobilities and the Urban Condition.* New York: Routledge, 2001.

Horan, Thomas A. *Digital Places: Building Our City of Bits.* Washington, D.C.: Urban Land Institute, 2000.

Institute for Civil Infrastructure Systems (ICIS). "Bringing Information Technology to Infrastructure." Workshop materials, New York University Institute for Civil Infrastructure Systems, July 2001.

Kotkin, Joel. *The New Geography: How the Digital Revolution Is Reshaping the*

American Landscape. New York: Random House, 2000.

Mitchell, William J. *City of Bits: Space, Place, and the Infobahn*. Cambridge: MIT Press, 1995.

Mitchell, William J. *E-topia: "Urban Life, Jim—But Not As We Know It."* Cambridge: MIT Press, 1999.

National Research Council. *Embedded Everywhere: A Research Agenda for Networked Systems of Embedded Computers* (Washington, D.C.: National Academy Press, 2001).

Nilles, Jack M. *Managing Telework: Strategies for Managing the Virtual Workforce*. New York: Wiley, 1998.

Silberman, Steve. "The Intelligent Web." *Wired*, July 2001, 114–27.

4. THE RURAL REBOUND OF THE 1990S AND BEYOND

Beale, Calvin L., and Kenneth M. Johnson. "The Identification of Recreational Counties in Nonmetropolitan Areas of the USA." *Population Research and Policy Review* 17 (1998): 37–53.

Brown, David L., et al. "Continuities in Size of Place Preference in the United States, 1972–1992." *Rural Sociology* 62 (1997): 408–28.

Cook, Peggy J., and Karen L. Mizer. *The Revised ERS County Typology: An Overview* (RDRR-89). Washington, D.C.: Economic Research Service, U.S. Department of Agriculture, 1994.

Frey, William H., and Kenneth M. Johnson. "Concentrated Immigration, Restructuring, and the Selective Deconcentration of the U.S. Population." In Paul J. Boyle and Keith H. Halfacree, eds., *Migration into Rural Areas: Theories and Issues*. London: Wiley, 1998.

Fuguitt, Glenn V., and Tim B. Heaton. "The Impact of Migration on the Nonmetropolitan Population Age Structure, 1960–1990." *Population Research and Policy Review* 14 (1995): 215–32.

Fuguitt, Glenn V., et al. "Elderly Population Change in Nonmetropolitan Areas: From the Turnaround to the Rebound." Paper presented at the annual meeting of the Western Regional Science Association, Monterey, Calif., 1998.

Johnson, Kenneth M. "Recent Population Redistribution Trends in Nonmetropolitan America." *Rural Sociology* 54 (1989): 301–26.

Johnson, Kenneth M., and Calvin L. Beale. "The Recent Revival of Widespread Population Growth in Nonmetropolitan Areas of the United States." *Rural Sociology* 59 (1994): 655–67.

Johnson, Kenneth M., and Glenn V. Fuguitt. "Continuity and Change in Rural Migration Patterns, 1950–1995." *Rural Sociology* 65:1 (2000): 27–49.

Johnson, Kenneth M., et al. "Local Government Fiscal Burden in Nonmetropolitan America." *Rural Sociology* 60 (1995): 381–98.

Klier, Thomas H., and Kenneth M. Johnson. "Effect of Auto Plant Openings on Net Migration in the Auto Corridor, 1980–1997." *Economic Perspectives* 24:4 (2000): 14–29.

McGranahan, David A. "Natural Amenities Drive Rural Population Change." *Agricultural Economic Report*, no. 781. Washington, D.C.: Economic Research Service, U.S. Department of Agriculture, 1999.

Radeloff, Volker C., et al. "Human Demographic Trends and Landscape Level Forest Management in the Northwest Wisconsin Pine Barrens." *Forest Science* 47:2 (2001): 229–41.

Schachter, Jason. "Geographic Mobility: 1999–2000." *Current Population Reports* (P20-538). Washington, D.C.: U.S. Census Bureau, U.S. Department of Commerce, 2001.

Stewart, Sue I., and Daniel J. Stynes. "Toward a Dynamic Model of Complex Tourism Choices: The Seasonal Home Location Decision." *Journal of Travel and Tourism Marketing* 3:3 (1994): 69–88.

Wear, David N., and P. Bolstad. "Land-Use and Change in Southern Appalachian Landscapes: Spatial Analysis and Forecast Evaluation." *Ecosystems* 1 (1998): 575–94.

Wear, David N., M. G. Turner, and R. J. Naiman. "Land Cover along an Urban-Rural Gradient: Implications for Water Quality." *Ecological Applications* 8:3 (1998): 619–30.

5. FARMLAND IN THE AGE OF THE INTERNET

American Farmland Trust. Organizational information available at www.farmland.org.

California Department of Conservation, Division of Land Resource Protection. "California Farmland Conservancy Program." Updated October 11, 2001. Available at www.consrv.ca.gov/dlrp/cfcp/index.htm.

Commonwealth of Pennsylvania, Department of Agriculture. "Pennsylvania Continues to Lead Nation in Farmland Preservation." Available at http://sites.state.pa.us/PA_Exec/Agriculture/bureaus/farmland_protection/press_release.htm.

Harford County Government. "Agricultural Land Preservation Program." Available at www.co.ha.md.us/ag_preservation.html.

Land Trust Alliance. Organizational information available at www.lta.org.

Marin Agricultural Land Trust. Organizational information available at www.malt.org/preserve/threat.html.

Missouri Precision Agriculture Farming Center. Organizational information available at www.fse.missouri.edu/mpac/index.htm.

Monterey County, California. "Monterey County Twenty-First Century General Plan Update." Available at www.co.monterey.ca.us/gpu/.

New York City Department of Environmental Protection, Bureau of Water Supply. Land Acquisition and Stewardship Program. "Watershed Conservation Easements." Available at www.ci.nyc.ny.us/html/dep/html/news/easement.html.

Pitkin County Government. "Building Permits." Available at www.pitkingov.com/sitepages/pid7.php.

U.S. Department of Agriculture, National Agricultural Statistics Service. "Farm Computer Usage and Ownership" (July 30, 2001). Available at http://usda.mannlib.cornell.edu/reports/nassr/other/computer/empc0701.txt.

U.S. Department of Agriculture, Natural Resources Conservation Service. "1997 National Resources Inventory, National Results" (January 9, 2001). Available at www.nrcs.usda.gov/technical/NRI/1997/national_results.html.

U.S. Department of Agriculture, Natural Resources Conservation Service. "1997 National Resources Inventory, Summary Report" (January 9, 2001). Available at www.nrcs.usda.gov/technical/NRI/1997/summary_report/body.html.

West Bay Real Estate. "Marin County Cities." Available at http://westbayre.com/marin.htm.

6. INTERNET USE IN A HIGH-GROWTH, AMENITY-RICH REGION

Beyers, William. "Trends in Producer Services Growth in the Rural Heartland." In *Economic Forces Shaping the Rural Heartland* (Kansas City, Mo.: Federal Reserve Bank of Kansas City, April 1996).

Beyers, William, and David P. Lindahl. "Lone Eagles and High Fliers in the Rural Producer Services." *Rural Development Perspectives* 11:3 (1996): 2–10.

Bureau of Economic Analysis, *Regional Economic Information System 1969–1996* (CD-ROM) (Washington, D.C.: U.S. Department of Commerce, 1999).

Charny, Ben. "More U.S. Households Online Than Not." *ZDNet News*, August 17, 2000. Available at www.zdnet.com/zdnn/stories/news/0,4586,2616761.html.

Davis, Judy, for Parsons Brinckerhoff Quade & Douglas, "Consequences of the Development of the Interstate Highway System for Transit." Transportation Research Board, National Research Council, Transit Cooperative Research Program. *Research Results Digest* 21 (August 1997): 3. Available at http://nationalacademies.org/trb/publications/tcrp/tcrp_rrd_21.pdf.

Dean Runyan Associates (DRA). *Oregon Travel Impacts, 1991–2001* (2002).

Deschutes Basin Land Trust. Organizational information available at www.deschuteslandtrust.org.

Deschutes County, Oregon. "Draft Deschutes County Final RPS Report" (June 2, 1999).

Deschutes Resources Conservancy. Organizational information available at www.deschutesrc.org/.

Elstrom, Peter. "Craig McCaw: The Prophet of Telecom." *Business Week,* September 29, 1998.

Johnson, Kenneth M., and Calvin L. Beale. "Recreational Counties in Non-metropolitan America" (revised May 14, 2002). Available at www.luc.edu/depts/sociology/johnson/p99webr.html.

Johnson, Kenneth M., and Calvin L. Beale. "The Rural Rebound." *Wilson Quarterly* 12 (1998): 16–27.

Kenworthy, Tom. "Peaks, Valleys Define Today's West." *USA Today,* May 18, 2001, 3A.

McGranahan, David A., Food and Rural Economics Division, Economic Research Service, U.S. Department of Agriculture. "Natural Amenities Drive Rural Population Change." *Agricultural Economic Report,* no. 781 (1999).

Murata, Toshihiko, and Patricia Gwartney. "1999 Pilot Survey of Oregon Telecommunications." In *Oregon Economic and Community Development Department's Oregon Household Telecommunications Survey, Methodology and Results* (winter 2000), available at http://darkwing.uoregon.edu/~osrl/telecomoedd/frmtelecom.htm.

Niles, John S. *Beyond Telecommuting: A New Paradigm for the Effect of Telecommunications on Travel.* Washington, D.C.: U.S. Department of Energy, Office of Energy Research, September 1994, 1–2. Available at www.lbl.gov/ICSD/Niles.

Oregon Birders Online. Organizational information available at www.cyberdyne.com/~lucyb/obol.html.

Oregon Water Resources Department. "The Deschutes Ground Water Work Group" (December 13, 1999). Available at http://powder.wrd.state.or.us/programs/deschutes/overview.shtml.

"PGE Donation Protects Essential Steelhead Habitat in Deschutes Basin." Available at www.portlandgeneral.com/about_pge/corporate_info/news/archives/pge_donation_protects_steelhead_habitat_deschutes_basin.asp.

Phillips, Spencer. *Windfalls for Wilderness: Land Protection and Land Value in the Green Mountains.* Washington, D.C.: Wilderness Society, 2000. Available at www.wilderness.org/newsroom/pdf/windfallsforwilderness.pdf.

Rasker, Ray, and Ben Alexander. *The New Challenge: People, Commerce, and the Environment in the Yellowstone to Yukon Region.* Washington, D.C.: Wilderness Society, October 1998. Available at www.wilderness.org/ccc/northrockies/y2y_report.pdf.

Salant, Priscilla, Lisa Carley, and Don Dillman. "Lone Eagles among Washington's In-Migrants: Who Are They and Are They Moving to Rural Places?" *Northwest Journal of Business and Economics* (Bellingham: Wash.: Center for Economic and Business Research, Western Washington University, 1997). Available as of October 2001 at www.ac.wwu.edu/~cebr/loneagle.pdf.

"Testimony in Support of H.R. 1787 by Ron Nelson, Chaiman, Deschutes Basin

Resources Conservancy, April 6, 2000." Available at www.house.gov/resources/ 106cong/water/00apr06/nelson.htm.

Texas A&M Real Estate Center. "Crook County, OR: Population and Components of Change." Available at http://recenter.tamu.edu/Data/popc/pc41013.htm.

Texas A&M Real Estate Center. "Deschutes County, OR: Population and Components of Change." Available at http://recenter.tamu.edu/Data/popc/pc41017.htm.

U.S. Department of Agriculture, Economic Research Service. "Natural Amenities and Rural Growth." Available at www.ers.usda.gov/topics/view.asp?T=104024.

U.S. Department of Commerce, National Telecommunications and Information Administration. *Falling through the Net: Defining the Digital Divide.* Revised November 1999. Available at www.ntia.doc.gov/ntiahome/digitaldivide/index.html.

Wild Birds Unlimited of Bend. Organizational information available at www.wbu.com/bend/.

Williams, Gerald, and Stephen Mark. *Establishing and Defending the Cascade Range Forest Reserve from 1885 to 1912: As Found in Letters, Newspapers, Magazines, and Official Reports.* Portland, Ore.: U.S. Forest Service, Pacific Northwest Region, and U.S. Park Service, Crater Lake National Park, September 1995. Available at http://fs.jorge.com/archives/Cascade_Range/Introduction.html.

Yuskavitch, Jim, and Leslie D. Cole. *The Insider's Guide to Bend and Central Oregon.* Helena, Mont.: Falcon, 1999.

7. RURAL DEVELOPMENT AND BIODIVERSITY

Clawson, Mark R. "An Investigation of Factors That May Affect Nest Success in CRP Lands and Other Grassland Habitats in an Agricultural Landscape." Master's thesis, Montana State University, 1996.

Coughenour, Michael B., and Francis J. Singer. "Elk Population Processes in the Yellowstone National Park under the Policy of Natural Regulation." *Ecological Applications* 6:2 (1996): 573–93.

Goodman, Daniel. "Viability Analysis of the Antelope Population Wintering near Gardiner." National Park Service report, Yellowstone National Park, Wyo., 1996.

Hansen, Andrew J., and Jay J. Rotella. "Biophysical Factors, Land Use, and Species Viability in and around Nature Reserves. *Conservation Biology* (2002).

Hansen, Andrew J., Jay J. Rotella, and Matthew L. Kraska. "Dynamic Habitat and Population Analysis: A Filtering Approach to Resolve the Biodiversity Manager's Dilemma." *Ecological Applications* 9:4 (1999): 1459–76.

Hansen, Andrew J., et al. "Ecology and Socioeconomics in the New West: A Case Study from Greater Yellowstone." *BioScience* 52:2 (2002): 151–68.

Hansen, Andrew J., et al. "Spatial Patterns of Primary Productivity in the Greater Yellowstone Ecosystem." *Landscape Ecology* 15 (2000): 505–22.

Marzluff, John M., F. R. Gehlbach, and David A. Manuwal. "Urban Environments: Influences on Avifauna and Challenges for the Avian Conservationist." In John M. Marzluff and Rex Sallabanks, *Avian Conservation: Research and Management.* Washington, D.C.: Island Press, 1998.

Oechsli, Lauren M. "Ex-urban Development in the Rocky Mountain West: Consequences for Native Vegetation, Wildlife Diversity, and Land-Use Planning in Big Sky, Montana." Master's thesis, Montana State University, 2000.

Powell, Scott, et al. "Rural Residences and Biodiversity: How Where We Live Influences Native Species." *Conservation Biology* (in review).

Power, Thomas M. "Ecosystem Preservation and the Economy of the Greater Yellowstone Area." *Conservation Biology* 5:3 (1991): 395–404.

Rasker, Ray, and Andrew J. Hansen. "Natural Amenities and Population Growth in the Greater Yellowstone Region." *Human Ecology Review* 7:2 (2000): 30–40.

Rotella, Jay J., Mark L. Taper, and Andrew J. Hansen. "Correcting Nesting-Success Estimates for Observer Effects: Maximum-Likelihood Estimates of Daily Survival Rates with Reduced Bias." *Auk* 11 (2000): 92–109.

Schullery, Paul. *Searching for Yellowstone: Ecology and Wonder in the Last Wilderness.* Boston: Houghton Mifflin, 1997.

Scott, J. Michael, et al. "Nature Reserves: Do They Capture the Full Range of America's Biological Diversity?" *Ecological Applications* 11:4 (2001): 999–1007.

Shepard, Brad, et al. "Status and Risk of Extinction for Westslope Cutthroat Trout in the Upper Missouri River Basin, Montana. *North American Journal of Fisheries Management* 17 (1997): 1158–72.

Skagen, Susan K., E. Muths, and R. D. Amans. *Towards Assessing the Effects of Bank Stabilization Activities on Wildlife Communities on the Upper Yellowstone River, U.S.A.* Fort Collins, Colo.: U.S. Geological Survey, Midcontinent Ecological Science Center, 2001.

Varley, John D., and Paul D. Schullery. *Yellowstone Fishes: Ecology, History, and Angling in the Park.* New York: Stackpole, 1998.

Yellowstone National Park. *Yellowstone's Northern Range: Complexity and Change in a Wildland Ecosystem.* Mammoth Hot Springs, Wyo.: National Park Service, 1997.

8. THE GREEN INTERNET

Advisory Planning Board of the National Biodiversity Information Center. *Strategic Planning Document—Environment and Natural Resources.* Washington, D.C.: National Science and Technology Council's Committee on Environment and Natural Resources, 1994. Available in July 2000 at www.whitehouse.gov/White_House/EOP/OSTP/NSTC/html/enr/enr-3a.html.

Allen, Craig D., and David D. Breshears. "Drought-Induced Shift of a Forest-

Woodland Ecotone: Rapid Landscape Response to Climate Variation." *Proceedings of the National Academy of Sciences USA* 95 (1998): 14839–42.

Black, Craig C., et al. *Loss of Biological Diversity: A Global Crisis Requiring International Solutions.* Washington, D.C.: National Science Foundation, National Science Board, 1989.

Bloch, Erich, et al. *Impact of Emerging Technologies on the Biological Sciences.* Arlington, Va.: National Science Foundation, 1995.

Carlton, James T. "Pattern, Process, and Prediction in Marine Invasion Ecology." *Biological Conservation* 78 (1996): 97–106.

Dobson, Andrew, Alison Jolly, and Daniel I. Rubenstein. "The Greenhouse Effect and Biological Diversity." *Trends in Ecology and Evolution* 4 (1989): 64–68.

Edwards, James L., et al. "Interoperability of Biodiversity Databases: Biodiversity Information on Every Desktop." *Science* 289 (2000): 2312–14.

Environment Australia. *The Darwin Declaration.* Canberra: Australian Biological Resources Study, Department of the Environment, Environment Australia, 1998.

Godown, Mandaline E., and A. Townsend Peterson. "Preliminary Distributional Analysis of U.S. Endangered Bird Species." *Biodiversity and Conservation* 9 (2000): 1313–22.

Holt, Robert D. "The Microevolutionary Consequences of Climate Change." *Trends in Ecology and Evolution* 5 (1990): 311–15.

Inouye, David W., et al. "Climate Change Is Affecting Altitudinal Migrants and Hibernating Species." *Proceedings of the National Academy of Sciences USA* 97 (2000): 1630–33.

Joseph, Leo. "Preliminary Climatic Overview of Migration Patterns in South American Austral Migrant Passerines." *Ecotropica* 2 (1996): 185–93.

Joseph, Leo. "Towards a Broader View of Neotropical Migrants: Consequences of a Re-examination of Austral Migration." *Ornitologia Neotropical* 8 (1997): 31–37.

Kaiser, Jocelyn. "Searching Museums from Your Desktop." *Science* 284 (May 7, 1999): 888.

Karl, Thomas R., et al. "Indices of Climate Change for the United States." *Bulletin of the American Meteorological Society* 77 (1996): 279–92.

Krishtalka, Leonard, and Philip H. Humphrey. "Fiddling While the Planet Burns." *Museum News* 77:2 (2000): 29–35.

O'Brien, Karen, and Diana Liverman. "Climate Change and Variability in Mexico." In Jesse C. Ribot, Antonio Rocha Magalhaes, and Stahis S. Panagides, eds. *Climate Variability, Climate Change and Social Vulnerability in the Semi-Arid Tropics.* Cambridge: Cambridge University Press, 1996.

Organisation for Economic Co-Operation and Development (OECD). *Final Report of the OECD Megascience Forum Working Group on Biological Informatics.* Paris: OECD Publications, 1999. Available at www.oecd.org/ehs/icgb/biodiv8.htm.

Parmesan, Camille. "Climate and Species' Range." *Nature* 382 (1996): 765–66.

Pennisi, Elizabeth. "Taxonomic Revival." *Science* 289 (2000): 2306–8.

Peters, Robert L., and Joan D. S. Darling. "The Greenhouse Effect and Nature Reserves." *BioScience* 35 (1985): 707–17.

Peters, Robert L., and John Peterson Myers. "Preserving Biodiversity in a Changing Climate." *Issues in Science and Technology 1991–1992*, 8(2): 66–72.

Peterson, A. Townsend. "Predicting Species' Geograpic Distributions Based on Ecological Niche Modeling." *Condor* 103:3 (2001): 599–605.

Peterson, A. Townsend, L. G. Ball, and K. M. Brady. "Distribution of Birds of the Philippines: Biogeography and Conservation Priorities." *Bird Conservation International* 10 (2000): 149–67.

Peterson, A. Townsend, L. G. Ball, and K. C. Cohoon. "Predicting Distributions of Mexican Birds Using Ecological Niche Modelling Methods." *Ibis* 144:1 (2002): E27–E32 (published online only).

Peterson, A. Townsend, and David A. Vieglais. "Predicting Species Invasions Using Ecological Niche Modeling." *BioScience* 51 (2001): 363–71.

Peterson, A. Townsend, Adolfo G. Navarro-Siguenza, and Hesiquio Benitez-Diaz. "The Need for Continued Scientific Collecting: A Geographic Analysis of Mexican Bird Specimens." *Ibis* 140 (1998): 288–94.

Porter, Warren P., Srinivas Budaraju, Warren Stewart, and Navin Ramankutty. "Calculating Climate Effects on Birds and Mammals: Impacts on Biodiversity, Conservation, Population Parameters, and Global Community Structure." *American Zoologist* 40 (2000): 597–630.

President's Council of Advisors on Science and Technology (PCAST). *Teaming with Life: Investing in Science to Understand and Use America's Living Capital.* Washington, D.C.: PCAST Panel on Biodiversity and Ecosystems, 1998.

Raven, Peter H., and Edward O. Wilson. "A Fifty-Year Plan for Biodiversity Surveys." *Science* 258 (1992): 1099–1100.

Sanchez-Cordero, Victor, and Enrique Martinez-Meyer. "Museum Specimen Data Predict Crop Damage by Tropical Rodents." *Proceedings of the National Academy of Sciences USA* 97 (2000): 7074–77.

Systematics Agenda 2000. New York: Charting the Biosphere, 1994.

Wilson, Edward O. "Integrated Science and the Coming Century of the Environment." *Science* 279 (1998): 2047–48.

9. BIRDSOURCE: USING BIRDS, CITIZEN SCIENCE, AND THE INTERNET AS TOOLS FOR GLOBAL MONITORING

Barrow, Mark V., Jr. *A Passion for Birds: American Ornithology after Audubon.* Princeton, N.J.: Princeton University Press, 1998.

Bildstein, Keith L. "Long-Term Counts of Migrating Raptors: A Role for Volunteers in Wildlife Research." *Journal of Wildlife Management* 62 (1998): 435–45.

BirdSource (sponsored by the Cornell Lab of Ornithology and the National Audubon Society). Organizational information available at www.birdsource.org.

Bonney, Rick, and André A. Dhondt. "Project FeederWatch." In Karen C. Cohen, ed., *Internet Links to Science Education: Student Scientist Partnerships*. New York: Plenum, 1997.

Carson, Rachel. *Silent Spring*. Boston: Houghton Mifflin, 1962.

Cordell, H. Ken, ed. *National Survey on Recreation and the Environment*. Washington, D.C.: U.S. Forest Service, 1997.

Hochachka, Wesley M., and André A. Dhondt. "Density-Dependent Decline of Host Abundance Resulting from a New Infectious Disease." *Proceedings of the National Academy of Sciences USA* 97 (2000): 5303–06.

Hochachka, Wesley M., et al. "Irruptive Migration of Common Redpolls." *Condor* 101 (1999): 195–204.

James, Francis C., C. E. McCulloch, and David A. Wiedenfeld. "New Approaches to the Analysis of Population Trends in Land Birds." *Ecology* 77 (1996): 13–27.

Kennedy, E. Dale, and Douglas W. White. "Bewick's Wren." In Alan Poole and Frank Gill, eds., *The Birds of North America*, no. 315. Washington, D.C.: American Ornithologists' Union, and Philadelphia: Academy of Natural Science, 1997.

Lewenstein, Bruce V., Rick Bonney, and Dominique Brossard. "Citizen Science: Measuring Scientific Knowledge and Attitudes in a New Type of Public Outreach Project." Texas A&M University, Center for Science and Technology Policy and Ethics, Discussion Paper Series #98-1, College Station, Tex., 1998.

Lewenstein, Bruce V., Rick Bonney, and Dominique Brossard. "Measuring Scientific Knowledge and Attitudes in Citizen Science." In Jon Miller, ed., *Perceptions of Biotechnology: Public Understanding and Attitudes*. Cresskill, N.J.: Hampton, 2002 (in press).

National Audubon Society. "CBC History." Available at www.birdsource.org/cbc/hist.htm.

National Audubon Society. "What Is the Status of the IBA Program?" Available at www.audubon.org/bird/iba/prog_status.html. See also various linked Web pages available at www.audubon.org/bird/iba/index.html.

National Research Council. *Biological Basis for the Endangered Species Act*. Washington, D.C.: National Academy Press, 1995.

The Nature Conservancy. "East Maui Watershed Partnership Wins Achievement Award." Press release (n.d.). Available at http://nature.org/wherewework/northamerica/states/hawaii/preserves/art2363.html.

Peterjohn, Bruce G., and John R. Sauer. "Population Status of North American Grassland Birds from the North American Breeding Bird Survey, 1966–1996." *Studies in Avian Biology* 19 (1999): 27–44.

Rosenberg, Kenneth V., James D. Lowe, and André A. Dhondt. "Effects of Forest

Fragmentation on Breeding Tanagers: A Continental Perspective." *Conservation Biology* 13 (1999): 568–83.

Rosenberg, Kenneth V., et al. *A Land Manager's Guide to Improving Habitat for Scarlet Tanagers and Other Forest-Interior Birds*. Ithaca, N.Y.: Cornell Lab of Ornithology, 1999.

Sherony, Dominic F., Brett M. Ewald, and Steve Kelling. "Inland Fall Migration of Red-Throated Loons." *Journal of Field Ornithology* 71 (2000): 310–20.

Temple, Stanley A., John R. Cary, and Robert E. Rolley. *Wisconsin Birds: A Seasonal and Geographical Guide*. Madison: University of Wisconsin Press, 1997.

Trumbull, Deborah J. *The New Science Teacher: Cultivation of Good Practice*. New York: Teachers College Press, 1999.

Trumbull, Deborah J., et al. "Thinking Scientifically during Participation in a Citizen-Science Project." *Science Education* 84 (2000): 265–75.

Vickery, Peter D., and James R. Herkert, eds. "Ecology and Conservation of Grassland Birds of the Western Hemisphere." *Studies in Avian Biology* 19 (1999).

Wells, Allison C. "Lab Awarded $2.25 Million for Citizen Science Online." *Birdscope: News and Views from Sapsucker Woods* (newsletter of the Cornell Lab of Ornithology) (Autumn 2001): 4.

Wells, Jeffrey V., et al. "Feeder Counts as Indicators of Spatial and Temporal Variation in Winter Abundance of Resident Birds." *Journal of Field Ornithology* 69 (1998): 577–86.

10. CONSERVATION ADVOCACY AND THE INTERNET

Aridjis, Homero. "El Silencio de las Ballenas" (The Silence of the Whales). *La Reforma* (Mexico), February 4, 1995.

Bremer, Juan Ignacio. "To All the Children of Mexico." Open letter published in *La Reforma* (Mexico), January 8, 2000, and in *El Universal* (Mexico), January 9, 2000.

Chacon, Richard. "Whales Sink Plans for Mexican Salt Plant." *Boston Globe*, April 2, 2000.

Dedina, Serge. *Saving the Gray Whale: People, Politics, and Conservation in Baja California*. Tucson: University of Arizona Press, 2000.

Dedina, Serge, and Emily Young. *Conservation and Development in the Gray Whale Lagoons of Baja California Sur, Mexico*. Final report to the U.S. Marine Mammal Commission, October 1995.

"El 007 Viene a Mexico." *La Reforma* (Mexico), March 12, 1997.

Friedman, Thomas W. "Responsive Governance in the 21st Century." *The Mansfield American-Pacific Lectures* at International House of Japan, Tokyo, September 21, 2000. Washington, D.C., Missoula, Mont., and Tokyo: Mansfield Center for Pacific Affairs, 2000. Available at www.mcpa.org/pro-grams/friedmancomplete.pdf.

González Torres, Jorge. "Ganaron las Ballenas" (The Whales Won). *El Universal* (Mexico), March 4, 2000.

Hardman, Chris. "Saving the Song of Sea Giants." *Américas* (April 2001): 38.

Hesse, Stephen. "Can Environmental Dialogue Prevent a Whale of a Disaster?" *Japan Times*, December 13, 1998.

"La UNESCO Analizará los Daños que Causan las Salineras a las Ballenas de México" (UNESCO Will Analyze the Damage Caused by Saltworks to the Whales of Mexico). *La Crónica*, August 23, 1999.

"Letter from the Chair of the World Heritage Committee to the President of Mexico concerning El Vizcaino." Ref: WHC/74/Mexico.nat/MR, March 7, 2000. Available at www.unesco.org/whc/nwhc/pages/news/main2.htm.

O'Donnell, Paul, Seth Stevenson, and Victoria S. Stefanakos. "A Whale of a Whale Problem." *Newsweek*, November 29, 1999.

"Sancionan Empresa $206 Mil 700" (Company Is Fined $206,700 [Pesos]). *La Reforma* (Mexico), December 8, 1999.

Secretaria de Medio Ambiente Recursos Naturales y Pesca (SEMARNAP). *San Ignacio Saltworks: Salt and Whales in Baja California.* SEMARNAP Working Papers, 1997.

Shea, Michael. "Beating Mitsubishi." *Campaigns and Elections Magazine.* (July 2000): 44.

Smith, James. "Activists Break New Ground to Help Shake Off Saltworks Project." *Los Angeles Times*, April 23, 2000, sec. A1, p. 10.

Spalding, Mark J. *Laguna San Ignacio: A Briefing Book.* Prepared for the Natural Resources Defense Council and the International Fund for Animal Welfare (January 1999).

United Nations Educational, Scientific and Cultural Organization (UNESCO). *Convention Concerning the Protection of the World Cultural and Natural Heritage.* Paris: UNESCO, 1972.

United Nations Educational, Scientific and Cultural Organization (UNESCO). *Report of the Mission to the Whale Sanctuary of El Vizcaino, Mexico, 23–28 August 1999.* Report presented to the Bureau of the World Heritage Committee, 23rd Extraordinary Session, Marrakesh, Morocco, November 26–27, 1999. Available at www.unesco.org/whc/archive/99-209-inf20.pdf.

I I. ENVISIONING RURAL FUTURES

Batty, Michael, et al. "Visualizing the City: Communicating Urban Design to Planners and Decision-Makers." Centre for Advanced Spatial Analysis Working Paper Series. London: University College of London, October 2000.

Bernard, Robert N. "Using Adaptive Agent-Based Simulation Models to Assist Planners in Policy Development: The Case of Rent Control." Paper presented at the annual conference of the Association of Collegiate Schools of Planning, Pasadena, Calif., November 5–8, 1998.

Forester, John. *The Deliberative Practitioner: Encouraging Participatory Planning Processes.* Cambridge: MIT Press, 1999.

Harris, Britton. "Computing in Planning: Professional and Institutional Requirements." *Environment and Planning B: Planning and Design* 26 (1999): 321–31.

Murdoch, Iris. *The Sovereignty of Good.* New York: Schocken, 1971.

Orton Family Foundation. "About the Foundation." Available at www.ortonfamilyfoundation.org/about.htm.

Orton, Lyman. "Bringing Citizens Back into Defining Their Town's Future." In *Vermont Country Store Catalog* (fall 2001). Available at www.vermontcountrystore.com/corp/editorials/bringing_citizens_back.asp.

Shenton, Edward. "The Happy Storekeeper of the Green Mountains." *Saturday Evening Post*, March 15, 1952. Available at www.vermontcountrystore.com/Corp/artHappy_Storekeeper.asp.

The Trust for Public Land. "Santa Fe, New Mexico—Wildlife, Mountains, Trails, and Historic Places Program." *Trust for Public Land Research Room* (June 24, 2001). Available at www.tpl.org/tier3_cdl.cfm?content_item_id=4527&folder_id=1365.

12. THE WATERSHED APPROACH, BIODIVERSITY, AND COMMUNITY PRESERVATION

Arendt, Randall. *Conservation Design for Subdivisions: A Practical Guide to Creating Open Space Networks.* Washington, D.C.: Island Press, 1996.

Born, Stephen M., and Kenneth D. Genskow. *Exploring the Watershed Approach: Critical Dimensions of State-Local Partnerships.* Final report of the Four Corners Watershed Innovators Initiative. Madison: University of Wisconsin Extension, 1999.

Hankin, Alan Lee, and the Secretary's Advisory Committee on Environmental Education. "Massachusetts Environmental Education Plan: Education to Protect, Restore, and Preserve Our Commonwealth." 2000.

Jewell, Paula M., et al. *The Massachusetts Watershed Initiative: Opportunities and Challenges in Reshaping Government* (Boston: Massachusetts Executive Office of Environmental Affairs, 1996).

Keystone Center. "Final Consensus Report of the Keystone Policy Dialogue on Biological Diversity on Federal Lands." Keystone, Colo.: The Keystone Center, 1991.

Lieberman, Gerald A., and Linda L. Hoody. *Closing the Achievement Gap: Using the Environment as an Integrating Context for Learning.* Poway, Calif.: Science Wizards, 1998.

Manik, Roy, and Lee A. Dillard. "Toxics Use Reduction in Massachusetts: The Blackstone Project." *Journal of Air and Waste Management Association* 40 (October 1990): 1368–71.

Massachusetts Department of Education. "Science and Technology/Engineering Curriculum Framework." 2001.

Massachusetts Department of Environmental Protection. "DEP's Proposed Total Maximum Daily Loads (TMDLs) Strategy to Improve the Water Quality of Massachusetts Rivers and Lakes." 1998.

Massachusetts Department of Environmental Protection, Division of Watershed Management. "Summary of Water Quality 2000." 2000.

Massachusetts Executive Office of Environmental Affairs (EOEA). "Community Preservation Vision." 2001.

Massachusetts Executive Office of Environmental Affairs (EOEA). *Exploring Biodiversity: A Workbook.* Braintree, Mass.: MacDonald and Evans, 2000.

Massachusetts Executive Office of Environmental Affairs (EOEA). "Planning for Growth: Implementation of Executive Order 385 within Agencies of the Commonwealth." 1996.

Massachusetts Executive Office of Environmental Affairs (EOEA). *The State of Our Environment.* Wellesley, Mass.: Susan Nappi Creative, 2000.

Massachusetts Executive Office of Environmental Affairs (EOEA), Division of Fisheries and Wildlife, Natural Heritage and Endangered Species Program. "BioMap: Guiding Land Conservation for Biodiversity in Massachusetts." 2001.

Massachusetts Executive Office of Environmental Affairs (EOEA), Ten Mile River Watershed Team Action Planning Subcommittee. "Five-Year Watershed Action Plan." 2001.

Massachusetts Executive Office of Environmental Affairs (EOEA), "Watershed Initiative." 2001.

Massachusetts General Laws. Chap. 144 of the Acts of 1997.

Massachusetts Legislature. *Legislative Record* (1989), 106S.

Massachusetts Watershed Initiative Steering Committee. "The Massachusetts Watershed Approach and Its Implementation: Status Report." 1995.

Noss, Reed F., and Allen Y. Cooperrider. *Saving Nature's Legacy.* Washington, D.C.: Island Press, 1994.

Ponting, Clive. *A Green History of the World: The Environment and the Collapse of Great Civilizations.* New York: Penguin, 1991.

Trust for Public Land. *Greenprint Gallery 2000.* San Francisco: Trust for Public Land, 2000. Available at www.tpl.org.

13. NATURAL AMENITIES AND LOCATIONAL
    CHOICE IN THE NEW ECONOMY

Baird, Justus. "New Economy Is Environmentalists' New Friend: Why City Hall and the Business Community Need Environmentalists in the New Economy." *CEC Newsletter* (December 2000). Available at www.cechouston.org/newsletter/nl_12-00/presletter.html.

Chattanooga News Bureau. "Chattanooga: America's Talking." Available at www.chattanooga-chamber.com/newsandinfo.headline.php3?id=23.

City of Austin. "Smart Growth Initiative." Available at www.ci.austin.tx.us/smart-growth/.

Deibel, Mary. "States Raise Stakes in Jobs Hunt." Scripps Howard News Service. Available at www.jrnl.net/news/00/May/jrn106220500.html.

Drabenstott, Mark, and Tim R. Smith. "The Changing Economy of the Rural Heartland." In *Economic Forces Shaping the Rural Heartland*. Kansas City, Mo.: Federal Reserve Bank, 1996.

ECONorthwest. "Quality of Life and Its Impact on Louisiana's Economy." Available at www.riversentinel.net/first.htm.

Engler, John. "Governor's Address" (January 31, 2001). Available at www.nga.org/governors/1,1169,C_SPEECH^D_1203,00.html.

Florida, Richard. "Competing in the Age of Talent: Environment, Amenities, and the New Economy." Report prepared for the R. K. Mellon Foundation, Heinz Endowments, and Sustainable Pittsburgh. Pittsburgh: Carnegie Mellon University, January 2000. Available at www2.heinz.cmu.edu/~florida/talent.pdf.

Florida, Richard. "Place and the New Economy." Champions of Sustainability lecture broadcast on WQED (Pittsburgh), August 27, 2000. Transcript available at www.heinz.cmu.edu/~florida.

Geddes, Pete. "Economy and Ecology in the Next West." *Journal of Forestry* 96: 8 (August 1998). Available at www.free-eco.org/pub/980800pg.html.

Gottlieb, Paul. "Amenities as an Economic Development Tool: Is There Enough Evidence?" *Economic Development Quarterly* (August 1994): 270–85.

Henton, Doug, and Kim Walesh. "Linking the New Economy to the Livable Community." San Francisco: James Irvine Foundation, April 1998. Available at www.irvine.org/frameset3.htm.

Hirschhorn, Joel. *Growing Pains: Quality of Life in the New Economy*. Washington, D.C.: National Governors Association, June 2000.

Hodges, Jim. "Governor's Address" (January 17, 2001). Available at www.nga.org/governors/1,1169,C_SPEECH^D_1113,00.html.

"In the Fast Lane: Delivering More Transportation Choices to Break Gridlock." Washington, D.C.: National Governors Association, November 2000.

IntelliQuest.com. "Environment Is of Greatest Concern to Residents of High Technology Communities" (June 20, 1998). Available at www.intelliquest.com/press/archive/release84.asp.

Johnson, Jerry D., and Ray Rasker. "Local Government: Local Business Climate and Quality of Life." *Montana Policy Review* 3:2 (1993): 11–19.

Johnson, Jerry D., and Ray Rasker. "The Role of Economic and Quality of Life Values in Rural Business Location." *Journal of Rural Studies* 11:4 (1995): 405–16.

Leavitt, Michael O. "Governor's Address" (January 16, 2001). Available at www.nga.org/governors/1,1169,C_SPEECH^D_1075,00.html.

Lerner, Steve, and William Poole. "Open Space Investments Pay Big Returns" (June 23, 1999). Available at www.tpl.org/tier3_cd.cfm?content_item_id=964&folder_id=765.

Marston, Ed. "Beyond the Revolution." *High Country News*, April 10, 2000.

New Mission for Brownfields: Attacking Sprawl by Revitalizing Older Communities. Washington, D.C.: National Governors Association, October 2000.

"The Northwest Faces Critical Choices." *Freeflow: Journal of the Pacific Rivers Council* (winter 1996).

Outdoor Recreation Coalition of America. "Economic Benefits of Outdoor Recreation." *State of the Industry Report, 1997.* Available at http://www.outdoorindustry.org.

Pachetti, Nick, and Alan Mirabella. "The Best Places to Live." *Money* (2000). Available at http://money.cnn.com/best/bplive/.

Pachetti, Nick, and Alan Mirabella. "The Best Places to Live: Best Big City." *Money* (2000). Available at www.money.com/money/depts/real_estate/bplive/portland.html.

Power, Thomas M. "Do Jobs Follow People or Do People Follow Jobs?" KUFM commentary (March 30, 1998). Available in 2001 at www.cas.umt.edu/econ/Power/kufm/1998/033098.htm.

Power, Thomas Michael. "The Economics of Wilderness Preservation in Utah." Testimony before the U.S. House of Representatives Committee on Resources and the Environment, June 29, 1995. Available at www.suwa.org/newsletters/1995/winter/insert1.html.

Rasker, Ray. "Entrepreneurs of the New West." Available in 2001 at www.yellowstonescience.com/views/rasker/entrepreneurs.html.

Rasker, Raymond. "A New Look at Old Vistas: The Economic Role of Environmental Quality in Western Public Lands." *University of Colorado Law Review* 65 (1994): 365–99.

Rauch, Jonathan. "The New Old Economy: Oil, Computers, and the Reinvention of the Earth." *Atlantic*, January 2001. Available at www.theatlantic.com/issues/2001/01/rauch.htm.

Rondinelli, Dennis A., and William J. Burpitt. "Do Government Incentives Attract and Retain International Investment? A Study of Foreign-Owned Firms in North Carolina." Kenan Institute of Private Enterprise, University of North Carolina at Chapel Hill, 1999.

"Urban Forests May Solve Houston's Problems," *The Forestry Source*, February 2001. Available at www.safnet.org/archive/houston201.htm.

Van Slambrouck, Paul. "Lifestyle Drives Today's Workers." *Christian Science Monitor*, September 11, 2000.

14. CONSERVATION PHILANTHROPY AND LEADERSHIP

Bielenberg, Warren. "Biographical Vignettes: Gilbert H. Grosvenor, 1875–1966." In *National Park Service: The First 75 Years.* Available at www.cr.nps.gov/history/online_books/sontag/grosvenor.htm.

Bryan, C.D.B. *The National Geographic Society: 100 Years of Adventure and Discovery*. New York: Abrams, 1987.

Chambers, S. Allen. "Thomas Jefferson Takes His Granddaughters to Natural Bridge: A Trip Attended with Disasters and Accidents." *Lynch's Ferry* 1 (fall 1998): 7.

Chandler, Alfred. *The Visible Hand*. Cambridge: Belknap Press of Harvard University Press, 1977.

Christen, Kris. "Biodiversity at the Crossroads." *Environmental Science and Technology* (March 1, 2000): 123A–28A.

Conservation International. "Conservation International Unveils Solution to Prevent Global Species Extinctions." Press release, December 9, 2001. Available at www.conservation.org/xp/CIWEB/newsroom/press_releases/2001/120901.xml.

Ford, Bill. Speech to the Coalition for Environmentally Responsible Economies (CERES) 2000 conference, April 14, 2000. Available at www.ford.com/en/our-company/environmentalinitiatives/environmentalactions/billfordspeaksatthe-ceres2000conference.htm.

George Perkins Marsh Online Research Center. Organizational information available at http://bailey.uvm.edu/specialcollections/gpmorc.html.

Grosvenor, Edwin S., and Morgan Wesson. *Alexander Graham Bell: The Life and Times of the Man Who Invented the Telephone*. New York: Abrams, 1997.

Jackson, Donald. *Thomas Jefferson and the Stony Mountains: Exploring the West from Monticello*. Champaign: University of Illinois Press, 1981.

Jefferson, Thomas, letter of instruction to Meriwether Lewis, June 20, 1803. Available at www.library.csi.cuny.edu/dept/history/lavender/jefflett.html.

Jefferson, Thomas. *Notes on the State of Virginia* (book online). Available at http://xroads.virginia.edu/~hyper/jefferson/ch05.html.

Jefferson, Thomas, to James Monroe, April 27, 1809. Cited in Garry Wills et al., *Thomas Jefferson: Genius of Liberty*. New York: Viking Studio with Library of Congress, 2000.

Jefferson, Thomas, to John Jacob Astor, May 24, 1812. Cited in Garry Wills et al., *Thomas Jefferson: Genius of Liberty*. New York: Viking Studio with Library of Congress, 2000.

Jefferson, Thomas, to John Trumbull, February 20, 1791. In Julian Boyd, ed., *The Papers of Thomas Jefferson*. Princeton, N.J.: Princeton University Press, 1950.

Jefferson, Thomas, to Maria Cosway, October 12, 1786. In Thomas Jefferson, *Thomas Jefferson: Writings*. Ed. Merrill D. Peterson. New York: Library of America, 1984.

Jefferson, Thomas, to William Caruthers, March 15, 1815. Cited in "Jefferson and the Natural Bridge." Available at www.monticello.org/resources/interests/natural_bridge.html.

Kellert, Stephen, and E. O. Wilson, eds. *The Biophilia Hypothesis*. Washington, D.C.: Island Press/Shearwater, 1993.

Levitt, James N. "Innovating on the Land: Conservation on the Working Landscape in American History." Paper prepared for Private Lands, Public Benefits: A Policy Summit on Working Lands Conservation, National Governors Association, March 16, 2001.

Library of Congress. "The West." In *Thomas Jefferson*. Exhibition at the Library of Congress, Washington, D.C., 2000. Available at www.loc.gov/exhibits/jefferson/jeffwest.html.

McCabe, Michael. "$100 Million Pledged for Coastal Open Space." *San Francisco Chronicle*, April 19, 2001, A18.

McNutt, Marcia. "How One Man Made a Difference: David Packard." Paper presented at the symposium Oceanography: The Making of a Science, Scripps Institution of Oceanography, La Jolla, Calif., February 8, 2000. Available at www.mbari.org/about/howonemanmadeadifference.pdf.

Mihaly, Mary. "All About University Circle: How the Circle Grew." Available as of November 2001 at www.universitycircle.org/go5.htm.

Miller, Charles A. *Jefferson and Nature: An Interpretation*. Baltimore: Johns Hopkins University Press, 1988.

Miller, Charles A. "The Rich Fields of Nature." In Gary Wills et al., *Thomas Jefferson: Genius of Liberty*. New York: Viking Studio with Library of Congress, 2000.

Monterey Bay Aquarium. Organizational information available at www.montereybayaquarium.org.

Monterey Bay Aquarium Research Institute. Organizational information available at www.mbari.org.

Morton, Carolyn, to Emma Morton, April 7, 1855. Cited in James C. Olson, *J. Sterling Morton*. Lincoln: University of Nebraska Press, 1942.

Morton, J. Sterling. "An Address by J. Sterling Morton on Arbor Day 1885." As reported in the *Nebraska City News*. Available on the National Arbor Day Foundation Web site at www.arborday.org/arborday/morton1887.html.

National Audubon Society. "Audubon Society Awards Julie Packard '98 Audubon Medal." Available at www.audubon.org/news/release98/jpackard.html.

The Nature Center at Shaker Lakes. "History of the Nature Center." Available at www.shakerlakes.org/history.htm.

Olson, James C. *J. Sterling Morton*. Lincoln: University of Nebraska Press, 1942.

Roberts, Ann Rockefeller. *Mr. Rockefeller's Roads*. Camden, Maine: Down East, 1990.

Smith, Robert J. "CPC Case Study: Natural Bridge of Virginia." Center for Private Conservation, posted June 1, 1998. Available at www.cei.org/gencon/025,01355.cfm.

Webster, Donovan. "Welcome to Turner Country." *Audubon*, January/February 1999, 56.

Wells, Owen. "Message from the President." The Libra Foundation, Portland,

Maine, 2001. Available at www.librafoundation.org/messagepresident/welcome.html.

Winks, Robin. *Frederick Billings.* Oxford: Oxford University Press, 1991.

Winks, Robin. *Laurance S. Rockefeller: Catalyst for Conservation.* Washington, D.C.: Island Press, 1997.

The World Almanac and Book of Facts 2002. New York: World Almanac Books, 2002.

15. CONCLUSION

Altshuler, Alan A., and Robert D. Behn, eds. *Innovations in American Government: Challenges, Opportunities, and Dilemmas.* Washington, D.C.: Brookings Institution Press, 1997.

Clarke, Arthur C. *Profiles of the Future: An Inquiry into the Limits of the Possible.* New York: Bantam, 1958.

Foster, Charles H. W., and James N. Levitt. "Beginner's Mind: Innovation and Environmental Practice." BCSIA Discussion Paper 2001-7, Environment and Natural Resources Program, John F. Kennedy School of Government, Harvard University, June 2001.

"Idaho Gray Wolf Recovery: Initiative Leads to Increased Tribal Sovereignty and Cultural Rejuvenation." In *Honoring Nations: Tribal Governance Success Stories, 1999.* Honoring Contributions in the Governance of American Indian Nations, administered by the Harvard Project on American Indian Economic Development, Cambridge, Mass., 2000.

Land Trust Alliance. Information on 2000 National Land Trust Census available at www.lta.org/aboutlta/census.shtml.

Murray, W. H. *The Scottish Himalaya Expedition* (1951). Cited in Meredith Lee, "Geothe Quotation Identified," *Goethe News and Notes* 19:1 (spring 1998). Available at www.humanities.uci.edu/tpsaine/Commitment.htm.

National Parks Conservation Association. "About NPCA: A Primer on Federal Lands." Available at www.npca.org/about_npca/park_system/default.asp.

Sellars, Richard West. *Preserving Nature in the National Parks: A History.* New Haven, Conn.: Yale University Press, 1997.

The Trustees of Reservations. Organizational information available at www.thetrustees.org.

Wilson, Edward O. *Naturalist.* Washington, D.C.: Island Press, 1994.

ABOUT THE
CONTRIBUTORS

JAMES H. BEACH is assistant director for informatics at the University of Kansas (KU) Biodiversity Research Center. He co-leads the Specify Software Project, a specimen database application used by museums around the world (www.usobi.org/specify), and is the leader of the Lifemapper Project (www.lifemapper.org), a distributed computing initiative that archives the known localities and predicted distributions of species. Prior to coming to KU, Beach worked for Harvard University, the University of California at Berkeley, the U.S. National Science Foundation, and the U.S. Geological Survey. Beach received his B.S. in botany from Michigan State in 1976 and his Ph.D. from the University of Massachusetts–Amherst in 1983.

BOB DURAND was appointed secretary of the Massachusetts Executive Office of Environmental Affairs by Governor Paul Cellucci on December 10, 1998. Prior to being named secretary of environmental affairs, Durand was best known for having been a leading advocate for environmental issues during his fifteen-year career in the Massachusetts legislature and has been recognized for his outstanding leadership by several conservation organizations, including the Environmental League of Massachusetts, the Massachusetts Audubon Society, the Massachusetts Association of Conservation Commissions, and the National Trust for Historic Preservation. Raised in Hudson, Massachusetts, Durand is a 1975 graduate of Boston College.

JOHN W. FITZPATRICK is director of the Cornell Lab of Ornithology and is a professor of ecology and evolutionary biology at Cornell University. Prior to coming to Cornell, Fitzpatrick served as executive director of the Archbold Biological Station in Central Florida and as chair of the Department of Zoology at the Field Museum of Natural History in Chicago. He serves on the national governing boards of the Nature Conservancy, the National Audubon Society, and the American Ornithologist's Union (AOU) and is a recipient of the AOU's William Brewster Memorial Award. Dr. Fitzpatrick received his bachelor's degree, magna cum laude, from Harvard University in 1974 and his Ph.D. from Princeton University in 1978.

FRANK B. GILL is senior vice president and director of science of the National Audubon Society. Dr. Gill came to National Audubon after twenty-five years guiding the Ornithology Department at the Academy of Natural Sciences in Philadelphia. He is the author of an acclaimed textbook, *Ornithology*, has served as president of the American Ornithologists' Union, and was honored with the William Brewster Memorial Award, the most prestigious award in American ornithology. Dr. Gill received both his undergraduate degree (1963) and a Ph.D. in zoology (1969) from the University of Michigan.

RALPH E. GROSSI is president of American Farmland Trust (AFT), a national nonprofit organization working to stop the loss of productive farmland and to promote farming practices that lead to a healthy environment. Grossi was a cofounder of Marin Agricultural Land Trust (MALT) in Marin County, California. He has served as chair of the MALT board of directors, as president of the Marin County Farm Bureau, and as a member of AFT's board of directors. A third-generation Marin County farmer, Grossi is managing partner of Marindale Ranch. He was named *Progressive Farmer Magazine*'s 2001 "Man of the Year," and is a recipient of the Feinstone Environmental Award. Grossi graduated from California Polytechnic State University in 1971.

ANDREW J. HANSEN is associate professor of ecology at Montana State University in Bozeman. His research focuses on the effects of natural and human disturbance on vegetation and vertebrates; landscape ecology; forest/vertebrate habit relationships; human habitat selection; computer modeling of habitat dynamics; and adaptive management. Dr. Hansen received a B.S. in ecology from Western Washington University in 1978 and a Ph.D. in ecology from the University of Tennessee, Oak Ridge National Lab, in 1984.

JOEL S. HIRSCHHORN is director of the Natural Resources Policy Studies Division at the Center for Best Practices of the National Governors Association. Prior to coming to the National Governors Association, he was president of Hirschhorn & Associates, an environmental consulting practice, and he edited *Remediation: The*

Journal of Environmental Cleanup Costs, Technologies, and Techniques. He has also served as a senior associate at the Congressional Office of Technology Assessment and was a full professor of engineering at the University of Wisconsin, Madison. He received a Ph.D. in 1965 in materials engineering at Rensselaer Polytechnic Institute, with a minor in physics. He received a B.S. in 1961 and an M.S. in 1962 in metallurgical engineering from the Polytechnic Institute of Brooklyn.

KENNETH M. JOHNSON is a professor of sociology at Loyola University–Chicago. He is a demographer and sociologist specializing in U.S. demographic trends, focusing on how the U.S. population is redistributing itself, and on the implications such demographic change has for the people and institutions of the nation. Johnson has published widely on the "rural rebound"—the net positive migration of Americans to nonmetropolitan counties in the 1990s—with Calvin Beale, the senior demographer at the U.S. Department of Agriculture. Johnson received his undergraduate training at the University of Michigan and completed his graduate training at the University of North Carolina–Chapel Hill.

LEONARD KRISHTALKA is director of the Natural History Museum and Biodiversity Research Center and a professor of ecology and evolutionary biology at the University of Kansas (KU), Lawrence. At KU, and as a member of the Advisory Committee for Biological Sciences at the National Science Foundation, Krishtalka has helped lead an international initiative for the knowledge networking of biodiversity information in order to meet grand challenges in research, education, and public policy in the twenty-first century. He helped form the North American Biodiversity Information Network and has received the Special Merit Award for Excellence in Writing from the Council for the Advancement and Support of Education. Prior to coming to KU, Krishtalka served on the staff of the Biological Sciences Directorate of the National Science Foundation, the Carnegie Museum of Natural History, and the University of Pittsburgh. He earned his B.S. (1969) and M.S. (1971) degrees from the University of Alberta, Edmonton, Canada, majoring in zoology, paleontology, and anthropology, and completed his doctoral studies in paleontology and evolutionary biology at KU, Lawrence, and Texas Tech University, Lubbock (Ph.D., 1975).

JAMES N. LEVITT is a fellow at the A. Alfred Taubman Center for State and Local Government at Harvard's John F. Kennedy School of Government and directs the Taubman Center's Internet and Conservation Project. Prior to coming to the Kennedy School, Levitt developed corporate strategy related to the emergence of the Internet and electronic commerce for Fortune 50–sized companies at GeoPartners Research Inc. He serves as a director of several conservation organizations, including the Massachusetts Audubon Society, the Quebec-Labrador

Foundation, and the Natural History Museum and Biodiversity Research Center at the University of Kansas. Levitt is a 1976 graduate of Yale College and earned his master's in public and private management from the Yale School of Management in 1980.

SHARON E. MCGREGOR serves as assistant secretary for Biological Conservation and Ecosystem Protection in the Massachusetts Executive Office of Environmental Affairs (EOEA). She works with Secretary of Environmental Affairs Bob Durand to promote the inclusion of biological conservation and ecosystem considerations in environmental and economic decision making. Within her management responsibility is the EOEA's Watershed Initiative, Water Policy, Growth Management and Land Acquisition, Forest Management, Wetlands Restoration, Environmental Education, and Geographic and Environmental Information Systems. McGregor earned a master in public administration from Harvard University's John F. Kennedy School of Government in 1997 and a B.S. in environmental science and policy from the University of New Hampshire, Durham in 1982.

WILLIAM J. MITCHELL is professor of architecture and media arts and sciences, head of the Media Arts and Sciences Program, and dean of the School of Architecture and Planning at the Massachusetts Institute of Technology (MIT). He has written several seminal books that examine the interaction of the built and electronic environments, including *City of Bits: Space, Place, and the Infobahn* and *E-Topia: Our Town Tomorrow*. Before coming to MIT, Mitchell was the G. Ware and Edythe M. Travelstead Professor of Architecture and director of the Master in Design Studies Program at the Harvard Graduate School of Design. He previously served as head of the Architecture/Urban Design Program at UCLA's Graduate School of Architecture and Urban Planning. He received a B.Arch. from the University of Melbourne in 1967, an M.E.D. from Yale University in 1969, and an M.A. from Cambridge University in 1977.

BRIAN H. F. MULLER is assistant professor of planning and design at the University of Colorado at Denver. His research interests include land use modeling, geographic information systems, the preservation of agricultural land and open space, and regional economic development. He received a B.A. in history and anthropology from Yale University in 1979 and a Ph.D. in city and regional planning from the University of California at Berkeley in 2001.

A. TOWNSEND PETERSON is associate professor in the Department of Ecology and Evolutionary Biology and curator of ornithology in the Natural History Museum at the University of Kansas. Peterson focuses his research on exploration of species' geographic distributions. These investigations include aspects of global climate change and its effects on species distributions, conservation biology, species invasions, and disease systems. He received his undergraduate degree in zoology from Miami University of Ohio in 1985 and completed a Ph.D. in evolutionary biology at the University of Chicago in 1990.

JOHN R. PITKIN is a consultant and president of Analysis and Forecasting Inc., a demographic research firm located in Cambridge, Massachusetts. Before becoming a consultant, he was research associate at the Joint Center for Urban Studies of Harvard University and Massachusetts Institute of Technology, where he studied housing demographics and population redistribution. Pitkin's current research activities include population and housing projections for the state of California for the University of Southern California's California Housing Futures project and the impacts of immigration on U.S. housing markets. He earned an A.B. in 1967 from Columbia University and has studied at Oxford University and the University of Helsinki.

WILLIAM E. ROPER is the director of Programs and Projects with the Orton Family Foundation. At the foundation, where he has worked since 1998, Roper oversees the development and delivery of planning tools for use by rural communities and place-based education in middle and high schools. From June 1997 to September 1998, Bill lived on the West Coast of Ireland with his wife and two daughters, spearheading national changes to Ireland's land use practices and regulations, forming Ireland's first land trust, and working on several local County Mayo projects. Prior to living in Ireland, Bill was a prominent land use attorney in Vermont, where he focused for fifteen years on encouraging sound economic and environmental development. He received a B.A. in 1977 from Williams College and a J.D. in 1983 from Vermont Law School.

JAY J. ROTELLA is associate professor of ecology at Montana State University in Bozeman. Rotella's research interests include avian population ecology and management, modeling and estimation of population parameters, effects of changes in land use on avian populations, and habitat relationships. He studies a variety of bird species in forests, grasslands, and wetlands. Rotella received a B.S. from the University of Vermont in 1982, an M.S. from Washington State University in 1985, and a Ph.D. from the University of Idaho in 1990.

S. JACOB SCHERR is a senior attorney with the Natural Resources Defense Council (NRDC) in its Washington, D.C., office, where he serves as director of NRDC's International Program. He also is president of Earth Summit Watch, which he founded to monitor national implementation of the commitments to sustainable development made at the 1992 Earth Summit. During his career with NRDC since

1976, he has worked extensively on a broad range of international environmental and nuclear issues. Prior to joining NRDC, he was a fellow at the American Society of International Law and a lecturer in international law at the University of Maryland School of Law. Scherr is a 1970 graduate of Wesleyan University. In 1974, he received his J.D. with highest honors from the University of Maryland Law School.

PETER R. STEIN is a general partner at the Lyme Timber Company in Lyme, New Hampshire, and is responsible for the development and structuring of large-scale timberland purchases and limited development projects in cooperation with regional and national land conservation organizations, as well as conservation advisory work with foundations, families, and corporate landowners. Prior to joining Lyme Timber Company in 1990, Stein was senior vice president of the Trust for Public Land, where he directed their conservation real estate acquisition activities in the Northeast and Midwest. His current and former board memberships include Island Press, Hubbard Brook Research Foundation, Land Trust Alliance, Appalachian Mountain Club, and Vermont Natural Resource Council. Stein and the Lyme Timber Company have been recognized by the Maine Coast Heritage Trust and by Friends of Acadia for their outstanding work on conservation initiatives in Maine. Stein received a B.A. from the University of California at Santa Cruz in 1975 and was a Loeb Fellow in Advanced Environmental Studies at Harvard University in 1981.

DAVID A. VIEGLAIS is a senior scientist at the University of Kansas (KU) Natural History Museum and Biodiversity Research Center, where he has led efforts to develop the Species Analyst, a biodiversity information network linking natural history collection and observation databases around the world. Prior to coming to KU, he served as a U.S. National Academy of Sciences postdoctoral fellow working on a joint Smithsonian/NASA project at the Kennedy Space Center in Florida investigating the terrestrial ecophysiological response to elevated atmospheric carbon dioxide. He has also served as director of research and development for a small technology company. Vieglais completed a B.Sc. in botany in 1989 and a Ph.D. in botany (ecophysiology) in 1992, both from the University of Queensland, Australia.

THOMAS J. VILSACK is governor of Iowa. First elected to that office in 1998, Vilsack became the first Democrat to be elected governor of Iowa in more than thirty years. He also served as chair of the Natural Resource Committee of the National Governors Association. In Iowa, Governor Vilsack and Lieutenant Governor Sally Pederson created the state's Clean Water Initiative to empower farmers to promote private land conservation and improve water quality conditions across the state.

Prior to becoming governor of Iowa, Vilsack served in the Iowa legislature and as mayor of Mount Pleasant, Iowa, and was active in the private practice of law. He graduated from Hamilton College in Clinton, New York, with a degree in history in 1972 and from Union University, Albany Law School, with a J.D. in 1975.

E. O. WILEY is a senior curator of fishes, University of Kansas (KU) Biodiversity Research Center; professor of ecology and evolutionary biology at KU; and a research associate of the National Museum of Natural History, Smithsonian Institution. Upon completion of his graduate studies, Wiley joined the scientific community at KU, where he has worked on a variety of projects, including empirical work on the systematics, evolution, and biogeography of fishes; theoretical aspects of systematics and taxonomy; and evolutionary theory. Wiley completed a B.S. in biology at Southwest Texas State Teachers College in 1966 and a Ph.D. in biology at the City University of New York in 1976, where he worked on the evolution and systematics of fossil and living gar fishes.

INDEX

355